Ball Milling Towards Green Synthesis
Applications, Projects, Challenges

RSC Green Chemistry

Editor-in-Chief:
Professor James Clark, *Department of Chemistry, University of York, UK*

Series Editors:
Professor George A Kraus, *Department of Chemistry, Iowa State University, Ames, Iowa, USA*
Professor Andrzej Stankiewicz, *Delft University of Technology, The Netherlands*
Professor Peter Siedl, *Federal University of Rio de Janeiro, Brazil*
Professor Yuan Kou, *Peking University, China*

Titles in the Series:
1: The Future of Glycerol: New Uses of a Versatile Raw Material
2: Alternative Solvents for Green Chemistry
3: Eco-Friendly Synthesis of Fine Chemicals
4: Sustainable Solutions for Modern Economies
5: Chemical Reactions and Processes under Flow Conditions
6: Radical Reactions in Aqueous Media
7: Aqueous Microwave Chemistry
8: The Future of Glycerol: 2nd Edition
9: Transportation Biofuels: Novel Pathways for the Production of Ethanol, Biogas and Biodiesel
10: Alternatives to Conventional Food Processing
11: Green Trends in Insect Control
12: A Handbook of Applied Biopolymer Technology: Synthesis, Degradation and Applications
13: Challenges in Green Analytical Chemistry
14: Advanced Oil Crop Biorefineries
15: Enantioselective Homogeneous Supported Catalysis
16: Natural Polymers Volume 1: Composites
17: Natural Polymers Volume 2: Nanocomposites
18: Integrated Forest Biorefineries
19: Sustainable Preparation of Metal Nanoparticles: Methods and Applications
20: Alternative Solvents for Green Chemistry: 2nd Edition
21: Natural Product Extraction: Principles and Applications
22: Element Recovery and Sustainability
23: Green Materials for Sustainable Water Remediation and Treatment
24: The Economic Utilisation of Food Co-Products
25: Biomass for Sustainable Applications: Pollution Remediation and Energy
26: From C-H to C-C Bonds: Cross-Dehydrogenative-Coupling
27: Renewable Resources for Biorefineries

28: Transition Metal Catalysis in Aerobic Alcohol Oxidation
29: Green Materials from Plant Oils
30: Polyhydroxyalkanoates (PHAs) Based Blends, Composites and Nanocomposites
31: Ball Milling Towards Green Synthesis: Applications, Projects, Challenges

How to obtain future titles on publication:
A standing order plan is available for this series. A standing order will bring delivery of each new volume immediately on publication.

For further information please contact:
Book Sales Department, Royal Society of Chemistry, Thomas Graham House, Science Park, Milton Road, Cambridge, CB4 0WF, UK
Telephone: +44 (0)1223 420066, Fax: +44 (0)1223 420247
Email: booksales@rsc.org
Visit our website at www.rsc.org/books

Ball Milling Towards Green Synthesis
Applications, Projects, Challenges

Edited by

Brindaban Ranu
University of Jadavpur, Jadavpur, Kolkata, India
Email: ocbcr@iacs.res.in

Achim Stolle
University of Jena, Jena, Germany
Email: achim.stolle@uni-jena.de

THE QUEEN'S AWARDS
FOR ENTERPRISE:
INTERNATIONAL TRADE
2013

RSC Green Chemistry No. 31

Print ISBN: 978-1-84973-945-0
PDF eISBN: 978-1-78262-198-0
ISSN: 1757-7039

A catalogue record for this book is available from the British Library

Published by The Royal Society of Chemistry,
Thomas Graham House, Science Park, Milton Road,
Cambridge CB4 0WF, UK

Registered Charity Number 207890

For further information see our web site at www.rsc.org

Printed and bound by CPI Group (UK) Ltd, Croydon, CR0 4YY

Foreword

Ball milling creates stress on substances. Solids break, and the resulting particles are modified in shape and size. Defects are induced and surface areas enlarged. Consequently, ball milling is highly relevant for various technological fields including mineral processing, materials engineering, and biomass degradation. In most of such applications the milling process is used for a particle size reduction. However, the energy induced by the mechanical treatment has other effects as well. On the molecular level, for example, the arrangements of chemical structures can change leading to products with altered properties. The potential of such mechanochemical approaches and the implications in utilizing them in modern organic synthesis are nicely illustrated by the excellent contributions collected in the book edited by Stolle and Ranu. Many of the discussed reactions are solvent-free leading to ecological and economical advantages over existing technologies providing the same products. Further benefits are recognized when comparing the energy efficiency of ball milling with other activation modes. Apparently, ball milling can be applied in a number of bond-forming processes, and various standard organic transformations (such as oxidations, reductions, and peptide formations) benefit from the use of this mechanochemical technique. Both low-molecular-weight compounds as well as polymers undergo specific chemical modifications in ball mills. Surprising observations have been made in both metal-catalyzed and organocatalytic C–C-bond formations including asymmetric versions thereof. Liquid-assisted grinding and kneading have proven advantageous for the preparation of coordination compounds. Being aware of the technological and process parameters is essential for achieving optimal results in synthetic transformations performed in ball mills.

Although mechanochemical activations have already been utilized for a long time, the advantages of applying ball mills in targeted organic synthesis

RSC Green Chemistry No. 31
Ball Milling Towards Green Synthesis: Applications, Projects, Challenges
Edited by Brindaban Ranu and Achim Stolle

Published by the Royal Society of Chemistry, www.rsc.org

has largely remained unrecognized until recently. The growing awareness of environmental implications of chemical processes and the search for greener solutions have led to a change, and today ball milling results are more present in the community than ever. Gaining a deeper understanding of the underlying mechanistic principles leading to mechanochemical activations and finding new reaction pathways resulting in products inaccessible by other means will further expand the synthetic ball milling opportunities.

This book will initiate new thought processes and promote the implementation of ball milling as a modern synthetic technique in existing lab structures. Experts in academia and industry as well as interested newcomers will benefit from the timely presentations collected by Stolle and Ranu, and I congratulate both editors and authors for their stimulating contributions.

Carsten Bolm
Aachen, Germany

Preface

During the last few years ball milling has emerged as a powerful tool in effecting various chemical reactions in a relatively green way by reducing the amount of solvent and operating at close to ambient temperature by application of mechanical energy. However, compared to other alternative forms of energy such as microwave and ultrasound, ball milling is still not familiar as a synthetic tool to a wide section of chemists and activity is limited to a comparatively small number of groups. Thus, the objective of this book is to create a general awareness of the importance of ball milling in chemical transformations among all section of readers including students, teachers and researchers. This book, entitled *Ball Milling Towards Green Synthesis: Applications, Projects, Challenges*, covers the current developments in the application of ball milling for various chemical transformations such as carbon–carbon and carbon–heteroatom bond formation, oxidation–reduction, organocatalytic reaction, dehydrogenative coupling, synthesis of peptide, polymeric materials, *etc.* highlighting its green aspects, scope and future prospects. This book also includes a chapter describing its origin, technological background and challenges.

We gratefully acknowledge the co-operation from all those eminent and active scientists who contributed chapters to this book and the support and guidance from the staff at the Royal Society of Chemistry.

We hope this book will be able to provide basic information regarding ball milling and its application in chemical synthesis and thus will be useful to a wide section of chemists including students and researchers associated with academy and industry.

Brindaban C. Ranu
Achim Stolle

RSC Green Chemistry No. 31
Ball Milling Towards Green Synthesis: Applications, Projects, Challenges
Edited by Brindaban Ranu and Achim Stolle

Published by the Royal Society of Chemistry, www.rsc.org

Contents

RSC Green Chemistry No. 31
Ball Milling Towards Green Synthesis: Applications, Projects, Challenges
Edited by Brindaban Ranu and Achim Stolle

Published by the Royal Society of Chemistry, www.rsc.org

CHAPTER 1

Carbon–Heteroatom Bond Forming Reactions and Heterocycle Synthesis under Ball Milling

BRINDABAN C. RANU,* TANMAY CHATTERJEE AND NIRMALYA MUKHERJEE

Department of Organic Chemistry, Indian Association for the Cultivation of Science, Jadavpur, Kolkata 700032, India
*Email: ocbcr@iacs.res.in; bcranu@gmail.com

1.1 Introduction

Chemical transformations involving mechanochemical (grinding) reactions using a mortar and pestle were initiated long ago during the early stage of evolution of chemistry. But due to variable and relatively low grinding strength and speed, limited chemical reactions were successfully carried out by grinding. To overcome this limitation, a mixer/shaker mill or a planetary mill[1] has been developed and these devices provide much higher energy and are more reliable than hand grinding. Mixing in such a mill is generally referred to as "milling". Mechanical grinding in these mills is aided by the milling balls and hence it is called ball milling. Mechanochemistry using ball milling has received intense attention in organic synthesis in recent decades and is now a fast growing branch. Mechanochemical reactions in a ball mill depend on several parameters like milling frequency, milling time,

RSC Green Chemistry No. 31
Ball Milling Towards Green Synthesis: Applications, Projects, Challenges
Edited by Brindaban Ranu and Achim Stolle

Published by the Royal Society of Chemistry, www.rsc.org

size and number of milling balls, and the material of milling balls and beakers. The first two have been found to be the most important parameters.[1] Basically two types of ball mills have been frequently used to perform organic transformations, *i.e.* mixer mill (MM) and planetary mill (PM) (for further details the reader is referred to Chapter 10).

The present chapter covers reactions involving carbon–heteroatom (C–N, C–O, C–S, C–Cl, C–Br, *etc.*) bond formation and synthesis of heterocycles under ball milling.

1.2 Carbon–Heteroatom Bond Forming Reactions under Ball Milling

1.2.1 C–N Bond Forming Reactions

Kaupp and co-workers have reported the quantitative formation of imines by ball milling a stoichiometric mixture of aldehyde and amine in less than 30 min (Scheme 1.1).[2,3]

All the reactions were carried out below 0 °C except for the condensations of 4-nitroaniline with 4-hydroxybenzaldehyde and 4-nitrobenzaldehyde, which were performed at 60 and 80 °C, respectively.

The same group also investigated the solid-state reactions of hydrazine–hydroquinone (1 : 1 complex) and of hydrazine hydrochloride with solid aldehyde, ketone, carboxylic acid, thiohydantoin and 4-nitrophenyl isothiocyanate and found that only the hydrazine hydroquinone complex provides quantitative addition, condensation, ring opening and ring closure (Schemes 1.2 and 1.3).[4]

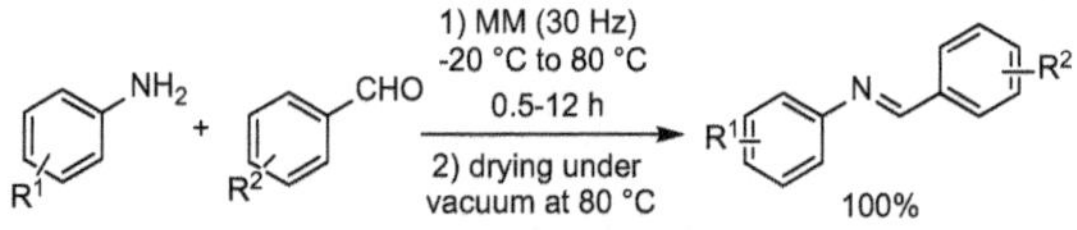

Scheme 1.1 Imine formation under mechanical milling.

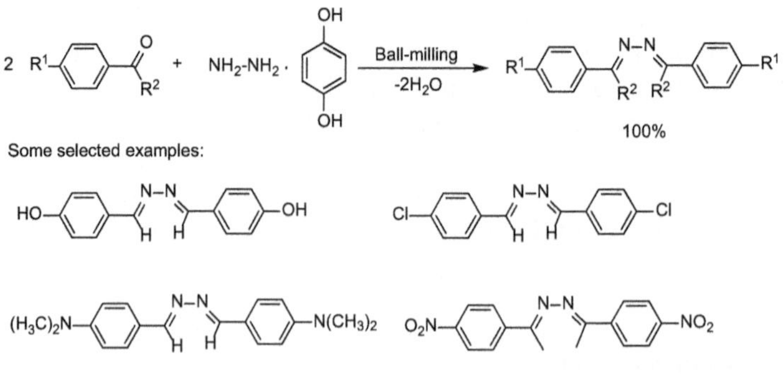

Scheme 1.2 Synthesis of azines from the solid state reaction between aldehyde or ketone and hydrazine–hydroquinone complex in a ball mill.

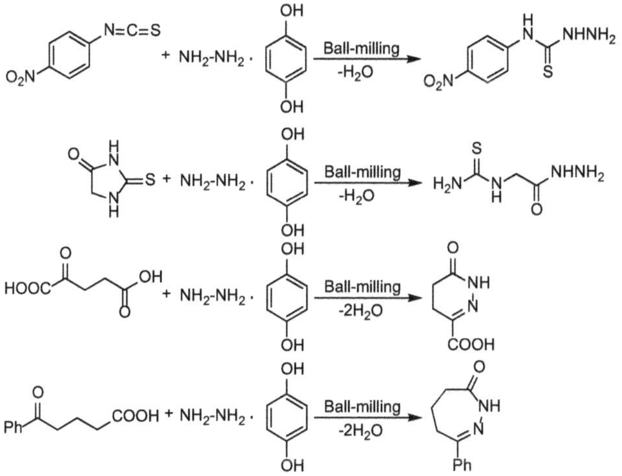

Scheme 1.3 Quantitative C–N bond forming reactions *via* condensation, ring opening and ring closure under ball milling.

Naimi-Jamal, Kaupp and co-workers employed "kneading ball-milling" for the stoichiometric quantitative preparation of synthetically versatile 2,4-dinitrophenylhydrazones from low-melting aldehydes and ketones.[5] Owing to the potential explosive nature of dry 2,4-dinitrophenylhydrazine, 50 wt% deionized water was added to wet the crystals. Water was used as an auxiliary to minimize the risk of explosive destruction. The stoichiometric reaction of 2,4-dinitrophenylhydrazine with aldehydes or ketones occurred rapidly in the kneading ball mill at 25–70 °C for 10–20 min, giving the desired hydrazones in 58–100% yields (Scheme 1.4). When these reactions are performed in solution, strong acid catalysts are required.[6] However, they occur much faster in the kneading ball mill in the absence of catalyst.

Similarly, aldehydes and ketones were stoichiometrically ball-milled with hydroxylamine hydrochloride at 25–140 °C for 10–120 min to provide the hydrated oxime salts in sticky form, which were then treated with a base to obtain the oximes in 75–100% yields (Scheme 1.5).[5] NaCl in water and CO_2 from carbonate were found to be the only wastes produced. Aldehydes were found to be more reactive than ketones towards hydroxylamine. The difference in reactivity of aldehydes and ketones can be utilized as a versatile method for selective protection of an aldehyde by oximation in the presence of a ketone.

Benzoylhydrazones were also obtained, by ball-milling benzhydrazide and solid aldehydes in a molar ratio of 1 : 1 for 1 h at 25–30 °C. However, the reaction of benzhydrazide with isatin required 3 h of ball milling for completion (Scheme 1.6).[7]

Liquid assisted grinding (LAG)[8] has been applied for the synthesis of imine from 5-aminosalicylic acid and vanillin or 2-hydroxy-1- naphthaldehyde in a ball mill for 5–30 min in the presence of a small amount of EtOH or EtOH–NEt_3 (Scheme 1.7).

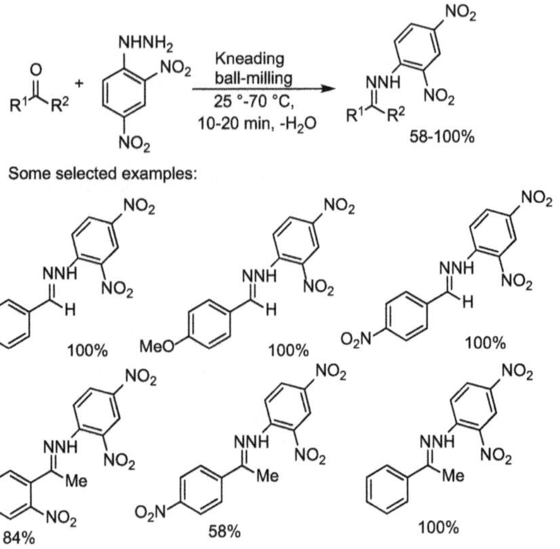

Scheme 1.4 Kneading ball-milling for the preparation of 2,4-dinitrophenylhydrazones.

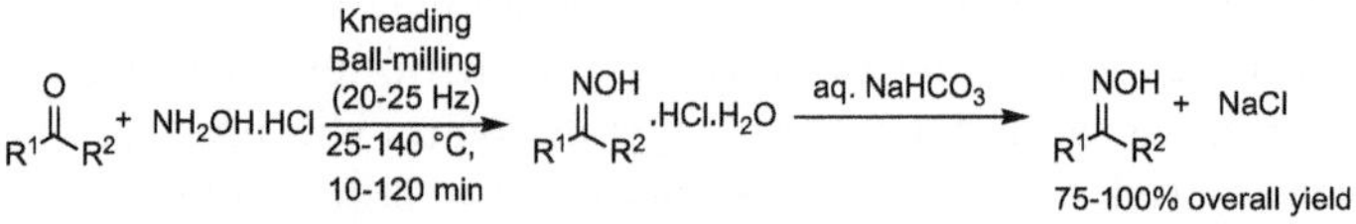

Scheme 1.5 Solvent-free kneading ball-milling of aldehyde or ketone with hydroxylamine hydrochloride.

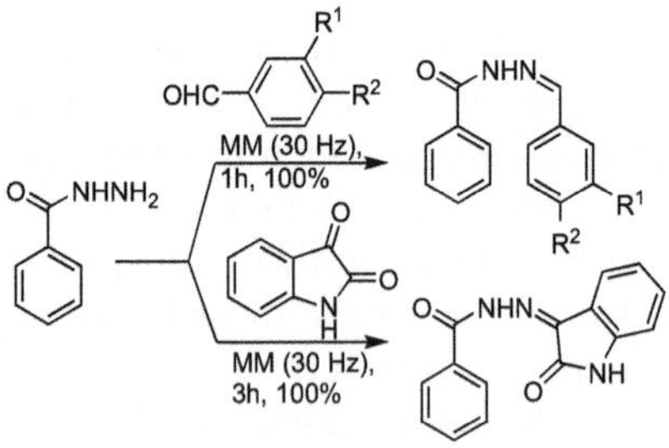

Scheme 1.6 Benzoylhydrazone formation by ball milling.

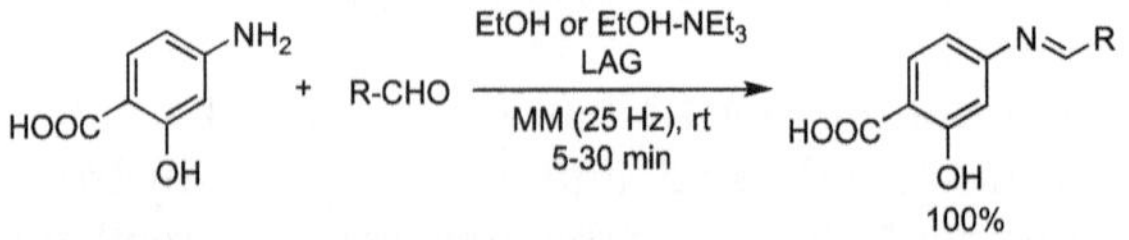

Scheme 1.7 Liquid assisted grinding (LAG) in a mechanical mill to form imine.

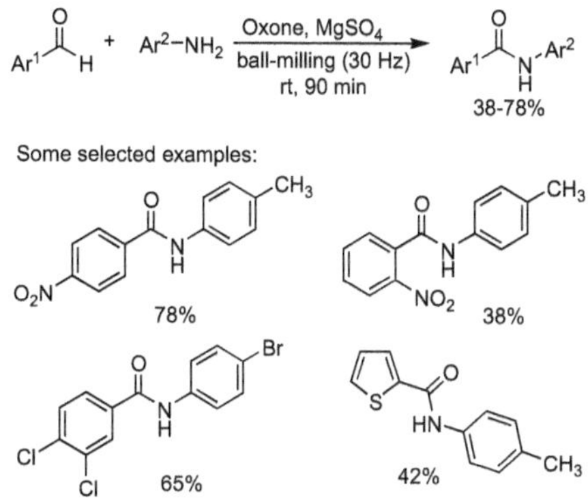

Scheme 1.8 Direct oxidative amidation of aldehydes with anilines.

Wang and Gao have reported direct oxidative amidation of aldehydes with anilines under ball milling (Scheme 1.8).[9] Oxone was used as an effective oxidant for the transformation of aldehydes into amides. Other oxidants like $K_2S_2O_8$ and I_2 failed to initiate the oxidative amidation.

$MgSO_4$ was found to be the best dehydrating agent for removing the water formed in the reaction. In addition, when the oxidative amidation reaction was performed in organic solvents such as acetonitrile or toluene, under identical reaction conditions, the yield of the corresponding product was comparatively lower than the reaction under solvent-free condition in a ball mill. Anilines bearing electron-donating and electron-withdrawing groups underwent facile reaction with aldehydes bearing electron-withdrawing groups only. Reactions with aldehydes containing electron-donating groups were unsuccessful, indicating the importance of the electronic character of aldehydes. Aliphatic anilines were also inert under the reaction conditions. The replacement of an aldehyde by the corresponding carboxylic acid did not afford the amides, thereby discarding the possibility of oxidation of aldehyde to carboxylic acid by Oxone followed by reaction with aniline to give amide.

A tentative mechanism for the amidation reaction of aldehydes with anilines based on the above results was proposed (Scheme 1.9). This direct oxidative amidation reaction may proceed *via* two pathways. In pathway A, the interaction of an aldehyde and aniline produces imine readily, which is then oxidized by Oxone to generate oxaziridine **1** as the key intermediate. Rearrangement of oxaziridine to amide is known for both photochemical and thermal reactions.[10] Initial cleavage of the N–O bond followed by migration of the substituent (hydrogen) *trans* to the nitrogen lone pair results in the formation of amide **2**. In pathway B, nucleophilic addition of aniline to aldehyde produces carbinolamine intermediate **3**, which is then oxidized by Oxone to form the amide. Even though control experiments showed that imine could be employed to perform the amidation directly, the exact mechanism of this reaction remains obscure and both pathways could be operational.

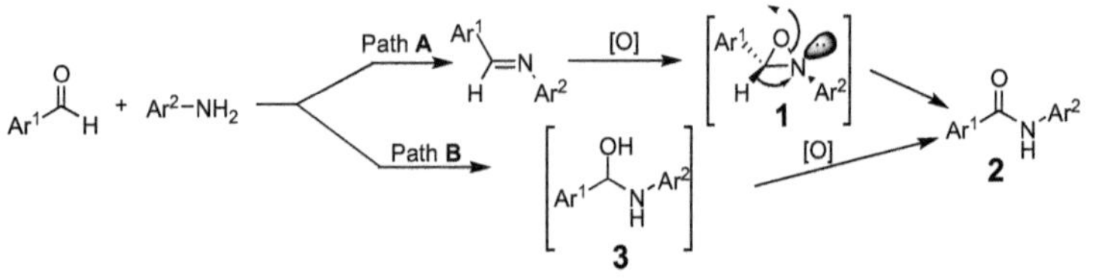

Scheme 1.9 Possible pathway for the direct oxidative amidation of aldehydes with anilines.

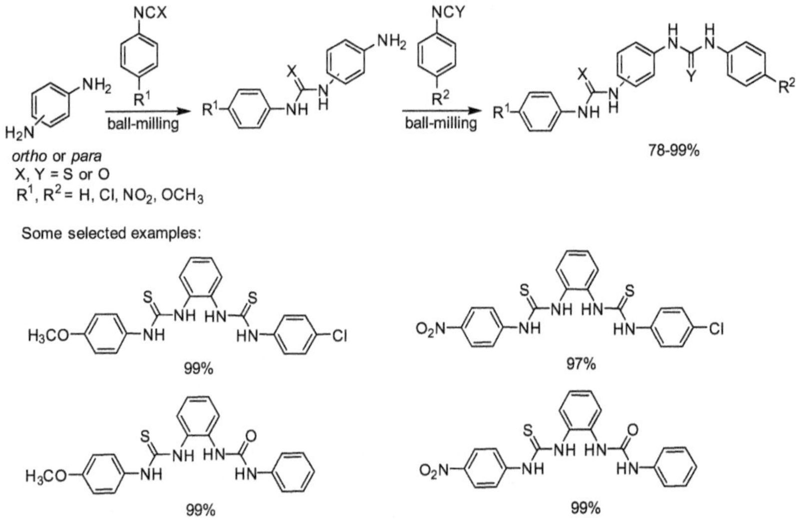

Scheme 1.10 Synthesis of symmetrical and non-symmetrical phenylenediamine(thio)urea derivatives by ball milling.

A new methodology has been developed for the desymmetrization of aromatic diamines to non-symmetrical thiourea derivatives by click-mechanochemistry in a ball mill.[11] Phenylenediamines (*o*-, *p*-) were desymmetrized through a one-pot mechanochemical click reaction sequence to furnish mono- and bis(thio)ureas or mixed thiourea–ureas (Scheme 1.10).

o-Phenylenediamine reacted selectively with either one or two equivalents of phenyl isothiocyanate to yield the non-symmetrical amino-thiourea or the symmetrical bis-thiourea in 95% and >99% yields respectively. The excellent control of the stoichiometric composition of the product in mechanochemical click-thiourea coupling demonstrates that it provides a facile and clean one-pot route to desymmetrization of aromatic diamines, and to the synthesis of symmetrical and non-symmetrical bis-(thio)ureas that are obtained in poor yields in solution.[12]

A chemoselective C–N bond formation has been achieved by the acylation of primary aliphatic amines using a vibrational ball mill in 10 min (120 min for aromatic amines). Azobenzene functionalized esters were employed to react with various amines (primary and secondary) in presence of a base,

N,N-dimethyl-4-aminopyridine (DMAP), under vibrational ball-milling to synthesize the corresponding amides (Scheme 1.11).[13]

Jin *et al.* have reported an efficient solvent and catalyst-free aza-Michael addition of chalcone to amine under the high-speed vibration ball-milling in a short reaction time (Scheme 1.12).[14] In general, excellent yields were obtained, eliminating the usual side reactions. It was also observed that the yields of the corresponding products obtained using anilines were lower compared to those using benzyl amine or piperidine.

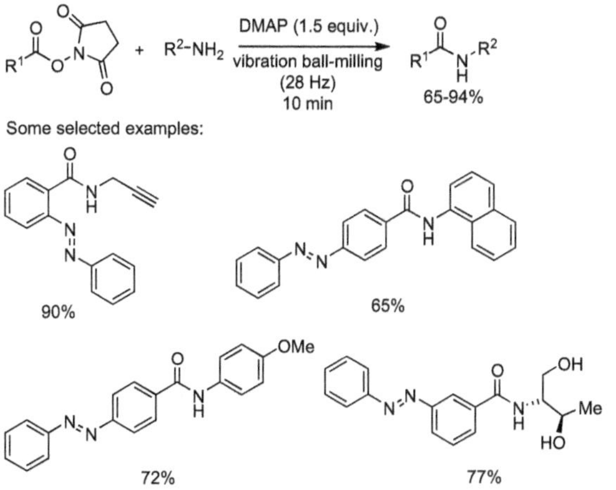

Scheme 1.11 C–N bond forming reaction of propargylamine with *N*-hydroxysuccinimidyl *p*-(phenylazo)benzoate under ball milling.

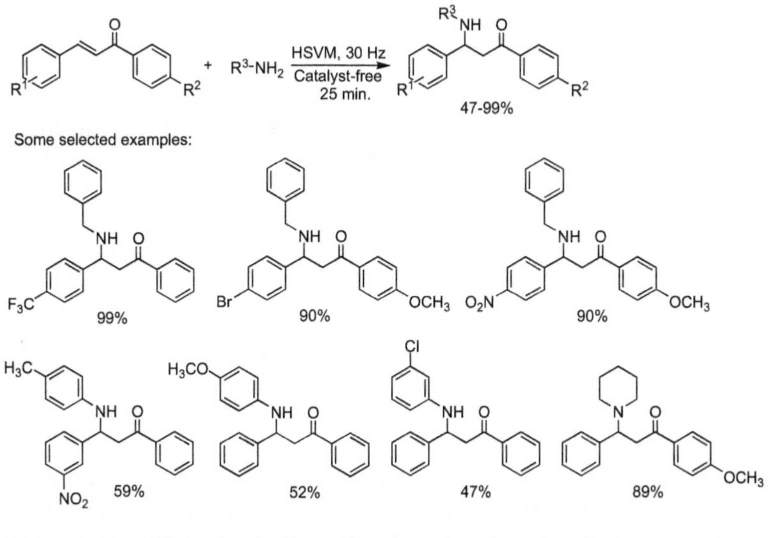

Scheme 1.12 Aza-Michael addition of different amines to chalcones under HSVM conditions.

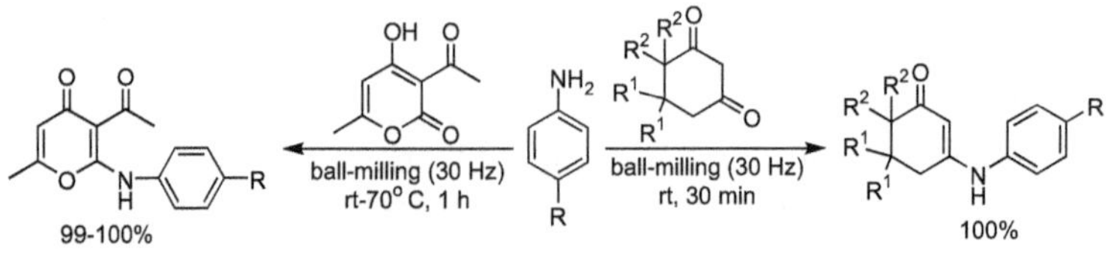

Scheme 1.13 Enamino ketone synthesis under ball milling.

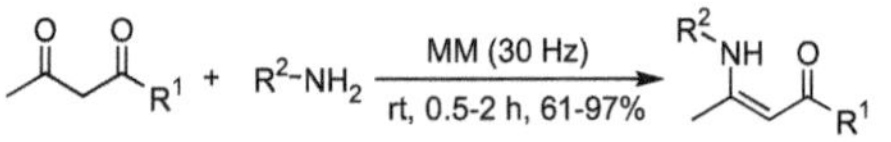

Scheme 1.14 Synthesis of enamino ketone and ester using aliphatic amine under ball milling.

Kaupp's group has reported quantitative synthesis of enamino ketones by the reaction of cyclic 1,3-dicarbonyl compounds such as 1,3-cyclohexanedione, dimedone and dehydroacetic acid with aniline derivatives without any catalyst under ball milling within 1 h, followed by drying at 0.01 bar at 80 °C (Scheme 1.13).[7]

Li and co-workers have reported the mechanochemical reaction of aliphatic primary amines with acyclic 1,3-dicarbonyl compounds such as 1,3-pentadione and ethyl acetoacetate in the absence of catalyst and solvent (Scheme 1.14).[15] A series of enamino ketones and esters were obtained in 61–97% yields by ball milling the mixtures of amines and 1,3-dicarbonyl compounds in a ratio of 1 : 1 in a mixer mill at 30 Hz for 0.5–2 h.

Stolle *et al.* have developed a solvent-free methodology for the synthesis of enamines by the addition of amines to dialkyl acetylene dicarboxylates or alkyl propiolates using a planetary ball mill at 800 rpm (13.3 Hz) (Scheme 1.15).[16] Fused quartz sand (SiO_2) was used as inert grinding auxiliary to facilitate the energy entry in the presence of liquid substrates by adsorbing them on the surface.[17] Significantly, reactions with several anilines and secondary alkyl amines were complete within five minutes with excellent yield of products. Besides the (*E*-/*Z*)-isomers, no other product was formed. Interestingly, addition of aniline or *p*-toluidine to dialkyl acetylene dicarboxylate produced the (*E*)-enamine as the major product whereas addition of the same amine to alkyl propiolate produces the (*Z*)-enamine as the major product.

Lamaty and co-workers reported the condensation of aldehydes with equimolar amounts of *N*-substituted hydroxylamines in a ball mill at a frequency of 30 Hz for 0.5–2 h to obtain various C-aryl and C-alkyl nitrones in 71–100% yields (Scheme 1.16).[18] Significantly, reactions were performed in the presence of air and moisture and the products were obtained pure.

Though urea is very unreactive toward alkylation, the reaction of urea with 4-bromobenzyl bromide under mechanical milling in the presence of NaOH produced di(4-bromobenzyl)urea with 41% conversion for a total milling time of 34 h (Scheme 1.17).[19]

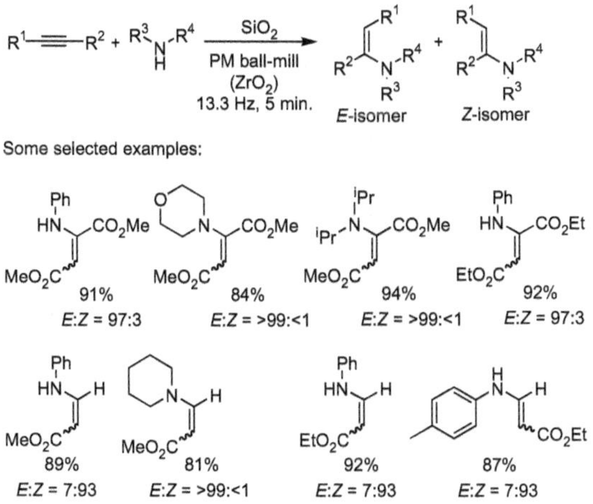

Scheme 1.15 Reaction of dialkyl acetylene dicarboxylates or alkyl propiolates with amines.

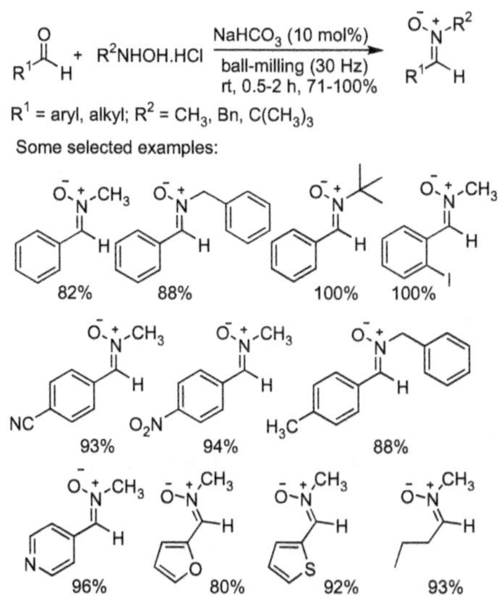

Scheme 1.16 Synthesis of nitrones by the reaction of aldehydes with *N*-substituted hydroxyl amines under ball milling.

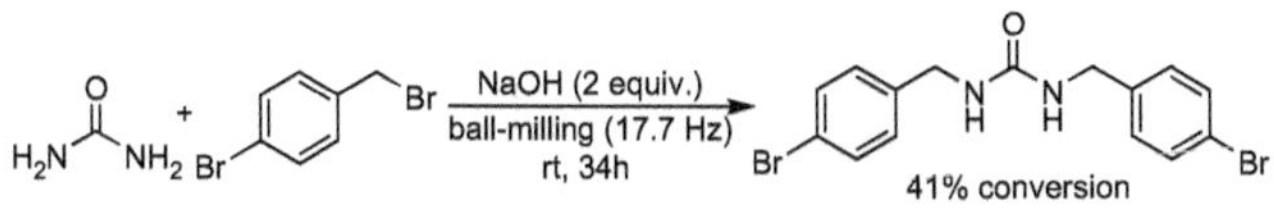

Scheme 1.17 *N*-Alkylation of urea with 4-bromobenzyl bromide under ball milling.

1.2.2 C–O Bond Forming Reactions

Mechanochemical reaction of pyrazolone derivatives and phenacyl bromide under ball milling afforded the corresponding pyrazolyl ether derivative quantitatively within 1 h *involving* the C–O bond formation. The products were obtained after washing the crude with sodium carbonate solution followed by drying at 0.01 bar at 80 °C under vacuum (Scheme 1.18).[20]

Mack and co-workers have reported the mechanochemical synthesis of dialkyl carbonates of various metal carbonates with the assistance of metal complexing reagents (Scheme 1.19).[21] The reaction of various metal carbonates including Li_2CO_3, Na_2CO_3, K_2CO_3 and Cs_2CO_3 with 4-bromobenzyl bromide gave poor results (0–18% yields). However, the addition of 2 equivalents of 18-crown-6 improved the yield from 2% to 74%. The same product was obtained as the major one in the reaction of cyclohexanone with *p*-bromobenzyl bromide using K_2CO_3 as the base in the presence of 18-crown-6.[22] Under the optimal conditions, other aliphatic halides such as benzyl bromide, (2-bromoethyl)benzene and benzyl chloride provided dialkyl carbonates in 58–67% yields.[21]

Recently, Ranu *et al.* have developed an efficient procedure for transesterification in a ball mill in the absence of any solvent, acid/base or metal catalyst (Scheme 1.20).[23] The reactions were carried out on the surface of basic alumina, which plays a dual role of a grinding auxiliary and a base. A wide variety of esters such as methyl, ethyl and allyl esters were transesterified with various alcohols including benzyl, cinnamyl, alkyl (primary, secondary) *etc.* Heteroaryl (pyridine, thiophene and furan) substituted

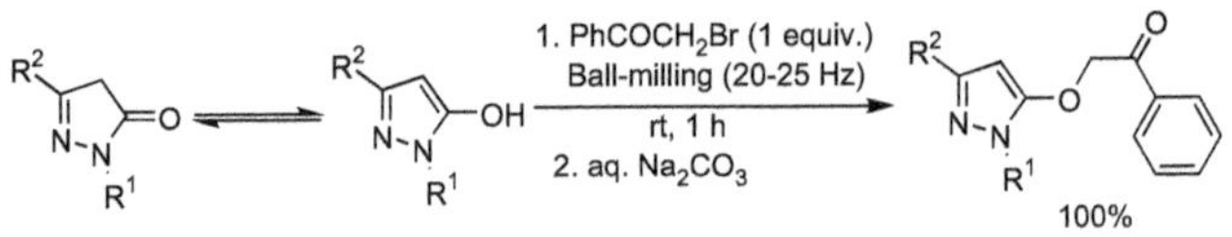

Scheme 1.18 Synthesis of pyrazolyl ether by the reaction of pyrazolone and phenacyl bromide *via* C–O bond formation under ball milling.

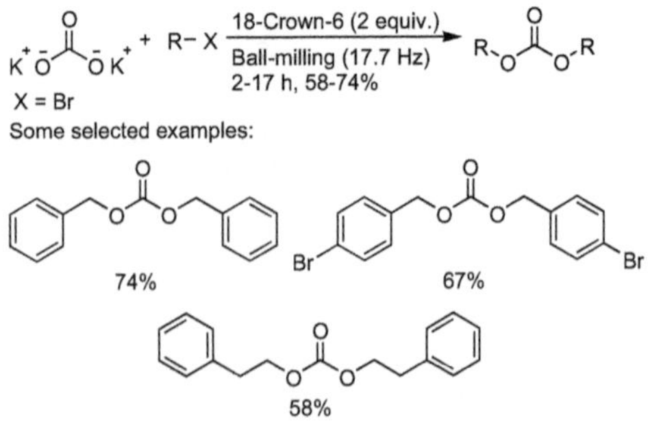

Scheme 1.19 Reaction between 4-bromobenzyl bromide and alkali salts with or without 18-crown-6 by ball milling.

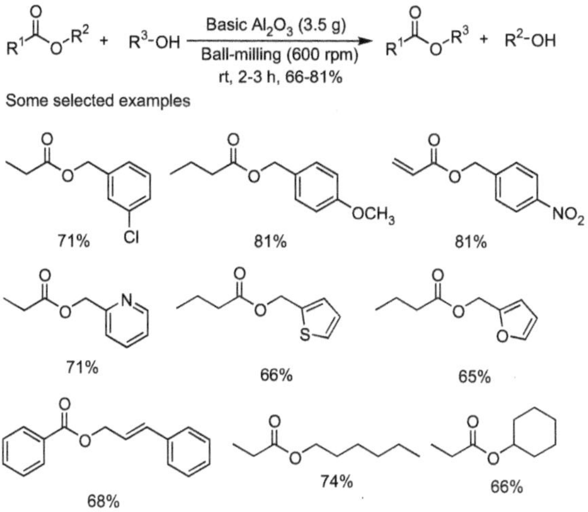

Scheme 1.20 Transesterification under ball milling.

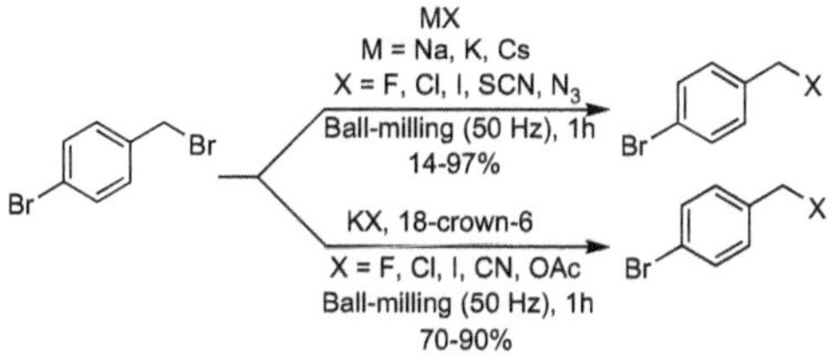

Scheme 1.21 Reaction between 4-bromobenzyl bromide and alkali salts with or without 18-crown-6 by ball milling.

methanols participated in the transesterification reaction without any difficulty.

1.2.3 C–X (X = F, Cl, I, SCN, OAc *etc.*) Bond Forming Reaction

Mack *et al.* have reported the heterogeneous nucleophilic additions of alkali salts of thiocyanate, azide, acetate and halides to 4-bromobenzyl bromide with or without 18-crown-6 by ball milling (Scheme 1.21).[24] It was found that thiocyanate and azide provided excellent yields of the nucleophilic addition products, whereas the potassium and sodium salts of fluoride, chloride, acetate and cyanide provided relatively low yields of the corresponding addition products. On the other hand, 18-crown-6 can complex with K^+ cation to increase the basicity and nucleophilicity of the employed nucleophiles, and thus the addition of 18-crown-6 led to a better conversion and increase in yield for all of the potassium salt nucleophiles including fluoride, acetate and cyanide,[25] which were previously unsuccessful.[24]

1.2.4 C–X (X = S, Se and Te) Bond Forming Reactions

Kaupp and co-workers have reported the quantitative formation of thiouronium salt from solid 2-mercaptobenzimidazole and phenacyl bromide under ball milling for 1 h (Scheme 1.22).[7]

Recently, Ranu *et al.* have developed a convenient and efficient procedure for the transition metal-, ligand- and solvent-free synthesis of aryl chalcogenides by employing a C–X (X = S, Se, Te) bond forming reaction between aryl diazonium tetrafluoroborates and diaryl dichalcogenides in the presence of KOH and neutral alumina as grinding auxiliary under ball milling (Schemes 1.23–1.25).[26] A wide variety of diaryl chalcogenides were synthesized within short reaction time (15–30 min) providing moderate to good yields (62–90%) of products.

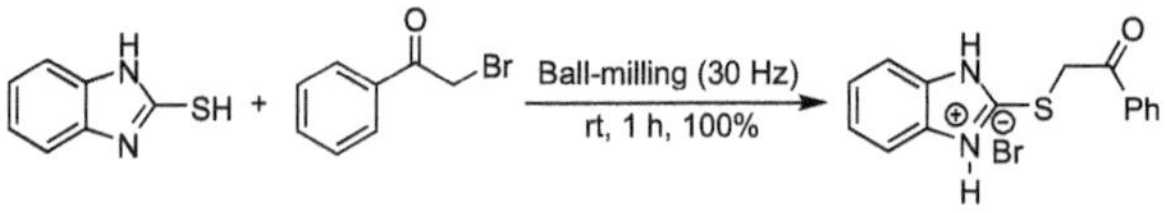

Scheme 1.22 Quantitative synthesis of thiouronium salt from 2-mercaptobenzimidazole and phenacyl bromide under ball milling.

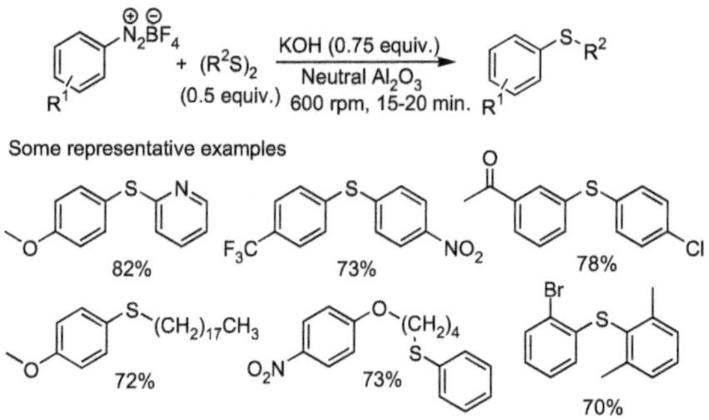

Scheme 1.23 Synthesis of diaryl or aryl-alkyl sulfides under ball milling.

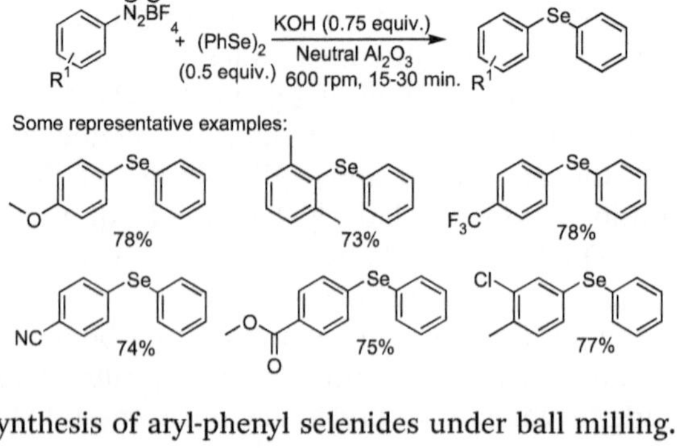

Scheme 1.24 Synthesis of aryl-phenyl selenides under ball milling.

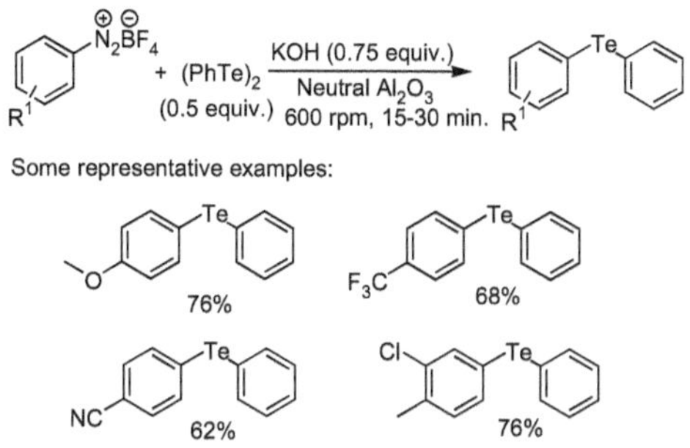

Scheme 1.25 Synthesis of aryl-phenyl tellurides under ball milling.

1.3 Synthesis of Heterocycles

1.3.1 Nitrogen-containing Heterocycles

1,2,3-Triazoles, five-membered nitrogen-containing heterocycles, have received considerable interest because of their useful applications as agrochemicals, dyes, corrosion inhibitors, photostabilizer, materials and in pharmaceutical industries.[27] Stolle and co-workers have reported the first ligand- and solvent-free mechanochemical synthesis of triazole from azides and alkynes using a planetary ball mill (Scheme 1.26).[28]

The applicability of this protocol to a broad variety of substrates enables easy access to a wide range of complex triazoles. Click polymerization in a ball mill without destroying the polymer backbone is also successful (Scheme 1.27).

Ranu *et al.* introduced a different protocol where 1,4-disubstituted-1,2,3-triazoles were synthesized by a simple one-pot three-component reaction of alkyl halide/aryl boronic acid, sodium azide and terminal alkynes on the surface of Cu/Al_2O_3 catalyst by ball-milling under solvent-free conditions (Scheme 1.28).[29] Usually no chromatographic separation/purification was required and no toxic organic solvent is used in the entire process. This protocol avoids handling of hazardous organo-azide and provides easy access to aryl-alkyl and aryl-aryl substituted 1,2,3-triazoles. From the X-band EPR spectrum and XPS study it was found that the copper was present in the +2 oxidation state throughout the reaction cycle. SEM and AFM analysis exhibited the uniform spherical morphology of the catalyst.

Another nitrogen-containing five-membered heterocycle, pyrazoline, can be effectively synthesized from chalcones and phenyl-hydrazines in the presence of $NaHSO_4 \cdot H_2O$ using high speed ball milling (HSBM) (Scheme 1.29).[30]

It was found that better yield of pyrazolines was obtained when the substrates contained electron-donating groups. The easy catalyst recovery and its reusability make this procedure more attractive. When other α,β-unsaturated ketones were treated with thiosemicarbazide/phenyl hydrazine in the presence of this catalyst the corresponding 2-pyrazoline derivatives

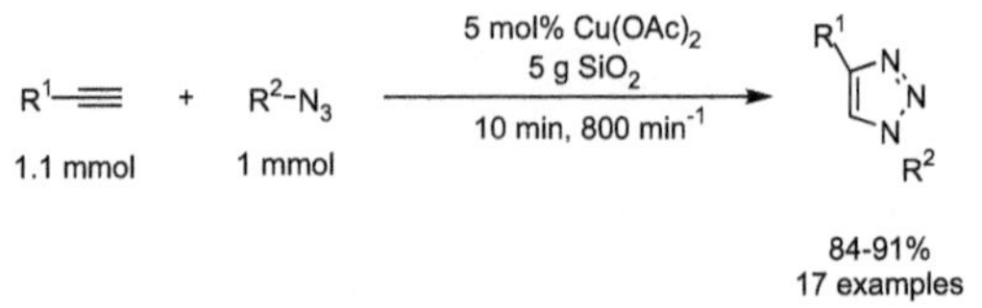

Scheme 1.26 Copper-catalyzed 1,3-dipolar cycloaddition of alkynes and azides forming 1,4-substituted-1*H*-1,2,3-triazoles.

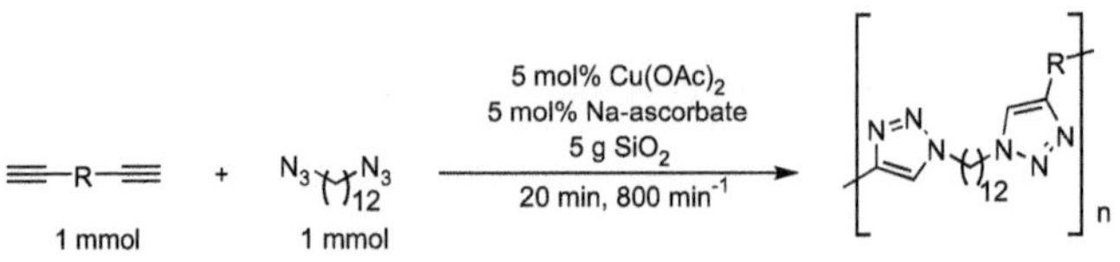

Scheme 1.27 Polymerization of 1,12-diazidododecane and bis-ethynyl compounds in ball mill.

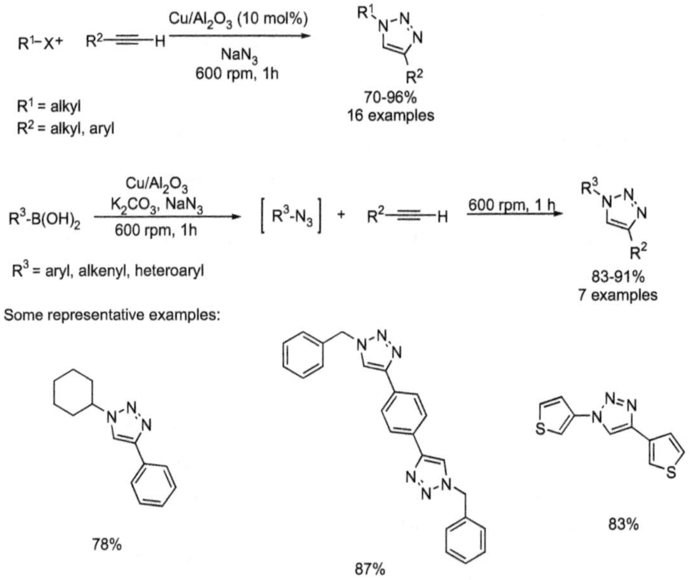

Scheme 1.28 Cu/Al_2O_3 catalyzed one-pot synthesis of 1,2,3-triazoles.

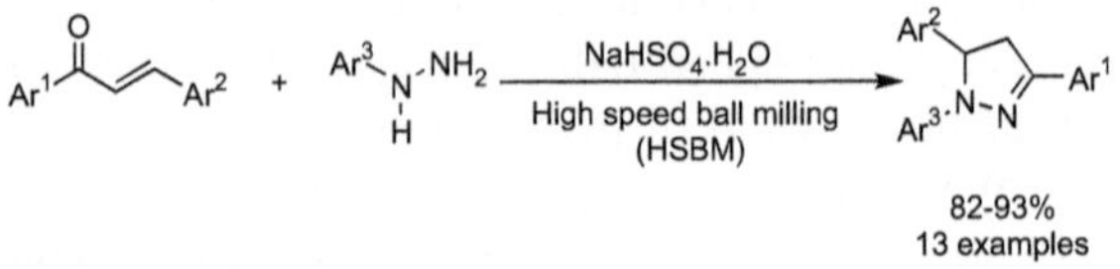

Scheme 1.29 Mechanochemical synthesis of 1,3,5-triaryl-2-pyrazolines catalyzed by $NaHSO_4 \cdot H_2O$.

were obtained in good yields (Scheme 1.30),[30] which demonstrates the general applicability of this protocol.

In 1999, Kaupp and co-workers reported a one-pot synthesis of highly substituted pyrroles (Scheme 1.31), which gives moderate yields in solution, but quantitative yields in solid–solid variant at much lower temparatures.[31] Mechanochemical reaction of primary or secondary enamine esters **4a–d** or the enamine ketone 7 with *trans*-1,2-dibenzoylethene (**5**) under ball milling

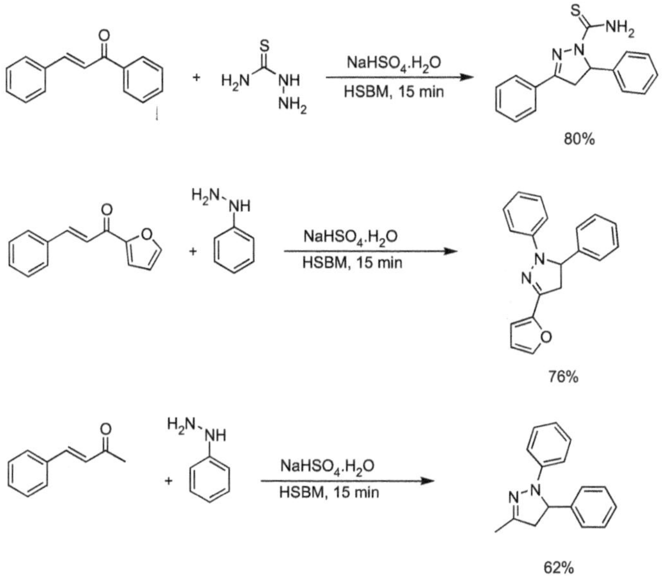

Scheme 1.30 Mechanochemical synthesis of pyrazoline derivatives from α,β-unsaturated ketone and phenylhydrazine/thiosemicarbazide.

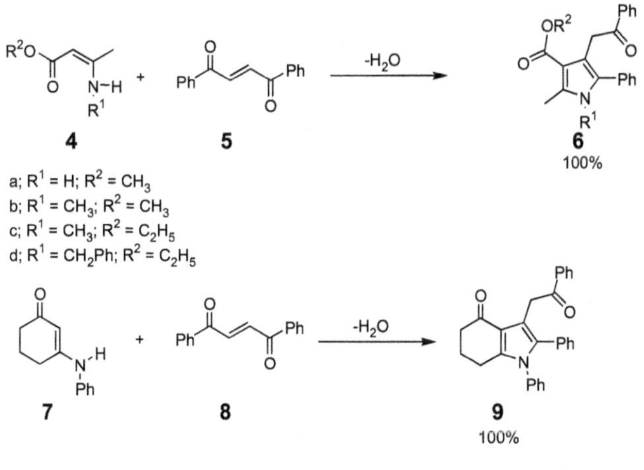

Scheme 1.31 One-pot synthesis of pyrrole derivatives and indoles under ball milling.

afforded pyrroles **6** or indole **9** quantitatively within 3 h despite the multistep course of the reaction.

A plausible reaction pathway is depicted in Scheme 1.32. In the first step, Michael addition followed by the hydrogen transfer produces **11**, which undergoes an imine/enamine rearrangement to give **12**. The generated amino group then interacts with the carbonyl function leading to the formation of a five-membered ring **13**, which provides pyrrole **14** by the elimination of water. An analogous mechanism was postulated for indole starting from **7**. Furthermore, the quantitative formation of thioorotic acid amide **17** from a cascade reaction between the thiohydantoin **15** and methyl amine shows the versatility of this method (Scheme 1.33).

Kaupp and co-workers have reported the quantitative formation of (*R*)-thiazolidine by ball-milling a stoichiometric mixture of L-cysteine and $(HCHO)_n$ for 1 h (Scheme 1.34).[32] It has been observed that a 1 : 1 mixture of phthalic anhydride and 4-toluidine when subjected to ball milling affords the corresponding imide in high yields. Interestingly, cyclic thiourea was obtained quantitatively when bifunctional isothiocyanates and amine were ball milled.

A series of heterocycles can be obtained by condensation of *o*-phenylenediamines with various 1,2-dicarbonyl compounds under ball

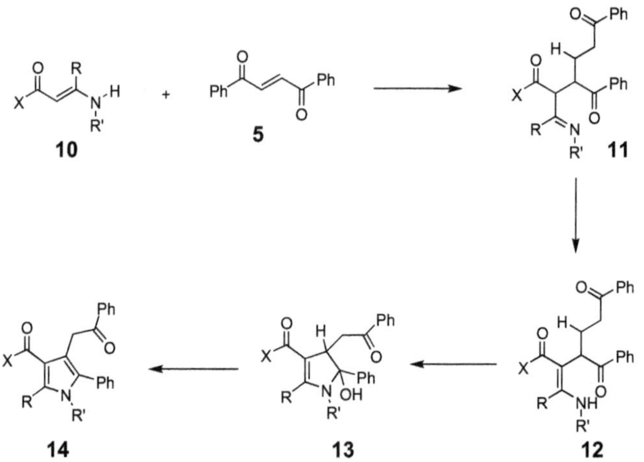

Scheme 1.32 Plausible mechanistic pathway of the cascade reactions to give pyrrole and indole derivatives.

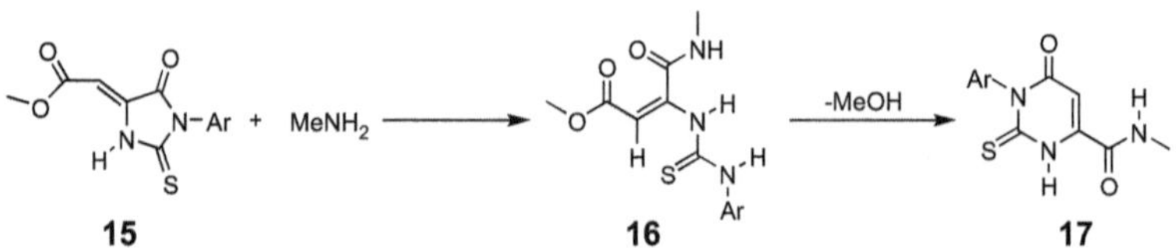

Scheme 1.33 Mechanochemical formation of thioorotic acid amide from a thiohydantoin.

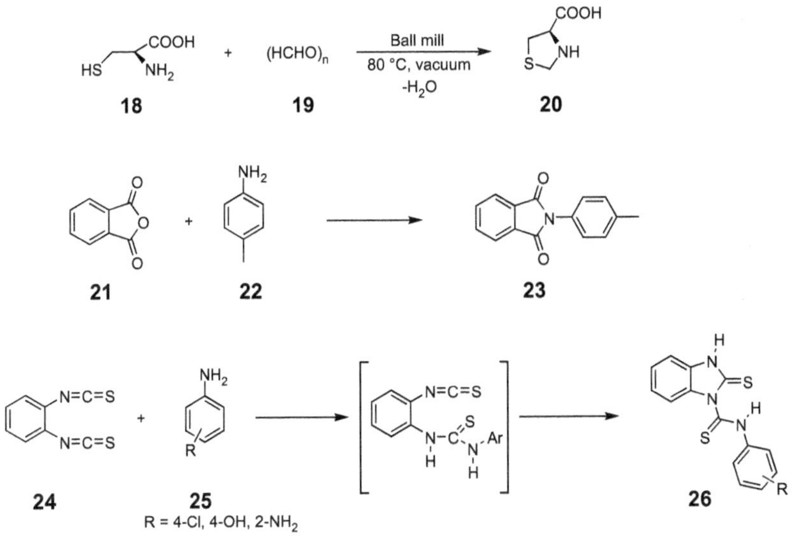

Scheme 1.34 Mechanochemical reactions of amines with carbonyl compounds and isothiocyanates.

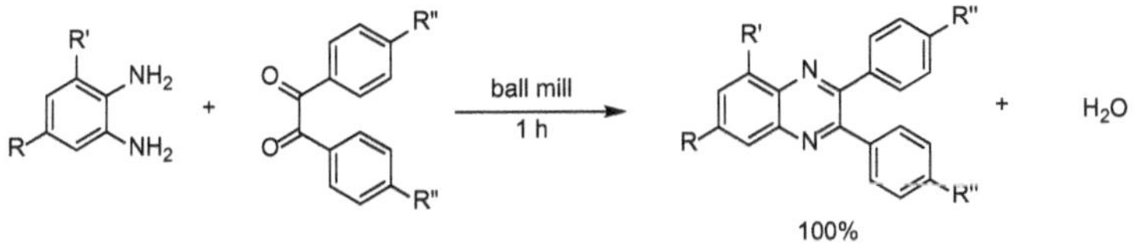

Scheme 1.35 Mechanochemical synthesis of quinoxaline derivatives.

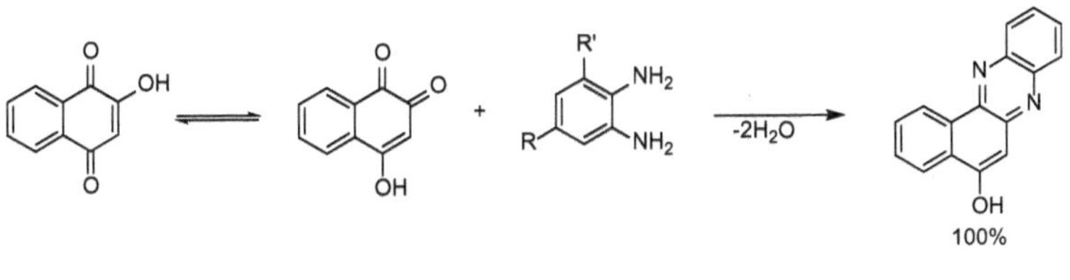

Scheme 1.36 Mechanochemical synthesis of benzo[*a*]phenzin-5-ol.

milling.[33] The condensation of substituted *o*-phenylenediamines and benzil produced the corresponding quinoxaline derivatives within 1 h (Scheme 1.35). Similarly, condensation between 2-hydroxy-1,4-naphthoquinone and *o*-phenylenediamine furnished the benzo[*a*]phenazin-5-ol within 15 min at 70 °C (Scheme 1.36). The yields are almost quantitative in all these reactions.

Kaupp and co-workers also investigated the solid state condensation of *o*-phenylenediamine with oxoglutaric acid, oxalic acid, 1,4-dihydroquinoxaline-2,3-dione and parabanic acid and found that in each case the corresponding products were obtained in high yields (Scheme 1.37).[33]

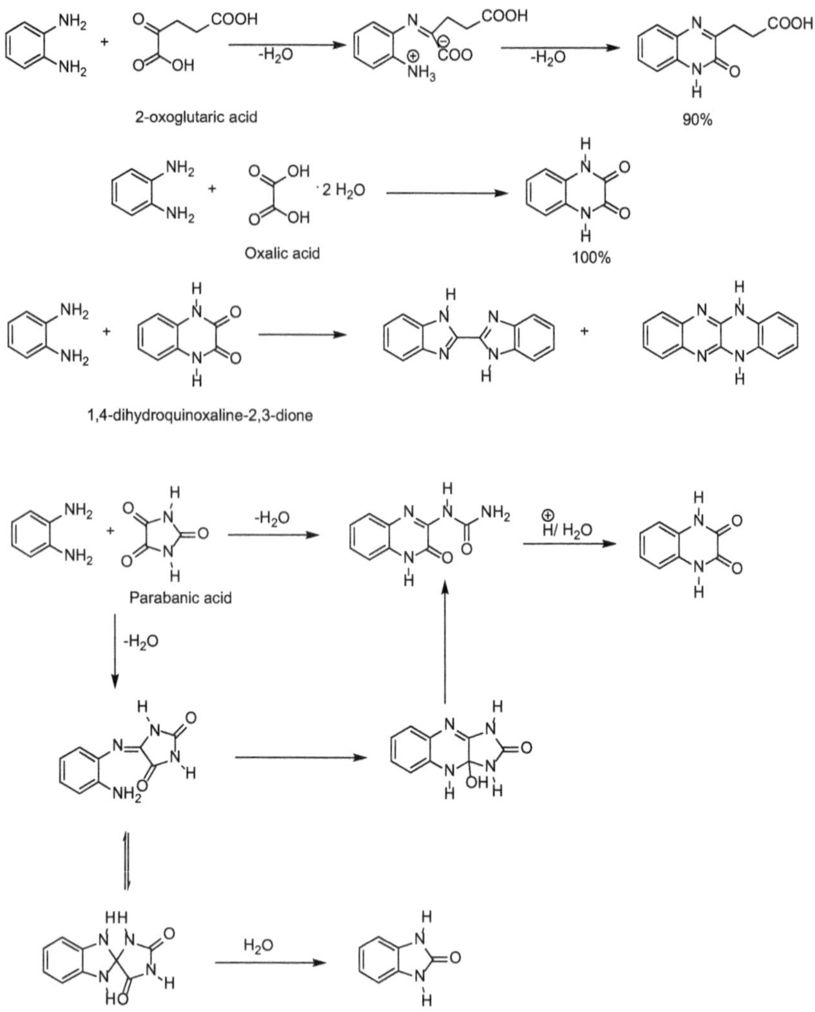

Scheme 1.37 Mechanochemical cascade condensation reaction.

Pyrimidine derivatives were synthesized by the direct condensation of an equimolar amount of an aldehyde, malononitrile, and thiourea/urea under ball milling within 40 min (Scheme 1.38).[34] This protocol offers a wide variety of substituted pyrimidine derivatives in pure form without further purification and in quantitative yields.

1.3.2 Oxygen-containing Heterocycles

In the last decade, much work has been done in the area of oxygen containing heterocycle synthesis under ball milling. It was observed that use of a ball mill not only allowed a net enhancement of the reaction yields but also

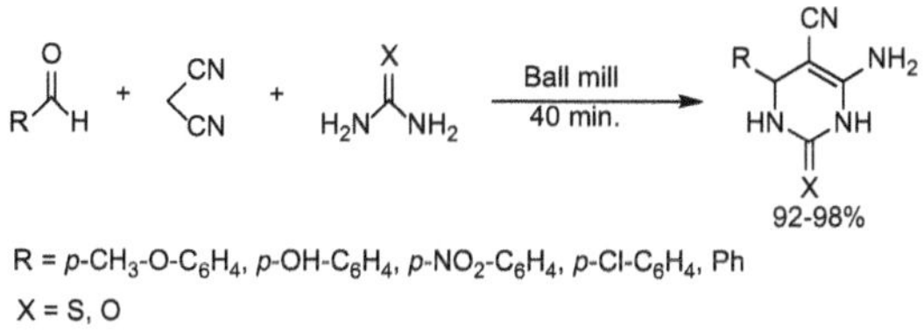

Scheme 1.38 Three component condensation in a ball mill forming substituted 2-thioxo and 2-oxo-1,2,3,4-tetrahydro-pyrimidine-5-carbonitriles.

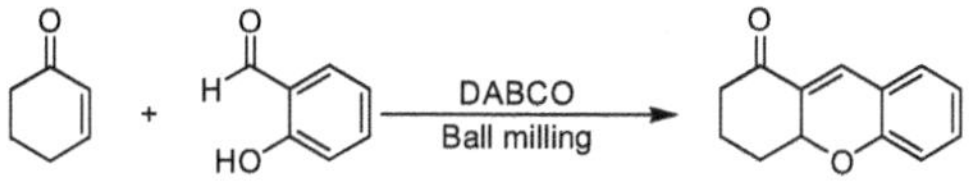

Scheme 1.39 One-pot domino oxa-Michael–aldol reaction to synthesize xanthone derivatives.

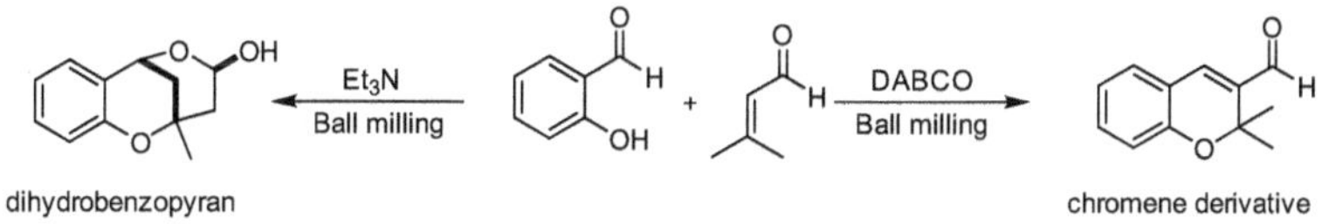

Scheme 1.40 Chemoselective base-catalyzed condensation of salicaldehyde with α,β-unsaturated aldehydes.

in some cases resulted in different chemoselectivities compared to conventional solution-phase synthesis. Xanthone and its isomers are six-membered oxygen-containing heterocycles and can be synthesized in a ball mill by a one-pot domino oxa-Michael–aldol reaction (Scheme 1.39).[35]

Bräse made a systematic study of the ball milling reaction to determine the influence of the ratio of the reactants, the rotational frequency of the mill, the number of balls, the milling time on the reactivity and chemoselectivity. The best yield was obtained using salicaldehyde and cyclohexenone in 1 : 2 molar ratio. It has been found that increasing the reaction time does not increase the yields linearly, probably because the high temperature generated during the milling procedure can be nullified by using water. Concerning the chemoselectivity, it has been observed that changing the base from DABCO to Et_3N increases the yield of dihydrobenzopyran, while use of DABCO gives the chromene derivative as a major product (Scheme 1.40).[35] The above experimental observation clearly established the fact that some parameters intrinsically linked to the ball-milling process largely influence the yield and sometimes chemoselectivity of a reaction.

Another heterocycle that contains a six-membered oxygen ring is naphthopyran, the derivatives of which have attracted considerable attention due to their interesting photochromic properties and biological applications.[36–38] An efficient solid state synthesis of naphthopyran was achieved by the cycloaddition reaction of propargylic alcohols with 2-naphthol in the presence of a Lewis acid catalyst ($InCl_3 \cdot 4H_2O$) under ball milling within 1 h (Scheme 1.41).[39] Other Lewis acids such as $ZnCl_2$ or $SnCl_4$ also catalyzed the cycloaddition reaction but the yields of the corresponding naphthopyrans were poor. Using this methodology several substituted naphthopyrans were synthesized. When propargyl alcohols bearing either a pyrazinyl group or a pyridyl group were used the expected naphthopyrans were not formed. This may happen due to the unwanted coordination of the pyrazinyl or pyridyl ligand to the indium metal.

Although there are several routes for the synthesis of flavone, containing a six-membered oxygen-containing ring, mechanically activated solid-state synthesis of flavones by high-speed ball milling (HSBM) received considerable attention as it avoids the use of solvent and provides waste-free products in almost quantitative yields. The synthesis of flavones was accomplished by cyclodehydration of 1-(2-hydroxyphenyl)-3-aryl-1,3-propanediones in the presence of $KHSO_4$ by HSBM within 15 min (Scheme 1.42).[40]

Kaupp and co-workers have reported the quantitative formation of six-membered oxygen containing heterocycles by Michael addition followed by rearrangement and cyclization under ball milling (Scheme 1.43).[41] Quantitative yields of products were obtained without using any catalyst in a waste free manner.

The two-component straightforward synthesis of 2,3,6,7-tetrahydro-4(5*H*)-benzofuranone derivatives under solvent-free ball-milling was initiated by $Mn(OAc)_3 \cdot 2H_2O$, which acts both as an oxidant and as a Lewis acid (Scheme 1.44).[42] In the standardized conditions oxidative radical reactions of 1,3-dicarbonyl compounds (dimedone, cyclohexane-1,3-dione) with (aza)chalcones catalyzed by $Mn(OAc)_3 \cdot 2H_2O$ afforded the products in moderate to high yields within 1 h without the need of any external base. This protocol is found to be widely applicable for a variety of chalcones. With the isolation of a key intermediate **30** (Scheme 1.45) after running the

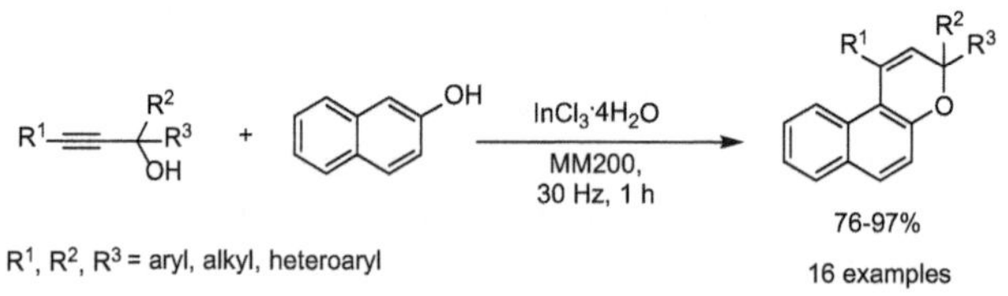

Scheme 1.41 Cycloaddition reaction for formation of naphthopyran derivatives under ball milling.

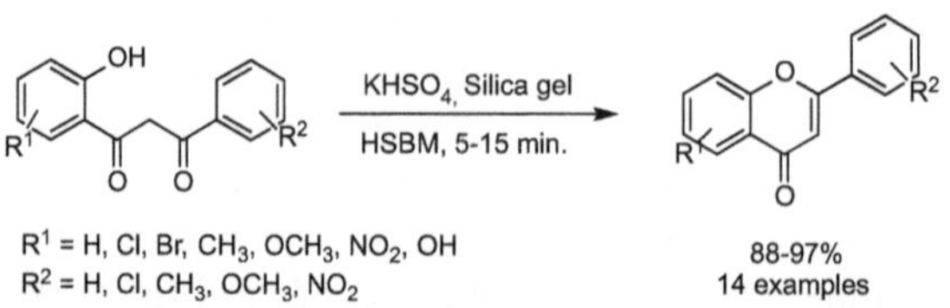

Scheme 1.42 Synthesis of flavones catalyzed by $KHSO_4$ under HSBM.

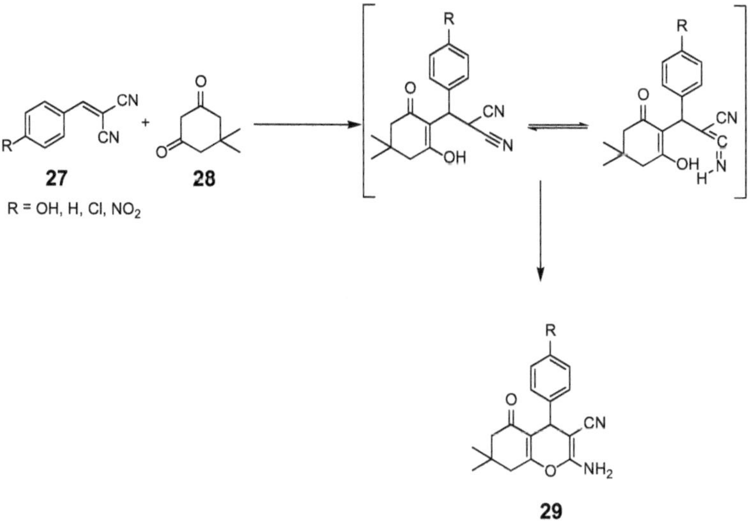

Scheme 1.43 Michael addition reaction followed by rearrangement and cyclization to give **29**.

Scheme 1.44 Manganese(III) catalyzed synthesis of 2,3,6,7-tetrahydro-4(5*H*)-benzofuranone derivatives under ball milling.

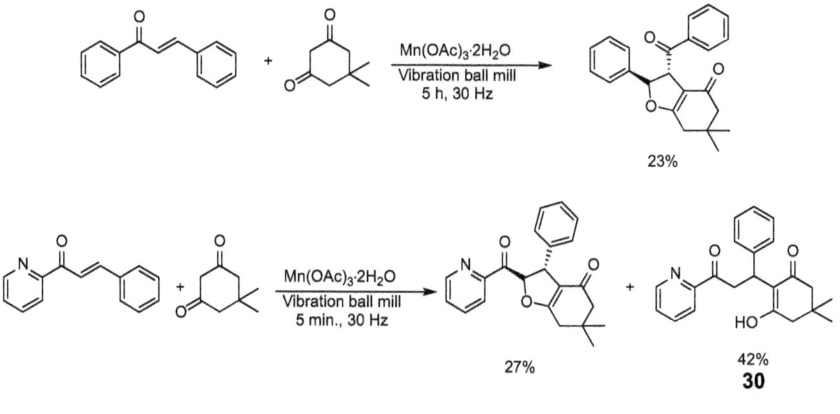

Scheme 1.45 Manganese(III) mediated oxidative radical reaction of 5,5-dimethyl-1,3-cyclohexanedione with (*E*)-1,3-diphenylprop-2-en-1-one and 1-(pyridin-2-yl)-enones.

reaction for 5 min, the reverse regioselectivity of the product in the case of 1-(pyridin-2-yl)-enones was established in comparison with the radical reaction of 1,3-dicarbonyl compounds and α,β-unsaturated compounds. The intermediate **30** is a Michael addition product, which was gradually converted into 2,3,6,7-tetrahydro-4(5*H*)-benzofuranone derivatives with prolonged reaction time. The reaction was initiated by the chelation of carbonyl and pyridinyl moieties of 1-(pyridin-2-yl)-enones with Mn(III), followed by 1,4 addition of 1,3-dicarbonyl compounds (Scheme 1.46). The enolization of intermediate **33** facilitates formation of radical **35** with loss of Mn(II) species. Oxidation of radical **35** affords the carbocation **36**, cyclization of which generates the final product.

1.3.3 Nitrogen–Oxygen-containing Heterocycles

Oxazolidinone, a five-membered heterocycle containing nitrogen and oxygen, is an important class of compounds that has applications as chiral auxiliaries, metal ligands, and pharmaceuticals (specifically as antibiotics).[43,44] A simple way to synthesize the oxazolidinone ring is by the reaction of an unactivated 2-alkyl or 2-aryl aziridine with carbon dioxide in the absence of any catalyst or solvent under high speed ball milling (Scheme 1.47).[45] When the reaction of 2-aryl substituted aziridine with CO_2

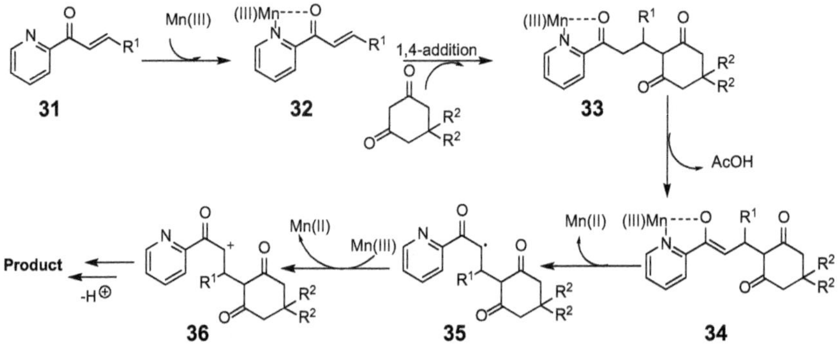

Scheme 1.46 Plausible reaction mechanism for the formation of 2,3,6,7-tetrahydro-4(5*H*)-benzofuranone derivatives.

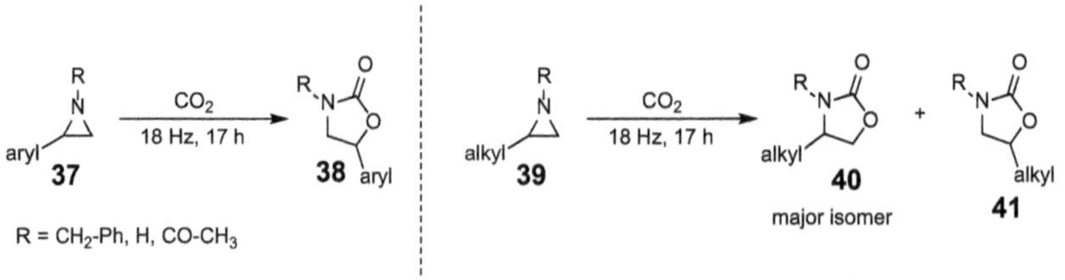

Scheme 1.47 Mechanical synthesis of oxazolidinone from aziridine and CO_2.

was performed in a conventional way the corresponding oxazolidinone **38** was obtained. However, when the same reaction was performed with an alkyl-substituted aziridine **39**, it afforded a mixture of oxazolidinones **40** and **41**. Interestingly, the reaction under high speed ball milling produced only one oxazolidinone **40** regioselectively within 17 h.

1.3.4 Boron-containing Heterocycles

Protection–deprotection is an important tool in organic synthesis as in many cases the molecules such as bi-or poly-functional alcohols, acids, amines, and thiols are required to be protected and subsequently deprotected to obtain the target molecule. Both protection and deprotection step should be simple and quantitative to make the synthetic procedure more economical and industrially suitable. Boronic acid reacts in the solid state with suitably substituted amines, amino acids, diols and polyhydroxy compounds like mannitol and inositol under ball milling to form the corresponding five- or six-membered boron containing heterocycles, which provides good protection to the functionalities.

1.3.4.1 Protection of Amines and Acid Functionality

Free amines were protected *via* the formation of six-membered diazaborinine by the reaction with phenyl boronic acid in solid state under ball milling (Scheme 1.48).[46] The formed diazaborinine is stable towards strong acid or strong base and it may be useful for electrophilic aromatic substitutions. The amine in anthranilic acid, with its free carboxylic functional group, was protected by reaction with boronic acid.

1.3.4.2 Protection of Aromatic 1,2-Diol Functionality

Suitably substituted diols are protected with boronic acid by the formation of corresponding five-membered heterocycles. For example, pyrocatechol

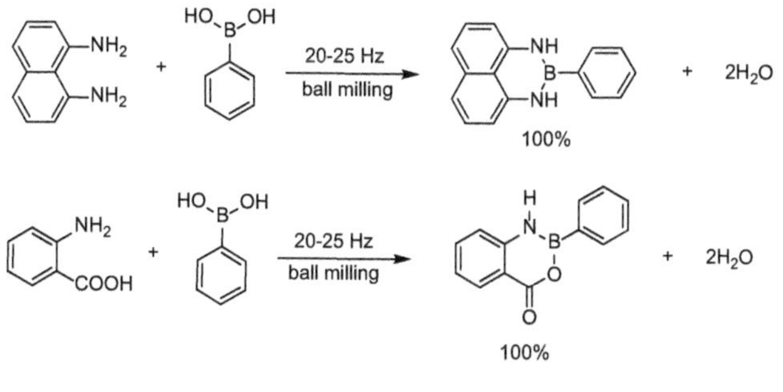

Scheme 1.48 Quantitative solid state formation of six-membered boron containing heterocycles.

and pyrogallol react with phenyl boronic acid when stoichiometric mixtures are ball milled at 80 °C (Scheme 1.49).[46] Deprotection also follows the same procedure by treatment with aqueous $NaHCO_3$ at 40 °C for 2 h.

1.3.4.3 Protection of Aliphatic 1,2-Diols

The cyclization reaction of aliphatic 1,2-diols with boronic acid is equally versatile and waste-free. Ball milling of a stoichiometric mixture of aliphatic 1,2-diols (pinacol or 2,2-dimethylpropane-1,3-diol) and boronic acids at room temperature provided the corresponding heterocycles quantitatively (Scheme 1.50).[46,47]

1.3.4.4 Protection of Polyhydroxy Functionality

Quantitative formation of five- and six-membered ring heterocycles in a waste-free manner by sugar alcohols and cyclic polyalcohols with phenyl boronic acid in the solid state eases the protection–deprotection step in carbohydrate chemistry.[46] For example, D-mannitol reacts with three molecules of phenyl boronic acid in a ball mill to give the corresponding product **42**, the structure of which has been further confirmed by X-ray crystal structure analysis (Scheme 1.51). Similarly, the racemic tris-borolic ester **43** with one five-membered and two six-membered rings was formed by the solid state reaction between cyclic hexol *myo*-inositol with phenyl boronic acid (1 : 3) at 95 °C (Scheme 1.51). Deprotection is performed by dissolving **43** in 0.01(N) HCl for 1 h at room temperature.

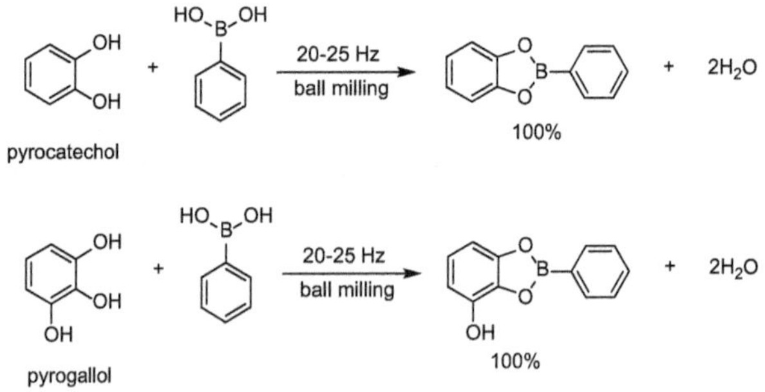

Scheme 1.49 Formation of five-membered boron containing heterocycles.

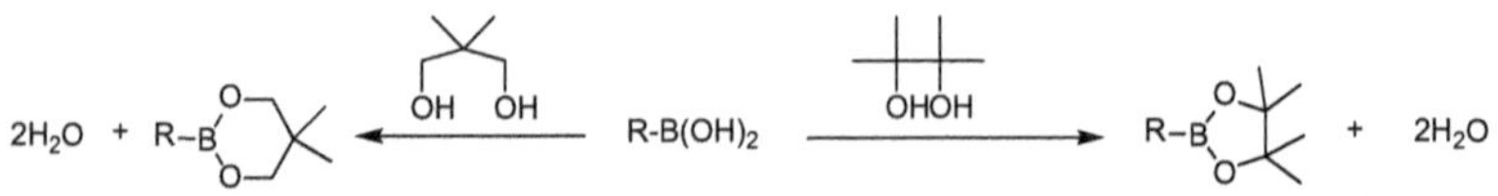

Scheme 1.50 Formation of five- and six-membered boron containing heterocycles.

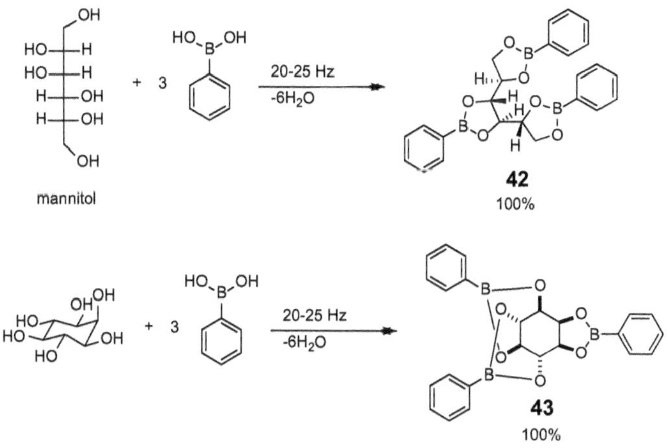

Scheme 1.51 Solid state reaction of polyhydroxy compounds with phenyl boronic acid under ball milling.

1.3.5 Fullerene Chemistry

During the last two decades, significant development has been made in the field of fullerene chemistry to study their chemical functionalizations and practical applications. Carbon allotropes are highly hydrophobic in nature. Thus, chemical reactions with fullerenes in solution are difficult to perform. Ball milling not only gives the opportunity to perform the reaction in the solid state but also provides an alternative where reactions furnish unexpected results that are not observed when the same reaction was performed in solution. Here comprehensive coverage on the oxygenation and cycloaddition reaction of fullerene is provided.

1.3.5.1 Heterocycle Formation by the Oxygenation of C_{60}

C_{60} was reported to be oxidized by high speed vibrational milling (HSVM) under oxygen at 1 atm. (Scheme 1.52).[48] In the formed poly-oxidized fullerene $C_{60}O_n$ containing the C–O–C and C = O bonds, n ranges from 1.4 to 12.4, depending on the milling time, amplitude and ball density. Interestingly $^1O_2$ was generated during the mechanochemical oxidation of C_{60} and it played a crucial role in this process. High speed ball milling of the poly-oxidized $C_{60}O$ under an Ar atmosphere afforded insoluble polymeric species. HPLC analysis revealed the co-existence of higher epoxides, $C_{60}O_n$ ($2 \leq n \leq 5$) and bare fullerene, as minor products possibly produced by the reverse reaction.

1.3.5.2 Heterocycle Formation by the Cycloaddition Reaction of C_{60}

Triazoline derivatives were obtained by the [2 + 3] cycloaddition of organic azide and C_{60} at high speed ball milling within 30 min (62–76%).[49] If the

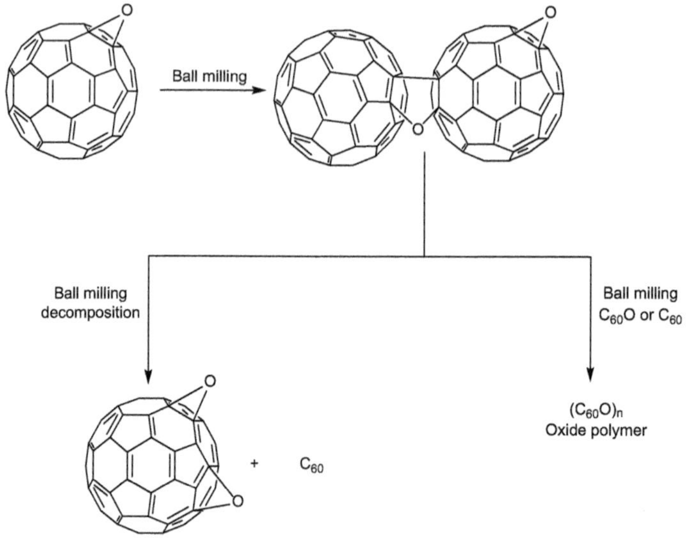

Scheme 1.52 Mechanochemical oxygenation of fullerene.

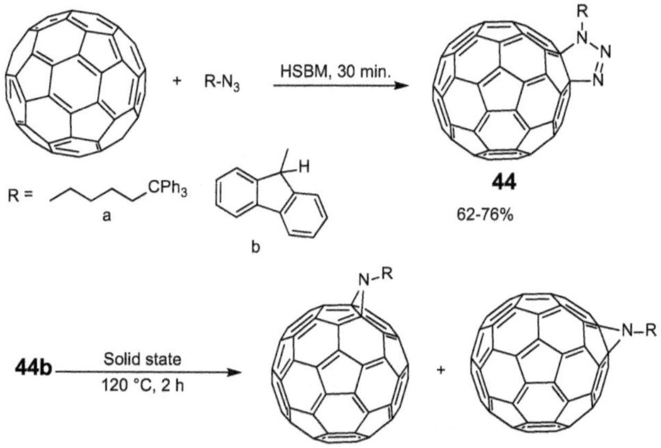

Scheme 1.53 Mechanochemical [2 + 3] cycloaddition reaction of fullerene and organic azide.

triazoline derivative was heated at 120 °C for 2 h in the solid state, the corresponding 5,6-open and 6,6-closed aza fullerenes were obtained (Scheme 1.53).

Mechanochemical [2 + 3] cycloaddition reaction of fullerene with diazo compound gives the corresponding C_{60}-fused-2-pyrazoline **46** within 30 min. 2-Pyrazoline **46** was obtained *via* isomerization of 1-pyrazoline **45**, which was formed directly by the 1,3-dipolar cycloaddition of C_{60} with ethyl diazoacetate generated in *situ* from the glycine ester hydrochloride and $NaNO_2$ (Scheme 1.54).[50]

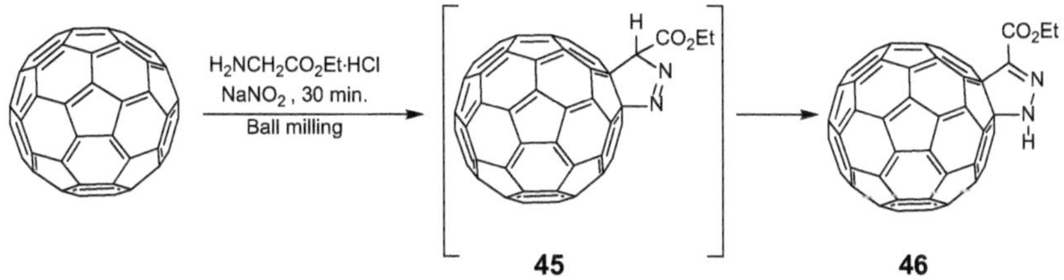

45 46

Scheme 1.54 Mechanochemical [2 + 3] cycloaddition reaction of fullerene with *in situ* generated diazo compound.

1.3.6 Macrocycle Formation

The use of mechanochemistry in the formation of metallomacrocycles and cages is not very common in structural supramolecular chemistry. However, ball milling provides a solvent-free route for the synthesis of macrocycles with better yield than any classical solution based method.

To synthesize complex molecular architectures based on boronic acids, Severin and co-workers reported a one-pot condensation of 4-formylphenylboronic acid **47**, pentaerythritol (**48**) and tris(2-aminoethyl)amine (tren) to afford a macrobicyclic structure.[51] Solid state structure determination of this macrobicyclic compound revealed a "collapsed" geometry with no accessible space inside the cage. To make larger and more expanded cage structures, the same condensation reaction was performed by exchanging the flexible tren building block for more rigid 1,3,5-tris(aminomethyl)-2,4,6-triethylbenzene (**49**). Mechanochemical condensation reaction between the boronic acid **47**, tetraol **48** and triamine **49** produced cage **50** quantitatively within 1 h at milling frequency 20 Hz. Furthermore, if a more elongated boronic acid, *i.e.* 4-(4-formylphenyl)phenylboronic acid (**51**) was used, it provided the corresponding cage **52** in good yield within 1 h (Scheme 1.55).[52] In comparison with solution phase, the yields of these reactions were always low (<40%) and the products also were always contaminated with significant amounts of an incomplete condensation product. The overall geometry and characterization of cages **50** and **52** were obtained from multinuclear NMR spectroscopy, mass spectrometry and single-crystal X-ray diffraction data.

In continuation with macrocycle preparation in a ball mill, a multicomponent polycondensation reaction of 4-formylbenzene boronic acid (**53**), t-$Bu_2Si(OH)_2$ (**54**) and diamines (4,4′-bis(aminomethyl)biphenyl (**55**) and (1*R*,2*R*)-1,2-diaminocyclohexane (**56**)) was performed under ball milling to produce the corresponding borasiloxane-based macrocycles **57** and **58** with an isolated yields of 85% and 65% respectively (Scheme 1.56).[53] Here, first the condensation of silane diols with boronic acids gave the cyclic borosiloxanes, and subsequent combination with the diamines produced the corresponding borosiloxane-based macrocycles.

Another class of macrocyclic compound, rotaxanes, has shown potential applicability as molecular actuators and switches within mesoscale molecular electronic devices. In recent times Chiu and co-workers have

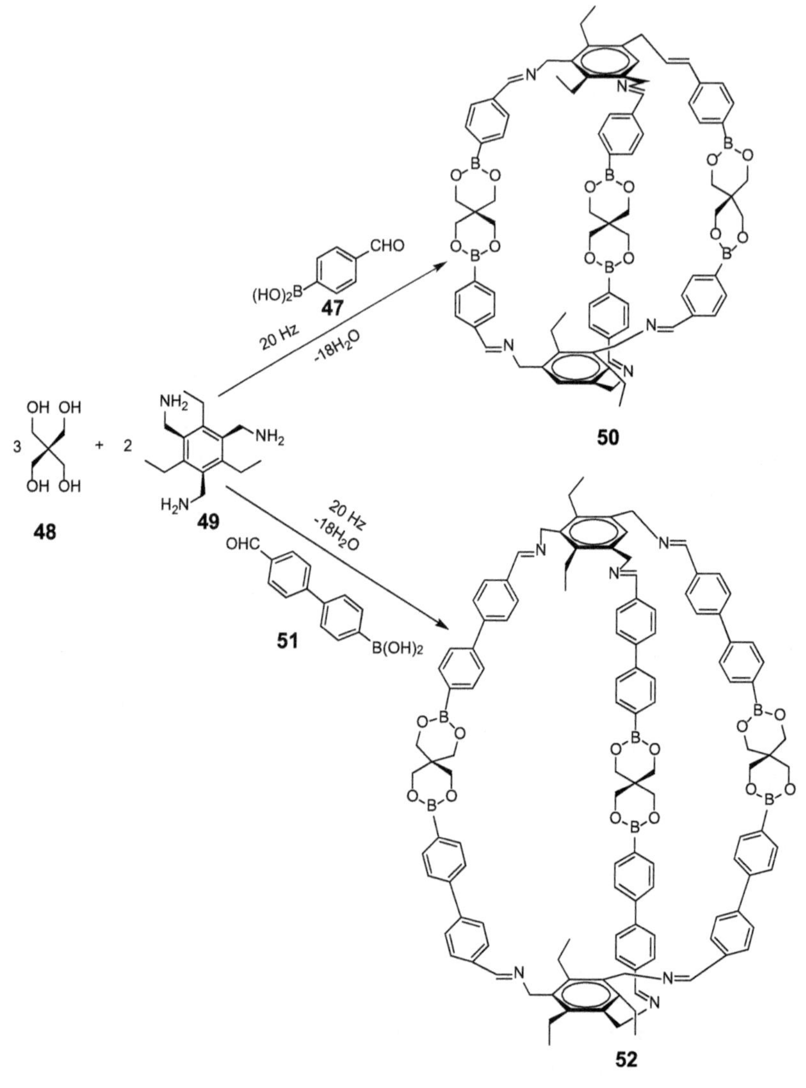

Scheme 1.55 One-pot multicomponent reaction for the synthesis of the cages **50** and **52** under ball milling.

developed several methods where rotaxanes can be synthesized by ball-milling in solid state with high yields. The solid state condensation of 1,8-diaminonaphthalene and benzaldehyde under ball milling provides an efficient, waste-free procedure for synthesizing interlocked molecules such as [2]- and [4]-rotaxanes due to the steric bulk and stability of the resulting dihydropyrimidine stopper units (Scheme 1.57).[54]

Furthermore, with rotaxane preparation, the smallest rotaxane was achieved in a ball mill by utilizing the Diels–Alder reaction of 1,2,4,5-tetrazine with the terminal alkyne units of a 21-crown-7 (21C7)-based

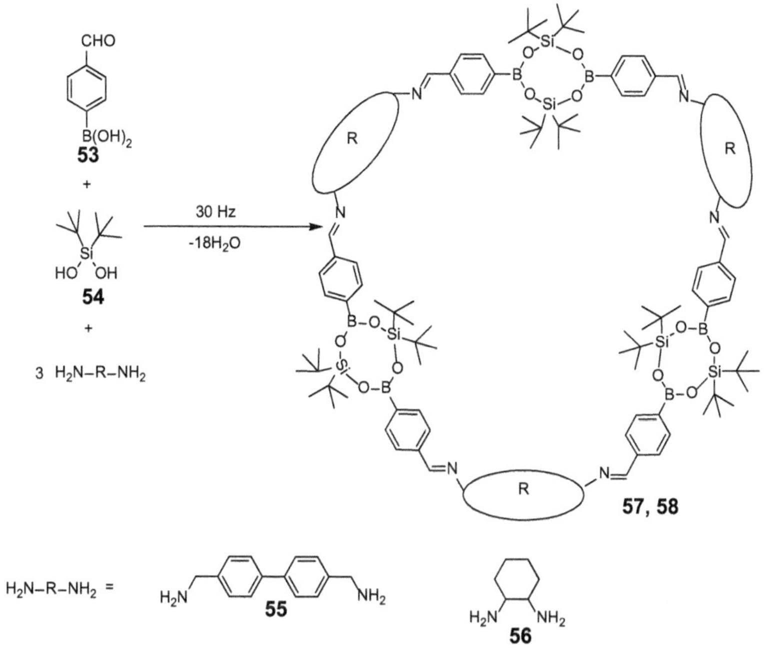

Scheme 1.56 Mechanochemical one-pot multicomponent polycondensation reaction for the synthesis of borasiloxane-based macrocycles.

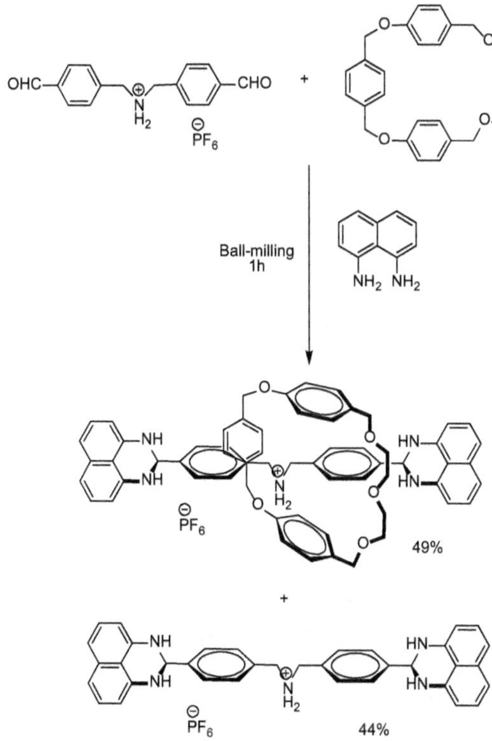

Scheme 1.57 Solid-state mechanochemical synthesis of [2]- and [4]-rotaxanes.

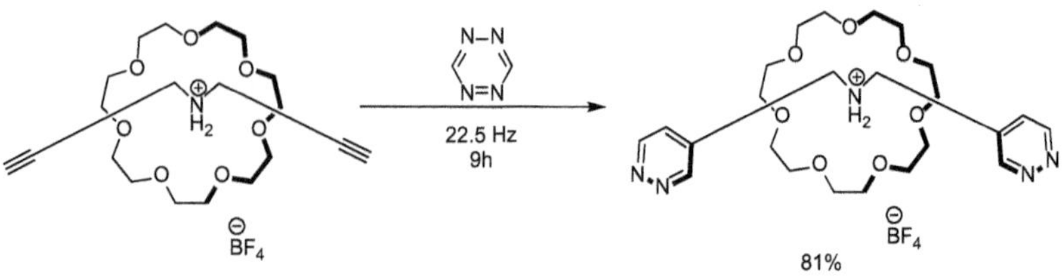

Scheme 1.58 Mechanochemical synthesis of smallest [2] rotaxane by utilizing the Diels–Alder reaction.

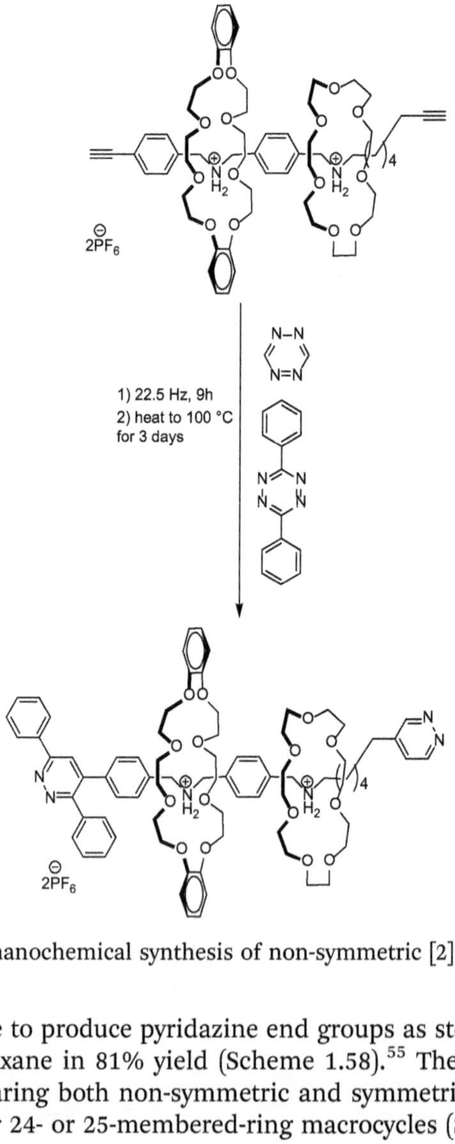

Scheme 1.59 Mechanochemical synthesis of non-symmetric [2] rotaxane.

[2]pseudorotaxane to produce pyridazine end groups as stoppers in a 21C7-containing [2]rotaxane in 81% yield (Scheme 1.58).[55] The same strategy is effective for preparing both non-symmetric and symmetric [2]rotaxanes incorporating either 24- or 25-membered-ring macrocycles (Scheme 1.59).[56]

1.4 Conclusion

In the present chapter, the efficacy of a ball mill for carbon–heteroatom bond formation and synthesis of heterocycles has been highlighted. In general, the ball milling mediated reactions possess several green aspects including solvent-free reaction, mild conditions, purification without chromatography and high yields. Certainly, ball milling can be used as an important tool in the synthesis of useful compounds and natural products, as an attractive alternative to conventional energy sources such as heating, microwave irradiation and sonication.

References

1. A. Stolle, T. Szuppa, S. E. S. Leonhardt and B. Ondruschka, *Chem. Soc. Rev.*, 2011, **40**, 2317.
2. (a) J. Schmeyers, F. Toda, J. Boy and G. Kaupp, *J. Chem. Soc., Perkin Trans. 2*, 1998, 989; (b) J. Schmeyers, F. Toda, J. Boy and G. Kaupp, *J. Chem. Soc., Perkin Trans. 2*, 2001, 132.
3. G. Kaupp, J. Schmeyers and J. Boy, *Chemosphere*, 2001, **43**, 55.
4. G. Kaupp and S. Schmeyer, *J. Phys. Org. Chem.*, 2000, **13**, 388.
5. J. Mokhtari, M. R. Naimi-Jamal, H. Hamzeali, M. G. Dekamin and G. Kaupp, *ChemSusChem*, 2009, **2**, 248.
6. R. L. Shriner, C. K. F. Hermann, T. C. Morrill, D. Y. Curtin and R. C. Fuson, *The Systematic Identification of Organic Compounds*, 8th edn., John Wiley & Sons, Inc., Hoboken, 2003, p. 656.
7. G. Kaupp, J. Schmeyers and J. Boy, *J. Prakt. Chem.*, 2000, **342**, 269.
8. (a) T. Friscic, A. V. Trask, W. Jones and W. D. S. Motherwell, *Angew. Chem., Int. Ed.*, 2006, **45**, 7546; (b) D. Braga and F. Grepioni, *Angew. Chem., Int. Ed.*, 2004, **43**, 4002.
9. J. Gao and G. W. Wang, *J. Org. Chem.*, 2008, **73**, 2955.
10. (a) E. Oliveros, M. Riviere, J. P. Malrieu and C. Teichteil, *J. Am. Chem. Soc.*, 1979, **101**, 318; (b) A. Lattes, E. Oliveros, M. Riviere, C. Belzecki, D. Mostowicz, W. Abramskj, C. Piccinni-Leopardi, G. Germain and M. V. Meerssche, *J. Am. Chem. Soc.*, 1982, **104**, 3929; (c) J. Aube, *Chem. Soc. Rev.*, 1997, **26**, 269; (d) S. Wenglowsky and L. S. Hegedus, *J. Am. Chem. Soc.*, 1998, **120**, 12468; (e) C. H. Leung, A. M. Voutchkova, R. H. Crabtree, D. Balcells and O. Eisenstein, *Green Chem.*, 2007, **9**, 976.
11. V. Strukil, D. Margetic, M. D. Igrc, M. Eckert-Maksic and T. Friscic, *Chem. Commun.*, 2012, **48**, 9705.
12. (a) S. J. Brooks, P. R. Edwards, P. A. Gale and M. E. Light, *New J. Chem.*, 2006, **30**, 65; (b) P. A. Gale, *Acc. Chem. Res.*, 2011, **44**, 216.
13. F. Ravalico, S. L. James and J. S. Vyle, *Green Chem.*, 2011, **13**, 1778.
14. L. Y. Jin, C. Y. Wen, X. F. Shuang, F. W. Ming, Y. W. Bin, J. J. Hong and G. J. Rong, *Sci. Chin. Chem.*, 2012, **55**, 1252.
15. W. Fan, J. Gao, J. Jia, L. Han, W. Sheng and Y. Li, *Chin. J. Org. Chem.*, 2010, **30**, 1732.

16. R. Thorwirth and A. Stolle, *Synlett*, 2011, 2200.
17. T. Szuppa, A. Stolle, B. Ondruschka and W. Hopfe, *ChemSusChem*, 2010, **3**, 1181.
18. E. Colacino, P. Nun, F. M. Colacino, J. Martinez and F. Lamaty, *Tetrahedron*, 2008, **64**, 5569.
19. D. C. Waddell, I. Thiel, A. Bunger, D. Nkata, A. Maloney, T. Clark, B. Smith and J. Mack, *Green Chem.*, 2011, **13**, 3156.
20. E. Abdel-Latif and M. A. Metwally, *Monatsh. Chem.*, 2007, **138**, 771.
21. D. C. Waddell, I. Thiel, A. Bunger, D. Nkata, A. Maloney, T. Clark, B. Smith and J. Mack, *Green Chem.*, 2011, **13**, 3156.
22. D. C. Waddell, I. Thiel, T. D. Clark, S. T. Marcum and J. Mack, *Green Chem.*, 2010, **12**, 209.
23. T. Chatterjee, D. Saha and B. C. Ranu, *Tetrahedron Lett.*, 2012, **53**, 4142.
24. P. Vogel, S. Figueira, S. Muthukrishnan and J. Mack, *Tetrahedron Lett.*, 2009, **50**, 55.
25. Y.-W. Dong, G.-W. Wang and L. Wang, *Tetrahedron*, 2008, **64**, 10148.
26. N. Mukherjee, T. Chatterjee and B. C. Ranu, *J. Org. Chem.*, 2013, **78**, 11110.
27. (a) M. Whiting, J. Muldoon, Y.-C. Lin, S. M. Silverman, W. Lindstrom, A. J. Olson, H. C. Kolb, M. G. Finn, K. B. Sharpless, J. H. Elder and V. V. Fokin, *Angew. Chem., Int. Ed.*, 2006, **45**, 1435; (b) H. C. Kolb and K. B. Sharpless, *Drug Discov. Today*, 2003, **8**, 1128; (c) M. J. Giffin, H. Heaslet, A. Brik, Y. C. Lin, G. Cauvi, C.-H. Wong, D. E. McRee, J. H. Elder, C. D. Stout and B. E. Torbett, *J. Med. Chem.*, 2008, **51**, 6263; (d) W. Q. Fann and A. R. Katrizky, in *Comprehensive Heterocyclic Chemistry II*, ed. A. R. Katrizky, C. W. Rees and E. F. V. Scriven, Elsevier Science, Oxford, 1996, vol. 4, p. 1.
28. R. Thorwirth, A. Stolle, B. Ondruschka, A. Wild and U. S. Schubert, *Chem. Commun.*, 2011, **47**, 4370.
29. N. Mukherjee, S. Ahammed, S. Bhadra and B. C. Ranu, *Green Chem.*, 2013, **15**, 389.
30. X. Zhu, Z. Li, C. Li, Q. Wu and W. Su, *Green Chem.*, 2009, **11**, 163.
31. G. Kaupp, J. Schmeyers, A. Kuse and A. Atfeh, *Angew. Chem., Int. Ed.*, 1999, **38**, 2896.
32. G. Kaupp, J. Schmeyers and J. Boy, *Tetrahedron*, 2000, **56**, 6899.
33. G. Kaupp and M. R. Naimi-Jamal, *Eur. J. Org. Chem.*, 2002, 1368.
34. M. O. M'hamed and O. K. Alduaij, *Phosphorus, Sulfur Silicon*, 2014, **189**, 235.
35. E. M. C. Gérard, H. Sahin, A. Encinas and S. Bräse, *Synlett*, 2008, 2702.
36. J. N. Moorthy, P. Venkatakrishnan, S. Samanta and D. K. Kumar, *Org. Lett.*, 2007, **9**, 919.
37. M. Frigoli and G. H. Mehl, *Angew. Chem., Int. Ed.*, 2005, **44**, 5048.
38. M. T. Stauffer, D. B. Knowles, C. Brennan, L. Funderburk, F.-T. Lin and S. G. Weber, *Chem. Commun.*, 1997, 287.
39. Y.-W. Dong, G.-W. Wang and L. Wang, *Tetrahedron*, 2008, **64**, 10148.
40. X. Zhu, Z. Li, Q. Shu, C. Zhou and W. Su, *Synth. Commun.*, 2009, **39**, 4199.

41. G. Kaupp, M. R. Naimi-Jamal and J. Schmeyers, *Tetrahedron*, 2003, **59**, 3753.
42. G.-W. Wang, Y.-W. Dong, P. Wu, T.-T. Yuan and Y.-B. Shen, *J. Org. Chem.*, 2008, **73**, 7088.
43. G. Zappia, E. Gacs-Baitz, G. D. Monache, D. Misiti, L. Nevola and B. Bott, *Curr. Org. Synth.*, 2007, 81.
44. X. E. Hu, *Tetrahedron*, 2004, **60**, 2701.
45. C. Phung, R. M. Ulrich, M. Ibrahim, N. T. G. Tighe, D. L. Lieberman and A. R. Pinhas, *Green Chem.*, 2011, **13**, 3224.
46. G. Kaupp, M. R. Naimi-Jamal and V. Stepanenko, *Chem. Eur. J.*, 2003, **9**, 4156.
47. M. Schnürch, M. Holzweber, M. D. Mihovilovic and P. Stanetty, *Green Chem.*, 2007, **9**, 139.
48. H. Watanabe, E. Matsui, Y. Ishiyama and M. Senna, *Tetrahedron Lett.*, 2007, **48**, 8132.
49. K. Komatsu, Y. Murata, G.-W. Wang, T. Tanaka, N. Kato and K. Fujiwara, *Fullerene Sci. Technol.*, 1999, 7, 609.
50. G.-W. Wang, Y.-L. Li, R.-F. Peng, Z.-H. Liang and Y.-C. Liu, *Tetrahedron*, 2004, **60**, 3921.
51. N. Christinat, R. Scopellit and K. Severin, *Angew. Chem., Int. Ed.*, 2008, **47**, 1848.
52. B. Icli, N. Christinat, J. Tönnemann, C. Schüttler, R. Scopelliti and K. Severin, *J. Am. Chem. Soc.*, 2009, **131**, 3154.
53. M. Pascu, A. Ruggi, R. Scopelliti and K. Severin, *Chem. Commun.*, 2013, **49**, 45.
54. S.-Y. Hsueh, K.-W. Cheng, C.-C. Lai and S.-H. Chiu, *Angew. Chem., Int. Ed.*, 2008, **47**, 4436.
55. C.-C. Hsu, N.-C. Chen, C.-C. Lai, Y.-H. Liu, S.-M. Peng and S.-H. Chiu, *Angew. Chem., Int. Ed.*, 2008, **47**, 7475.
56. C.-C. Hsu, C.-C. Lai and S.-H. Chiu, *Tetrahedron*, 2009, **65**, 2824.

CHAPTER 2

Carbon–Carbon Bond Forming by Ball Milling

KATHARINA JACOB, ROBERT SCHMIDT AND ACHIM STOLLE*

Institute for Technical Chemistry and Environmental Chemistry, Friedrich-Schiller University Jena, Lessingstr 12, D-07743 Jena, Germany
*Email: Achim.Stolle@uni-jena.de

2.1 Introduction

The formation of carbon–carbon bonds is an important field in organic chemistry. Essential compounds like diphenylacetylene or biphenyl derivatives can be produced that find application in medicine, drug design, electronics (LCDs (liquid crystal displays), semiconductors) and many more areas.

Mechanochemical techniques are widely common in inorganic comminution and alloying but in recent years they have received attention in organic syntheses, too. The application to organic reactions has been described in several reviews.[1–6] Thereby, the scope of organic reactions performed under mechanochemical conditions is manifold and reaches from metal catalyzed cross-coupling reactions to organocatalytic asymmetric reactions.

Bolm and co-workers have demonstrated for different carbon–carbon bond forming reactions, like Aldol or Baylis–Hillman reactions, the advantages of ball milling compared to microwave- or ultrasound-assisted synthetic routes.[6] Compared to classic syntheses, reactions performed with these techniques often require shorter reaction times and previously unknown molecular transformations have been reported.[7] Furthermore, reactions in ball mills can be performed solvent-free while in the case of

RSC Green Chemistry No. 31
Ball Milling Towards Green Synthesis: Applications, Projects, Challenges
Edited by Brindaban Ranu and Achim Stolle

Published by the Royal Society of Chemistry, www.rsc.org

classic, microwave- and ultrasound-assisted syntheses solvents are often required.

In this chapter various carbon–carbon-bond coupling reactions performed in ball mills will be discussed. In addition to cross-coupling reactions, metal-mediated reactions, condensation reactions, Wittig reaction and homo-coupling reactions are presented.

2.2 Cross-coupling Reactions

Reactions between carbon electrophiles and organometallic nucleophiles are known as cross-coupling reactions. They are catalyzed in almost all cases with palladium compounds. Some of these couplings have been realized under mechanochemical stress and are discussed in this chapter. The listing follows the names given for the typical combination of transition metal catalyst and organometallic nucleophile.

2.2.1 Sonogashira Cross-coupling

The coupling of terminal alkynes and aryl (or alkyl) halides (Br^-, I^-) under an air atmosphere (Scheme 2.1) was realized by Mack and co-workers using a high energy ball mill (SPEX CertiPrep 8000M mixer mill).[8] A minimal catalyst amount of 2.5 mol% [$Pd(PPh_3)_4$] was required to achieve high yields after 17 h of milling. Different ball materials were tested and tungsten carbide balls showed best results. Adding the ligands to less expansive Pd-species ($PdCl_2$, $Pd(OAc)_2$) did not work under milling conditions. A substrate screening showed moderate to high yields (20–95%) for different aryl halides coupled with trimethylsilyl acetylene or phenylacetylene. The absence of CuI led to considerably lower yields in the Sonogashira reaction. Application of copper balls for the reaction also furnished lower amounts of the desired product. When using a copper vessel combined with copper balls, similar results compared to the reaction with CuI (WC as grinding media) were reached.

Investigations in another mill type (planetary ball mill (PBM), Pulverisette 7, Fritsch GmbH) using $Pd(OAc)_2$ and DABCO as catalyst and basic component in the absence of Cu-species led to good conversions (47–99%) and high selectivities (<96%) to the desired products in only 20 min milling time

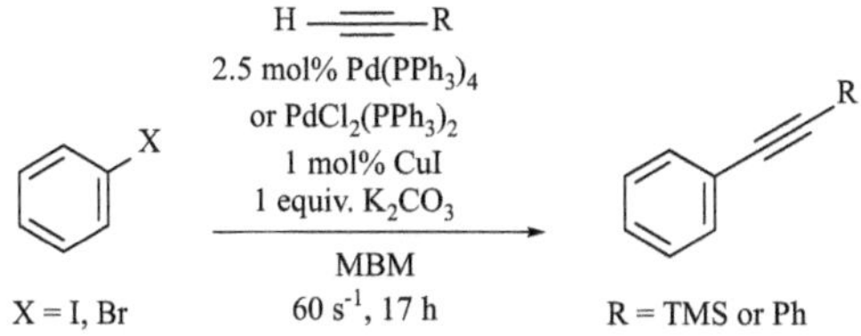

Scheme 2.1 Sonogashira cross-coupling of terminal alkynes with aryl halides in a mixer ball mill (MBM).

(Scheme 2.2).[9] The influence of the base and milling auxiliary was studied in this work. Additionally, the ratio of DABCO and phenylacetylene was varied. A minimum of 1.5 equiv. base was necessary to reach high conversion. Beside substituted phenylacetylene long-chained alkynes were successfully tested in the coupling (45–92% yield). Furthermore, 2-iodothiophene was converted under these conditions and double Sonogashira reactions were also possible.

The coupling of phenylacetylene and 4-iodoanisole accomplished in a planetary ball mill (Pulverisette 7, Fritsch GmbH) led to 57% isolated yield after 20 min milling.[10] As the addition of Cu caused lower selectivity, it was possible to dispense with this co-catalyst. Beside $Pd(OAc)_2$, $PdCl_2$ and $[Pd(PPh_3)_4]$ were tested successfully in this reaction.

2.2.2 Suzuki–Miyaura Cross-coupling

The cross-coupling of 2-bromonaphthalene (Scheme 2.3) gave moderate yields (21–53%) for different bases after 30 min milling in a planetary ball mill (Pulverisette 7, Fritsch GmbH).[11] Best results were obtained in the presence of K_2CO_3 and 2 equiv. of phenylboronic acid. Various aryl halides were tested in this reaction, giving 0–96% analytical yield in 30–50 min under an air atmosphere.

This work was re-considered by Gálvez *et al.* in analyzing the results with chemometric methods, like multi-linear regression or linear discriminant analysis.[12] By means of the regression equations and the Moran autocorrelation indices it was possible to predict the influence of electron-withdrawing or -donating groups and reaction time. The calculations were very close to the experimental results; consequently, time and costs can be saved.

In 2003 investigations of the reaction of 4-bromotoluene and phenylboronic acid revealed that $Pd(OAc)_2$ is a better catalyst than $PdCl_2$ and NEt_3 a significantly better base than Na_2CO_3 (Scheme 2.4).[13] After grinding

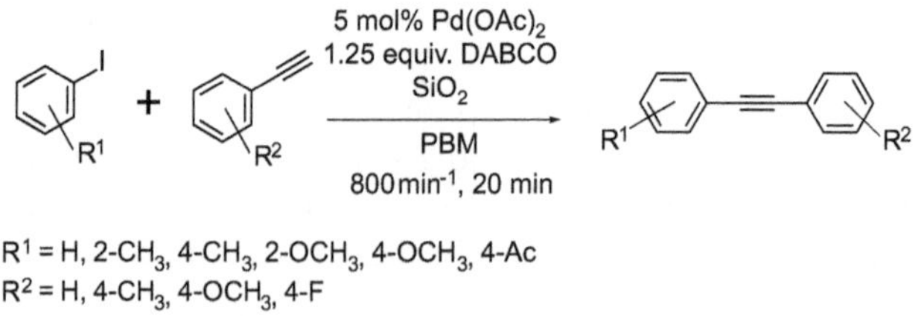

Scheme 2.2 Copper-free Sonogashira cross-coupling with DABCO as base.

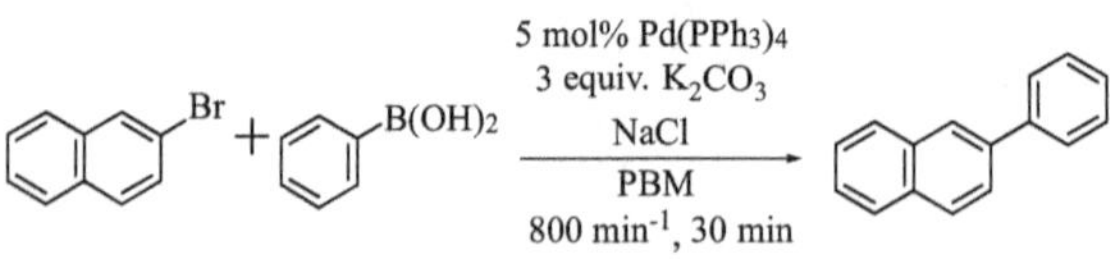

Scheme 2.3 Suzuki cross-coupling of phenylboronic acid with 2-bromonaphthalene.

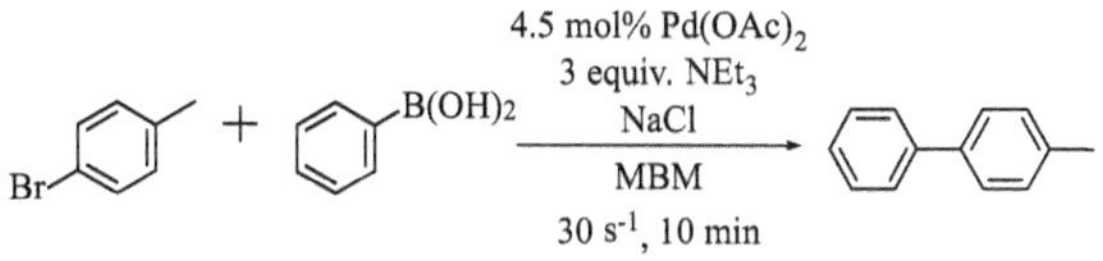

Scheme 2.4 Suzuki cross-coupling catalyzed by $Pd(OAc)_2$ in a mixer ball mill.

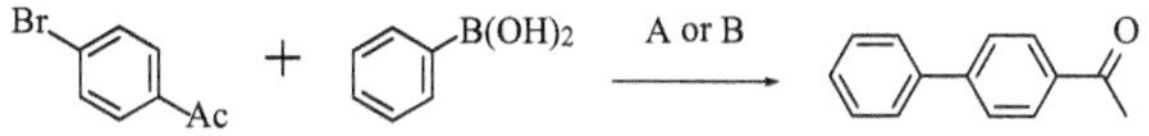

method A: 3.6 mol% $Pd(OAc)_2$, KF/Al_2O_3, PBM, 800 min^{-1}, 10 min
method B: Pd(II)-complex (0.1 mol% Pd), 2 equiv. K_2CO_3 or KOH, MBM, 1.5-7 h

Scheme 2.5 Synthesis of 4-phenylacetophenone under ball milling conditions.

the substrates in a mixer ball mill (MM200, Retsch GmbH) for 10 min the product was provided in 95% yield. Further studies show the versatility of this reaction procedure leading to >90% yield in 6 out of 13 reactions.

Schneider and Ondruschka studied the reaction of phenylboronic acid and 4-bromoacetophenone in the presence of KF/Al_2O_3 in a planetary ball mill (Scheme 2.5, method A, Pulverisette 7, Fritsch GmbH).[14] The catalytic species was $Pd(OAc)_2$. The *in situ* generation of base (KOH) occurred in the presence of small amounts of water. After 10 min milling at 800 min^{-1} 42% and >90% yield could be achieved using a content of 10 or 20–50 wt% KF, respectively. Further studies on the influence of KF loading and alumina modification can be found in the literature.[15] To investigate the scope of the Suzuki cross-coupling a huge number of substances was tested under optimized conditions. For aryl bromides and iodides yields of 18–98% (mostly >50%) could be achieved. Different boron-bearing compounds led to yields ranging from 9% to 98% for the reaction with 4-bromoacetophenone.[14]

In an advanced publication the Suzuki–Miyaura cross-coupling was evaluated with regard to yield, selectivity, batch size and line power consumption.[16] Therefore, two different ball mill types and microwave-assisted techniques were compared using curtate life cycle assessment.

Working with Pd(II)-complexes, embedded in a mesoporous silica framework, good results could be achieved in the Suzuki–Miyaura cross-coupling (Scheme 2.5, method B).[17] The catalyst preparation started with a suspension of $[PdCl_2(PhCN)_2]$ and tetraethoxysilane. Afterwards, pyridine or amine ligands were added. The authors utilized a mixer ball mill (MM200, Retsch GmbH) for the synthesis of 4-phenylacetophenone and, additionally, compared the results with those from stirred reactions with or without solvent. Under solvent-free conditions mostly two solid substances were mixed. In the solid phase a solid compound was mixed with a fluid by stirring to avoid a decrease of yield, because of the lower interaction with the catalyst. As basic compounds, K_2CO_3 or KOH were added. After a milling time of 1.5–7 h, analytical yields of 42–78% were accomplished for the reaction of phenylboronic acid and 4-bromoacetophenone.

Studies have shown that the conversion of poorly reactive aryl chlorides is complicated.[11,13–15] Cravotto *et al.* revealed the possibility for such a conversion.[18] For the coupling of phenylboronic acid with various aryl halides a Pd/chitosan (CS) catalyst in the presence of K_2CO_3 led to yields between 29% and 99% (Scheme 2.6). The use of aryl bromides and iodides led to amounts of 15–35% homo-coupling product, whereas aryl chlorides lead to <1%. It was also possible to convert heteroaryl chlorides in good rates. The formation of biphenyl could be slightly reduced when the coupling was carried out under an argon atmosphere. Reactions were carried out in a planetary ball mill (PM100, Retsch GmbH). In this publication the ability of catalyst recycling was shown. Within four reuses the yield decreased from 84% to 50%, caused by weight loss during work up.

Some of the desired boron compounds used in the Suzuki–Miyaura reaction (4,4,5,5-tetramethyl-2-phenyl-1,3,2-dioxaborolane and 5,5-dimethyl-2-phenyl-(1,3,2)dioxaborinane) were also prepared using the ball milling technique (Pulverisette 7, Fritsch GmbH, 560 min^{-1}, 1 h).[19]

Furthermore, it was possible to generate catalysts used in the Suzuki cross-coupling under ball milling conditions.[20] In a SPEX mill (SPEX CertiPrep 8000D High-Energy Shaker Mill) multi-walled carbon nanotube powder and $Pd(OAc)_2$ were mixed. This method involved no reducing agent in the following microwave-assisted reaction and the activity was increased to a TON of 7250 and a TOF of 217 500 h^{-1} if the catalyst amount was reduced drastically to 0.008 mol% instead of 0.5 mol%. Moreover, this catalyst proved to be very stable and suitable for recycling. It could be used, reaching total conversion, up to nine times. Table 1 provides a summary of different reaction conditions for Suzuki–Miyaura cross-coupling reactions.

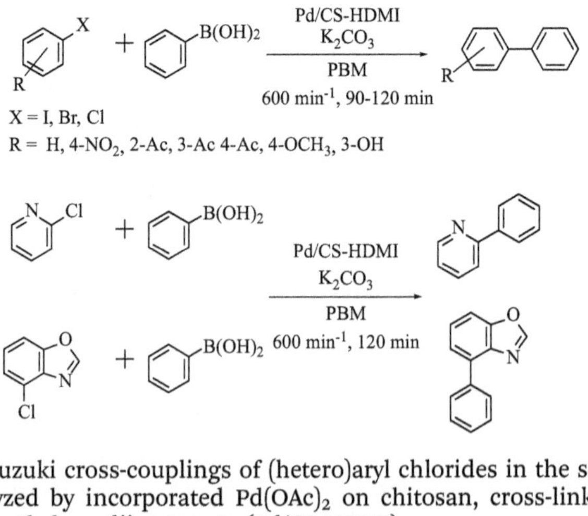

Scheme 2.6 Suzuki cross-couplings of (hetero)aryl chlorides in the solid-state catalyzed by incorporated $Pd(OAc)_2$ on chitosan, cross-linked with hexamethylene diisocyanate (Pd/CS-HDMI).

Table 2.1 Suzuki–Miyaura cross-couplings performed in ball mills.

Catalyst	Additive	Mill type[a]	Frequency	Time	Yield (%)	Reference
$[Pd(PPh_3)_4]$	K_2CO_3, NaCl	PBM	800 min^{-1}	30 min	53	11
$Pd(OAc)_2$	NEt_3, NaCl	MBM	30 s^{-1}	10 min	95	13
$Pd(OAc)_2$	KF/Al_2O_3	PBM	800 min^{-1}	10 min	95	14,15
Pd in MSU-X	K_2CO_3 or KOH	MBM	N.a.	1.5–7 h	42–78	17
Pd/CS-HDMI	K_2CO_3	PBM	600 min^{-1}	1.5–2 h	29–99	18

[a]MBM = mixer ball mill, PBM = planetary ball mill.

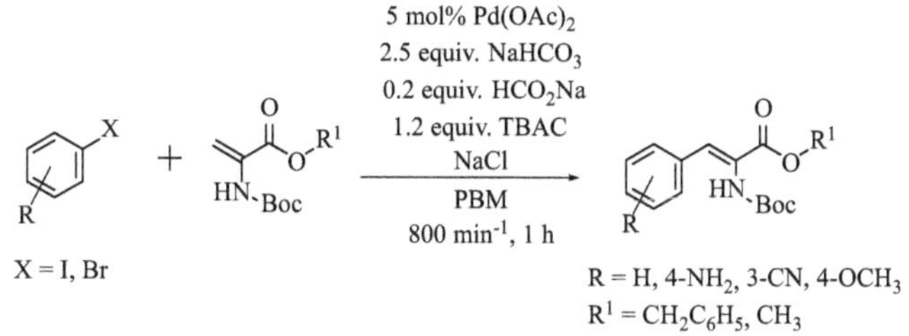

Scheme 2.7 Heck–Jeffery reaction yielding substituted dehydroalanines.

2.2.3 Mizoroki–Heck Reaction

In 2004 Frejd and co-workers published the first synthesis protocol for the Heck–Jeffery reaction under ball milling conditions (Scheme 2.7).[21] The formation of substituted dehydroalanines from aryl halides and alkenes was investigated using a planetary ball mill (Scheme 2.7, Pulverisette 7, Fritsch GmbH). Moderate yields (13–77%) were observed after 60 min milling for different aryl bromides and iodides under an argon atmosphere. For aryl chlorides no reaction was observed. Tetrabutylammonium chloride (TBAC) was important for the reaction and sodium formate acted as a good reductant for the Pd(II)-species.

The above-mentioned reaction was advanced by using 2-*tert*-butoxycarbonylaminoacrylic acid methyl ester as alkene (Scheme 2.8). In this reaction only aryl iodides led to the desired product. The syntheses were successful in the range of 29–85% yield. Applications of iodoanilines with electron-withdrawing groups had a negative effect. The coupling gave intermediates for indole derivatives. They could be generated by heating under reflux in acetic anhydride for 12 h.[22]

Declerck *et al.* investigated the model reaction of phenyl iodide and *tert*-butyl acrylate in a mixer ball mill (MM200, Retsch GmbH).[23] In the presence of a Pd-catalyst ($Pd(OAc)_2$), a reducing agent (HCOONa) and an inorganic base (K_2CO_3), best results (100% analytical yield) were achieved using polymer PEG-2000-OH (Scheme 2.9). In contrast to H_2 or $NaBH_4$, sodium formate was preferred as reducing agent. Compared to other syntheses methods no phosphine ligand was necessary in this reaction protocol. Furthermore various aryl iodides were tested under the best reaction conditions leading to 27–100% yield.

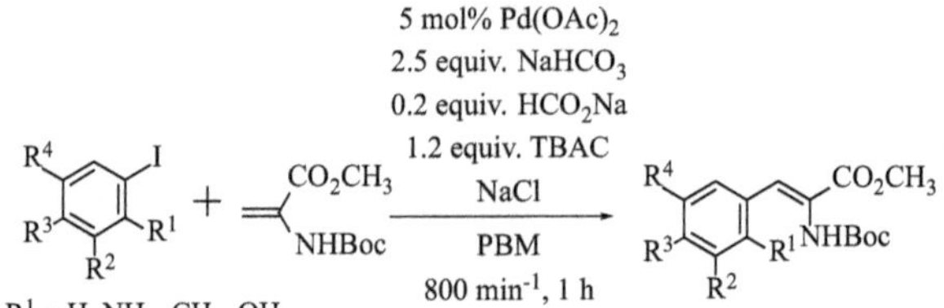

R^1 = H, NH_2, CH_3, OH
R^2 = H, NH_2, CH_3
R^3 = H, NH_2, CF_3, OH
R^4 = H, NH_2, CN, F, CO_2CH_3

Scheme 2.8 Mizoroki–Heck reaction of iodoanilines.

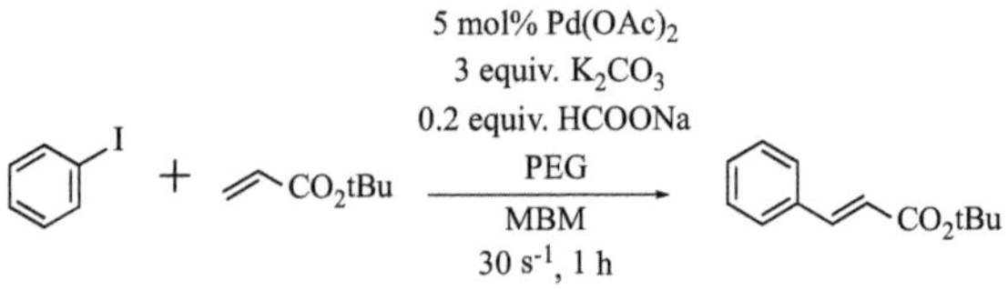

Scheme 2.9 Mizoroki–Heck reaction using ball milling in poly(ethylene glycol) (PEG).

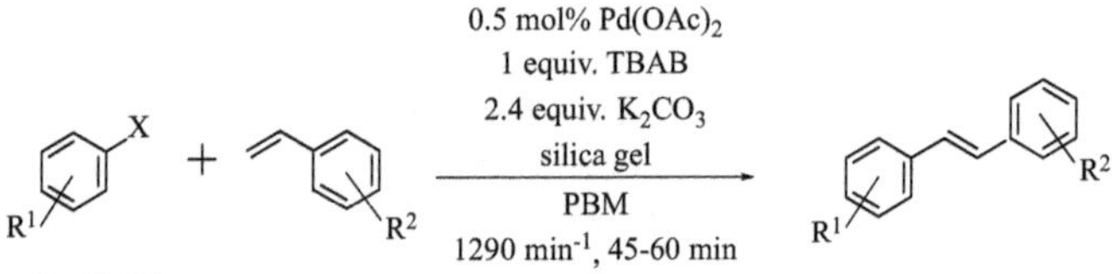

X = Br, Cl
R^1 = H, 4-$CO_2C_2H_5$, 4-CH_3, 4-OCH_3, 3-F, 3,4,5-$(F)_3$, 3-NO_2, 4-NO_2, 3-Ac, 4-CHO
R^2 = H, 3,4-$(OCH_3)_2$, 3,5-$(OCH_3)_2$, 3-OCH_3, 4-Cl

Scheme 2.10 Mizoroki–Heck reaction with silica gel as milling auxiliary in a planetary ball mill (PBM).

Su and co-workers have described the Mizoroki–Heck coupling using tetra-*n*-butylammonium bromide (TBAB), K_2CO_3 as base and silica gel as milling auxiliary in a planetary-centrifugal mill AGO-2 to synthesize (*E*)-stilbenes (Scheme 2.10).[24] Moderate to very good isolated yields (67–92%) were achieved after 45–60 min milling with a 5 min pause every 15 min. Without milling auxiliary only 19% yield of the desired stilbene was obtained.

2.3 Homo-coupling Reactions

2.3.1 Homo-coupling of Phenols

Binaphthols have a wide range of application. They are used in processes for dyes, pigments, fluorescent whiteners or as antioxidants or antiseptics.[25] Furthermore they are useful as chiral auxiliaries or catalysts.

The syntheses of 2,2′-dihydroxy-1,1′-binaphthyl *via* oxidative coupling of 2-naphthol was realized in a planetary ball mill (Pulverisette 7, Fritsch GmbH), giving 87% yield (Scheme 2.11, method A).[26] The substrate was ball milled with $FeCl_3 \cdot 6H_2O$ and NaCl for 60 min at ambient temperature.

This reaction was improved by Shayesteh and co-workers.[25] 2-Naphthol and $FeCl_3 \cdot 6H_2O$ were placed in stainless steel capsules and shaken with a vibration mill at 50 s^{-1} (Scheme 2.11, method B). After 8 min, 95% yield was achieved. The reaction was controlled with the help of TLC and UV spectroscopy. Additionally, the reaction method was compared to classic syntheses in solvents to emphasize the good reaction conditions (short time, room temperature).

2.3.2 Glaser Coupling

Investigations of Sonogashira reactions have noted that there is a small amount of the homo-coupling side product. Replacing the Pd-catalyst with a copper species, the product (1,4-diphenylbuta-1,3-diyne) was achieved after 10 min in 93% isolated yield (Scheme 2.12).[10] A mixer mill from Retsch was used (MM301) in this reaction. Extensive investigations of the reagent system showed that KF/Al_2O_3 is most suitable for the homo-coupling.[27] Thereby different modifications and prior treatment of the additives were examined. Furthermore the influence of KF amount and other bases were tested. A catalyst screening with various metal salts revealed that, beside Cu-catalysts, Ni- and Co-salts can promote the reaction while with Fe-salts no conversion could be observed.

The Glaser coupling also took place using a Pd-catalyst.[28] Application of $[Pd(PPh_3)_4]$ gave 74% yield and with $[PdCl_2(PPh_3)_2]$, and an excess of PPh_3, 90% homo-coupling product was generated. Reactions were performed in stainless steel or Teflon® milling vials, but the material showed no significant effect on the yield. Different phenylacetylene derivatives could be converted using this method in 62–82% yield (Scheme 2.13).

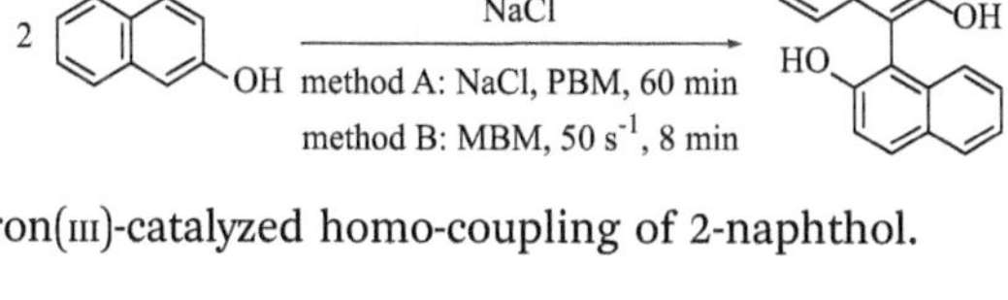

Scheme 2.11 Iron(III)-catalyzed homo-coupling of 2-naphthol.

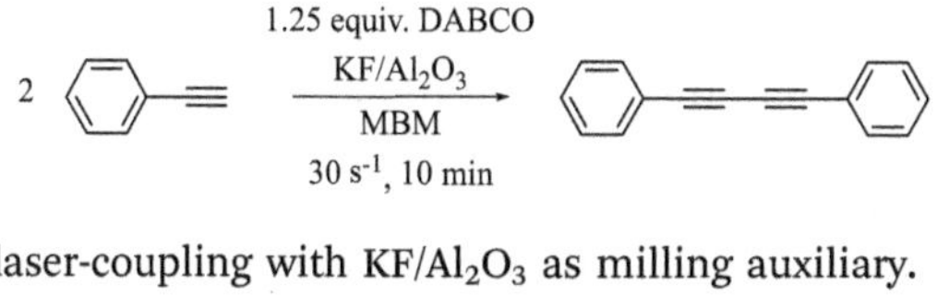

Scheme 2.12 Glaser-coupling with KF/Al_2O_3 as milling auxiliary.

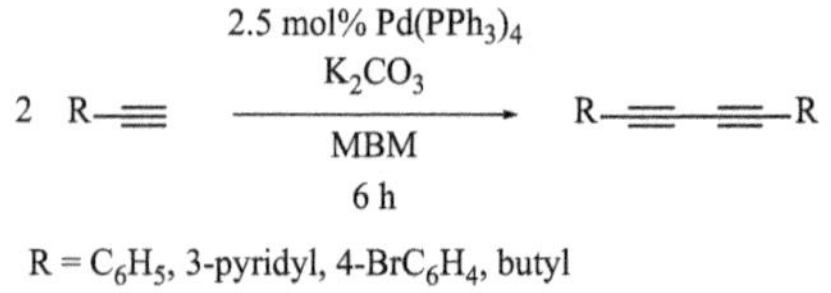

Scheme 2.13 Glaser-coupling catalyzed by $[Pd(PPh_3)_4]$.

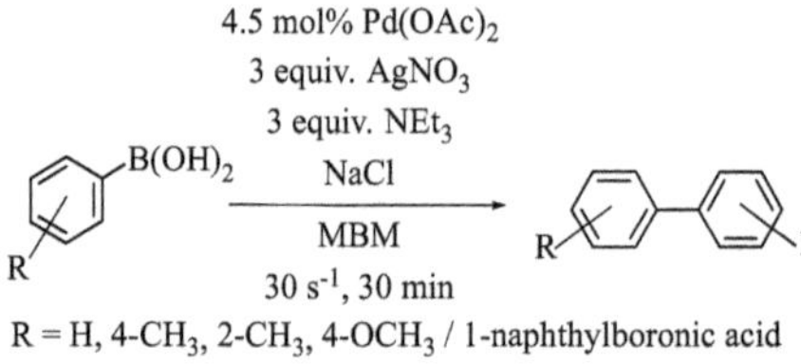

Scheme 2.14 Homo-coupling of boronic acids in a mixer ball mill (MBM).

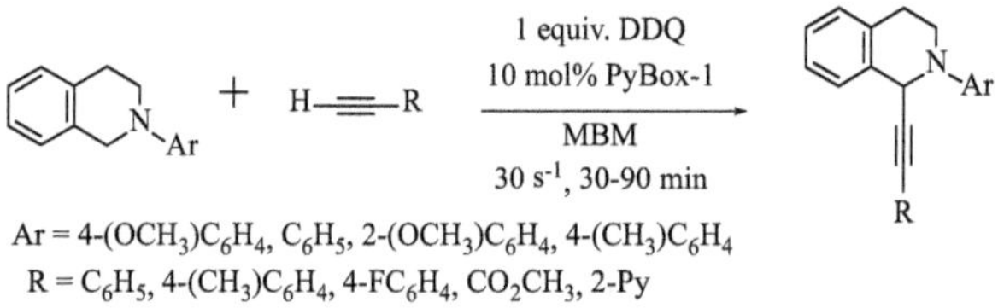

Scheme 2.15 Synthesis of optically active 1-alkynyltetrahydroisoquinoline derivatives.

2.3.3 Homo-coupling of Boronic Acids

The Suzuki reaction described by Nielsen and co-workers revealed a significant amount of the arylboronic acid homo-coupling product (13–28% analytical yield) after 30 min.[11] Similar results were detected by Cravotto *et al.* using Pd/CS-HDMI.[18] Higher yields of the biphenyl were obtained by Klingensmith and Leadbeater.[13] Biphenyls were synthesized from two molecules of phenylboronic acid using $AgNO_3$, NEt_3 and $Pd(OAc)_2$ as catalyst and base (Scheme 2.14). NaCl was added to make the reaction mixture less fluid. The substrates were treated in a mixer ball mill (MM200, Retsch GmbH) for 30 min and 19–72% of the desired product was obtained.

2.4 Miscellaneous

The syntheses of optically active 1-alkynyltetrahydroisoquinoline derivatives was realized with the ball milling technique using recoverable copper balls (Scheme 2.15).[29] In the investigations of chiral ligands PyBox led to good yields and the highest enantiomeric excess (74%). In the presence of 2,3-dichloro-5,6-dicyano-1,4-benzoquinone (DDQ) product yields of 60–77%

were achieved in the asymmetric cross-dehydrogenative coupling after 30–90 min milling.

2.5 Carbon–Carbon Bond Formation by Condensation Reactions

Condensations reactions are an important tool for the formation of carbon–carbon single and double bonds. Several reaction protocols have been developed and some famous name reactions like the Michael addition or Knoevenagel reaction belong to this group.

2.5.1 Aldol-type Reactions

Through the reaction of enolates and carbonyl compounds, one can gain access to α-hydroxy carbonyl compounds or α,β-unsaturated carbonyl compounds. These reactions, known as aldol reaction and aldol condensation, are valuable synthetic steps, widely used for syntheses of bioactive compounds, *e.g.* the aldol reaction is an important part in the synthesis of epothilone.[30]

The first aldol condensation in a mixer ball mill (MBM) was published by Raston and Scott.[31] Veratraldehyde, 4-phenylcyclohexanone and 1-indanone were directly milled with NaOH in various combinations – prior formation of the enolates was not necessary. The authors observed that reaction under solvent-free ball milling conditions was much more selective than in solution and single condensation products could be achieved in good yields. The self-condensation product was only formed in traces. In contrast, the solution based reaction protocols delivered mixtures of all possible products. The reaction proceeds in short milling times, 2 min of milling over a period of 10 min, yielding the crossed aldol products in 77–98%.

Waddell *et al.* observed that cyclohexanone undergoes an aldol type reaction in the presence of ethyl bromide and NaOH.[32] After 17 h milling, the aldol product was formed instead of the actually intended alkylated product. Owing to heating of the milling vessel by grinding, ethyl bromide was in the gas phase and therefore not available for the alkylation reaction. On this account cyclohexanone undergoes self-aldol condensation.

While Raston found the aldol condensation product, Kaupp *et al.* observed only an aldol addition product for the reaction of ninhydrin with dimedone (Scheme 2.16).[33] After 1 h milling in a mixer ball mill they obtained the aldol

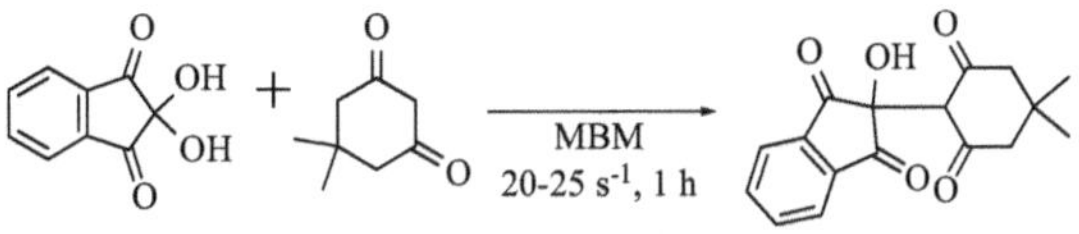

Scheme 2.16 Aldol condensation of ninhydrin with dimedone.

addition product in quantitative yield. A subsequent dehydration to achieve the double bond was not observed, likely due to the nature of the substrate.

Aluminium as catalyst for the self-aldol reaction of valeraldehyde was reported by Heintz and co-workers.[34] Non-spherical millimeter-sized flakes of bulk aluminium were stored together with the reactants in the milling beaker under a nitrogen atmosphere and milled for 24 h. Yields of >67% could be achieved. As without aluminium no reaction could be observed the authors suppose that bare aluminium surfaces were created while milling, leading to an acid-catalyzed aldol addition. The tentative mechanism includes an enolization of one carbonyl group and an activation step of the coupling partner on the fresh metal surface leading to an attack of the enol on the carbonyl group.

A reaction protocol for the asymmetric aldol reaction catalyzed by chiral amine macrocycle–metal(II) complexes was developed by Tanaka *et al.* for reaction in a planetary ball mill.[35] The scope of reaction was demonstrated applying several (hetero)cyclic ketones or aliphatic ketones with several substituted aromatic aldehydes. Yields ranged from 4% to 82%. Best results were obtained for the reaction of cyclohexanone and 4-nitrobenzaldehyde to give (1′*R*,2*S*)-2-[hydroxy(4-nitrophenyl)methyl]cyclohexanone (*anti* : *syn* = 60 : 40) (Scheme 2.17). The authors demonstrated that reaction in a ball mill was superior to protocols in solution or stirring of the neat liquids. Milling for 24 h in a planetary ball mill at 100 min^{-1} led to yields of 82% with 93% ee. In comparison, reaction in CH_2Cl_2 yielded 9% with 63% ee, while under neat-stirring conditions only 4% with 63% ee were obtained.

2.5.2 Knoevenagel Condensation

The Knoevenagel condensation is a long known and important carbon–carbon-coupling reaction for construction of α,β-unsaturated compounds. Traditionally, this reaction is performed in solvents like benzene or DMF and catalyzed by weak amine bases. Wada *et al.* investigated the Knoevenagel reaction of malononitrile or methyl cyanoacetate with aromatic aldehydes catalyzed by calcite and fluoride (Scheme 2.18).[36] The reaction was performed in a mixer ball mill and yields of 26–100% could be achieved within

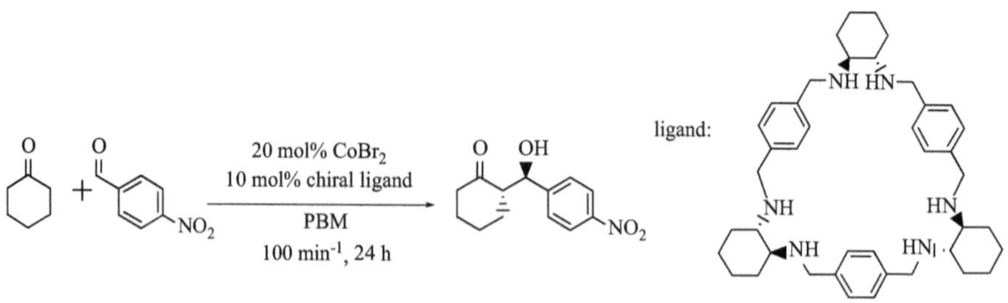

Scheme 2.17 $CoBr_2$-catalyzed aldol reaction of cyclohexanone and aromatic aldehyde.

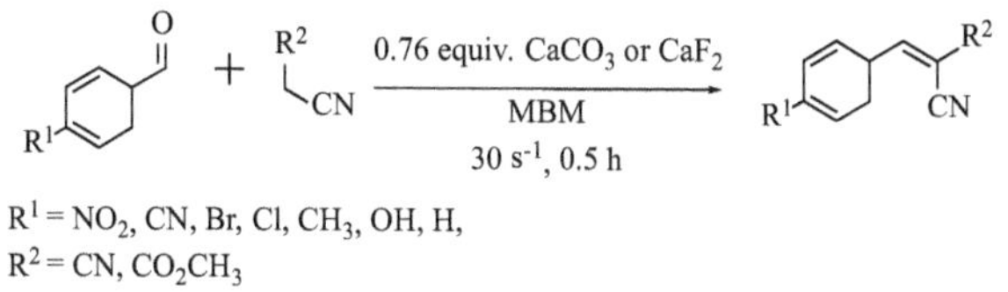

Scheme 2.18 Knoevenagel condensation of aldehydes with malononitrile or methyl cyanoacetate.

0.5 h milling at 30 Hz. Interestingly, applying the catalyst as fine powder did not promote the reaction, while adding the catalyst in 1–3 mm pieces led to good yields. The authors assumed that this effect is based on *in situ* created ionic species, acting as strong bases. Ondruschka *et al.* remarked that the missing activity of the powder is in fact based on a diluting effect, making the grinding process less effective.[37]

Another catalyst for Knoevenagel condensation with subsequent reduction of the double bond in a mixer ball mill was reported by Zhang *et al.*[38] Hantzsch-1,2-dihydropyridine catalyzed the reductive benzylization of malononitrile with substituted aldehydes. Thereby, 1,2-dihydropyridine acted on the one hand as catalyst for the Knoevenagel condensation and, on the other hand, as reductant for the double bond of the coupling product.

Beside catalyzed Knoevenagel reactions several authors have developed reaction protocols avoiding the use of a catalyst. In milling solid aldehydes with barbituric acid, Meldrum's acid, malononitrile or methyl cyanoacetate, quantitative yields were reported after 1 h milling in a mixer ball mill by Kaupp and co-workers.[39] By using stoichiometric amounts of reactants the product isolation is straightforward, no further clean-up was necessary and no further waste was produced. In addition to reactions on a low mmol-scale, the authors also reported a 200 g-scale reaction using a 2 l horizontal stirred media mill. The reaction proceeded at 1200 min^{-1} within 1 h. Based on the nature of the substrates, the reactions took place as solid-state reactions or as reactions in melt, *e.g.* reactions of Meldrum's acid at 50 °C proceeded in melt while reactions with barbituric acid occurred in solid state.

Ondruschka *et al.* further investigated the influence of reaction parameters in vibration and planetary ball mills for the reaction of aldehydes with malononitrile.[37] High yields were reported after milling the substrates at 800 min^{-1} for 1 h in a planetary ball mill or at 30 Hz in a mixer ball mill. With regard to energy efficiency they compared power consumption for reaction in ball mills and microwaves. They could show that performing the reaction in a planetary ball mill is very energy effective and that the power consumption per mol is 650 times lower in comparison to reaction in microwaves.

Although there are some example for performing Knoevenagel condensations in ball mills, the influence of process parameters, like rotation frequency, milling ball diameter or milling ball filling degree on the outcome of the reaction were rarely examined. Recently published results by Stolle *et al.* deliver insights into the influence of these parameters on the Knoevenagel condensation of vanillin and barbituric acid.[40,41] The results showed that milling balls with smaller diameter led to higher yields in shorter reaction times. A medium milling ball filling degree was also beneficial due to optimal energy entry. Under optimized conditions quantitative yields could be achieved within 10 min milling at 650 min^{-1}. The model reaction could also be scaled-up by a factor of 15, from 20 mmol to 300 mmol scale.

2.5.3 Michael Addition

The α,β-unsaturated carbonyl compounds, accessible by aldol or aldol-like reactions, are valuable synthetic intermediates as a further nucleophilic attack on the carbon–carbon double bond is possible. Such a reaction of α,β-unsaturated carbonyls with enolates or nucleophiles is known as the Michael addition. Zhang *et al.* developed a reaction protocol for the solvent-free Michael addition in a mixer ball mill. The reaction proceeded with the weak base K_2CO_3 and yields in the range 76–99% could be achieved within 10–60 min milling (Scheme 2.19). This protocol offered several advantages in contrast to reaction in solvents, *e.g.* milder reaction conditions, higher yields, shorter reaction times and no side reactions. With stronger bases like KF/Al_2O_3 intramolecular aldol addition and aldol condensation products were formed together with the Michael product. The authors also observed a strong influence of the rotation frequency. While at 20 Hz only 19% yield was achieved, yields of 47 and 99% could be observed at 30 Hz and 58 Hz (s^{-1}), respectively.[42]

Knoevenagel reactions and Michael additions are often important steps in cascade reactions allowing the synthesis of complex molecules. Kaupp *et al.* observed the Michael addition in a 1 : 2 mixture of aldehyde and dimedone, ball milled for 1 h at 50 °C (Scheme 2.20).[39] In this one-pot synthesis dimedone and aldehyde reacted to form the intermediate Knoevenagel product that further underwent Michael addition with another dimedone molecule. The reaction proceeded in an intermediate melt with subsequent crystallization of the product.

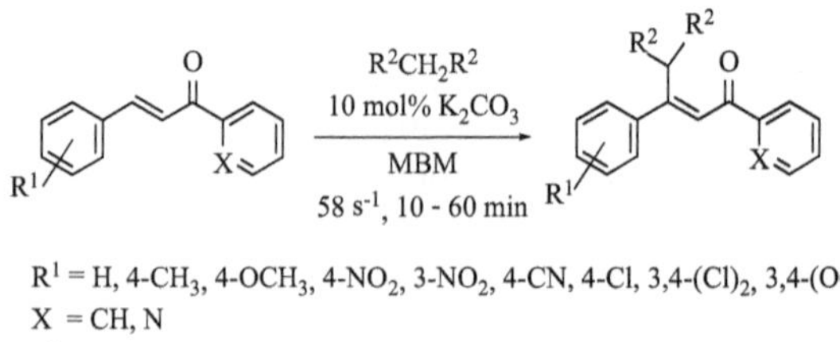

R^1 = H, 4-CH_3, 4-OCH_3, 4-NO_2, 3-NO_2, 4-CN, 4-Cl, 3,4-$(Cl)_2$, 3,4-(OCH_2O)-
X = CH, N
R^2 = $CO_2C_2H_5$

Scheme 2.19 Michael addition between diethyl malonate and (aza)chalcones catalyzed by K_2CO_3.

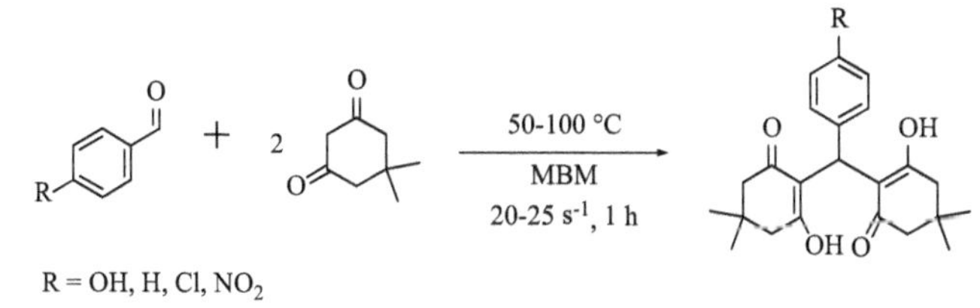

Scheme 2.20 Cascade reaction of dimedone with aromatic aldehydes.

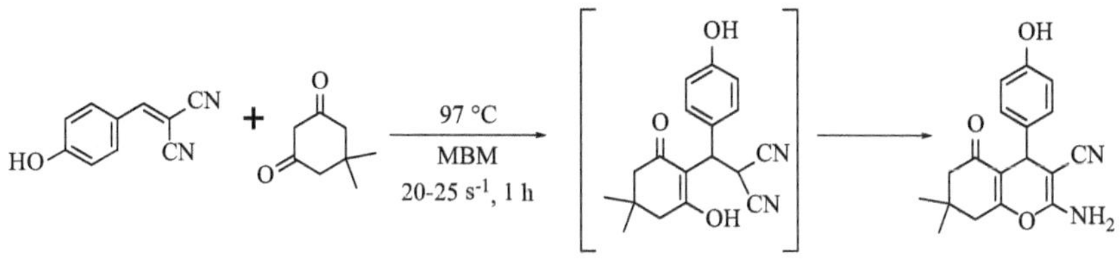

Scheme 2.21 Michael addition of dimedone and 2-[(4-hydroxyphenyl)methylene]-propanedinitrile as part of a cascade reaction.

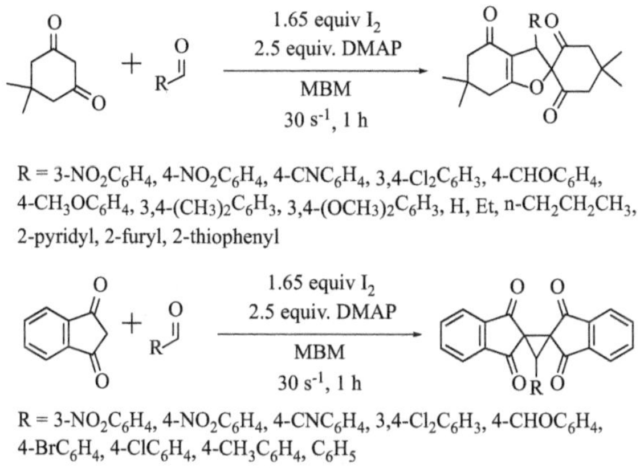

Scheme 2.22 Reaction sequence of Knoevenagel condensation and Michael addition forming dihydrofuran and cyclopropane derivatives.

Furthermore, dimedone underwent Michael addition with 2-[(4-hydroxyphenyl)-methylene]propanedinitrile followed by a rearrangement and cyclization step (Scheme 2.21).[39] 2-Amino-3-cyano-7,7-dimethyl-4-(4-hydroxyphenyl)-5-oxo-5,6,7,8-tetra-4*H*-chromene could be obtained in quantitative yields after milling at 97 °C in the melt for 1 h.

Dihydrofuran and cyclopropane structure motifs can be found in bioactive compounds and are interesting for the synthesis of pharmaceuticals. Wang and Gao developed a solvent-free reaction protocol to synthesize these structures by ball milling (Scheme 2.22).[43] This cascade reaction started with a Knoevenagel condensation of dimedone with aldehyde followed by an attack of the 1,3-dicarbonyl compound leading to the Michael addition product. Promoted by I_2 and DMAP the spiro dihydrofuran was formed selectively. A cyclopropane structure could be attained by using 1,3-indandion instead of dimedone. Yields of 77–92% for the dihydrofuran and 60–89% for the cyclopropane were achieved after milling at 30 Hz for 1 h.

Kaupp *et al.* published a Michael addition as part of a cascade reaction yielding pyrroles (Scheme 2.23).[44] Starting from primary or secondary amines and 1,2-dibenzoylethene a vinylogous Michael addition occurred as the first out of the four-step cascade giving quantitative yields by milling for 3 h at 20–30 Hz. Solution based methods delivered the products in yields of 46–81% only.

A three-component reaction yielding pyrano[2,3-*d*]pyrimidine-2,4(1*H*,3*H*)-diones was published by Mashkouri and co-authors (Scheme 2.24).[45] Reaction was accomplished by milling aldehyde, barbituric acid and malononitrile together in a mixer ball mill at 20–25 Hz for 15–70 min at 96 °C. Yields of 94–99% could be observed. Reactions without temperature control in the ball mill showed no conversion. The authors supposed that the reaction proceeded *via* an initial Knoevenagel condensation of malononitrile and aldehyde, followed by a Michael addition of barbituric acid to the Knoevenagel product. The reaction was completed with a tautomerization, cyclization and tautomerization sequence.

Gérard *et al.* developed a protocol for the solvent-free oxa-Michael–aldol condensation (Scheme 2.25). In a one-pot reaction they synthesized tetrahydroxanthenones by reacting salicylaldehyde with cyclohexanone in the presence of DABCO.[46] The authors investigated the influence of the overall milling ball mass and the milling time. The milling time showed an optimum at 48 h, longer or shorter reaction times led to decreased yields. The decrease of yield for milling times longer than 48 h was explained by the heat sensitivity of the product. Higher milling ball mass also led to higher yields.

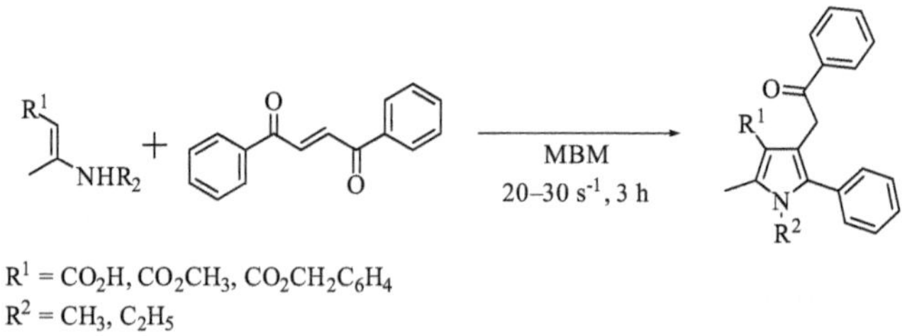

Scheme 2.23 Michael addition as part of a cascade reaction yielding pyrroles.

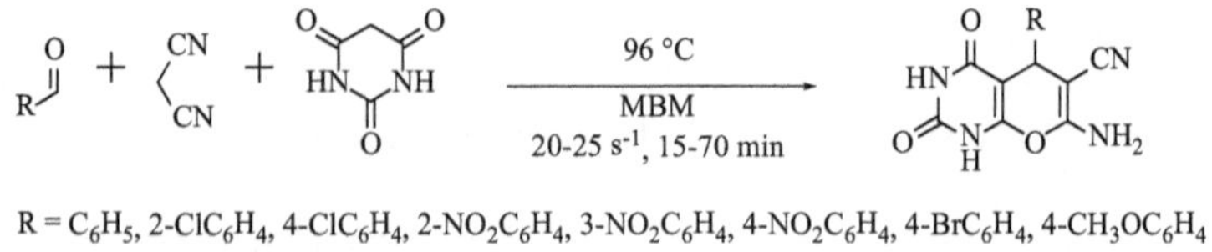

Scheme 2.24 Cascade reaction including Knoevenagel reaction and Michael addition.

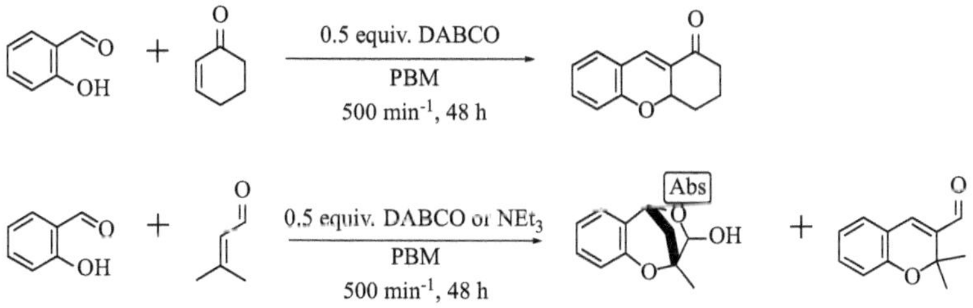

Scheme 2.25 DABCO catalyzed oxa-Michael–aldol reaction.

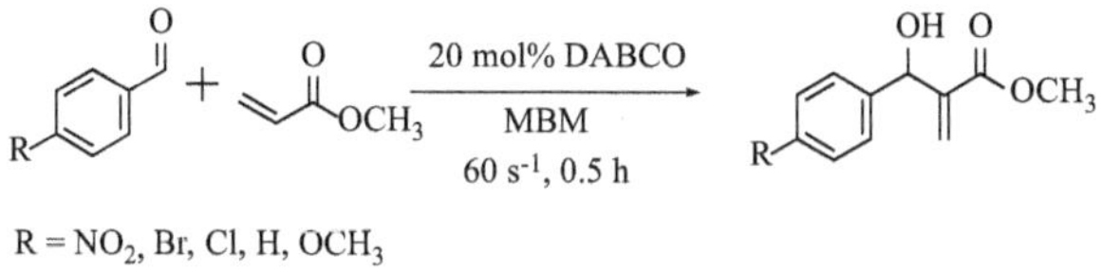

Scheme 2.26 Synthesis of α-hydroxy alkylated vinyl derivatives by Baylis–Hillman reaction.

The reaction of salicylaldehyde and prenal led to a mixture of two products since either oxa-Michael–aldol reaction took place or a vinylogous aldol reaction occurred followed by oxa-Michael addition giving dihydrobenzopyran. With NEt_3 or DABCO as base comparable yields were achieved for reaction in ball mills and solution. In contrast, on milling reactants with Na_2CO_3, only traces of the product could be detected while in solution 19% and 65% of the coupling products were formed.

2.5.4 Baylis–Hillman Reaction

α-Hydroxy alkylated vinyl derivatives can be obtained by the Morita–Baylis–Hillman reaction and are interesting molecules for synthesis of drugs, *e.g.* for antimalarial agents. Mack *et al.* reported high yields within 0.5 h by milling methyl acrylate, 4-substituted aromatic aldehydes and DABCO as catalyst (Scheme 2.26).[47] They observed a significant rate enhancement of the reaction in comparison to other solvent-free methods where reaction times of days are necessary.

2.5.5 Wittig Reaction

The Wittig reaction is of great interest and an industrially important method for the formation of double bonds, *e.g.* in the synthesis of vitamin A.[48] Shearouse *et al.* investigated the Wittig reaction under ball milling conditions (Scheme 2.27).[49] By reacting a premilled PPh_3-functionalized polystyrene with various aldehydes and ethyl bromoacetate or benzyl bromide, stilbenes could be obtained in high yields. Referring to the concept of liquid assisted grinding (LAG, see Chapter 7.3.3) the addition of small amounts of solvents was beneficial to the reaction. While without additional solvents a

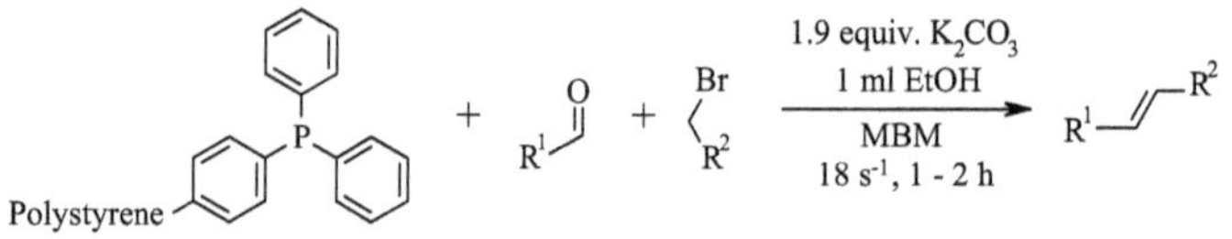

$R^1 = C_6H_5$, 4-$CH_3OC_6H_4$, 4-$NO_2C_6H_4$, 2-furyl, 2-naphthyl, 4-BrC_6H_4, $CHCHC_6H_5$, $CH_2CH_2C_6H_5$
$R^2 = CO_2C_2H_5$, $CH_2C_6H_5$

Scheme 2.27 Wittig reaction with functional resin.

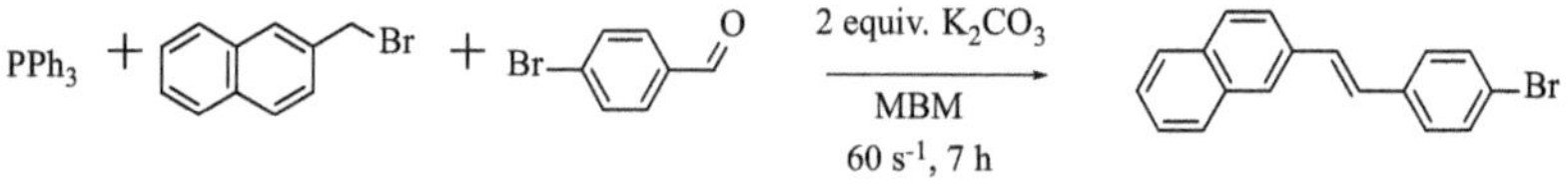

Scheme 2.28 One-pot Wittig reaction.

yield of 17% was achieved a rate enhancement could be observed for reactions with EtOH. Thereby, yields of 99% were gained within 2 h. Adding small amounts of solvents has furthermore the advantage that the (*E*)/(*Z*) ratio of the product was tunable by varying the solvent. Using the resins offered several advantages like simple work up. By suspending the mixture in EtOAc and subsequent filtration, the pure product could be obtained.

Balema *et al.* reported the Wittig reaction in a mixer ball mill under helium atmosphere.[50] (Benzyl)triphenylphosphonium chloride was milled with aromatic aldehydes, ketones and K_2CO_3. Samples were milled for 7 h in a glove box, affording the Wittig product in 70–93% yield. To avoid heating while milling, forced air cooling of the milling beakers of the Spex 8000 mill was employed. Beside application of preformed phosphorus yields the Wittig reaction can also be performed as one-pot reaction by using PPh_3, an organic halogenide, and an organic carbonyl compound (Scheme 2.28). In contrast to reactions in solution, the Wittig reaction of (benzylidene)- or (2-naphthylene)triphenylphosphoranes and aromatic carbonyl compound preferably yields the more thermodynamically stable (*E*)-stilbenes instead of (*Z*)-stilbenes or mixtures with nearly equal content of (*Z*)- and (*E*)-isomers.

Shearouse *et al.* also published details of a Horner–Wadsworth–Emmons reaction for the synthesis of α,β-unsaturated esters (Scheme 2.29).[51] The milling vessels were equipped with triethyl phosphonoacetate, 1.5 equiv. aldehyde and 2 equiv. K_2CO_3 and the reactants milled at 18 Hz for 6 h. Yields of 50–96% could be achieved with (*E*) : (*Z*)-ratio of 99 : 1 in almost all cases. As aldehyde was added in excess, after milling 4 equiv. Oxone® were added to the reaction mixture, which was then further milled for 12 h to remove the unreacted aldehyde. This strategy may offer an interesting starting-point for work up of reaction mixtures in ball mills. The aldehyde was oxidized to the carboxylic acid, which can be simply removed by acid–base extraction.

Boc-protected unsaturated amino esters were synthesized by Baron and co-workers.[52] A Horner–Wadsworth–Emmons reaction between aminophosphoryl acetate (1.1 equiv.), aldehyde (1 equiv.) and a base (2 equiv.) gave the

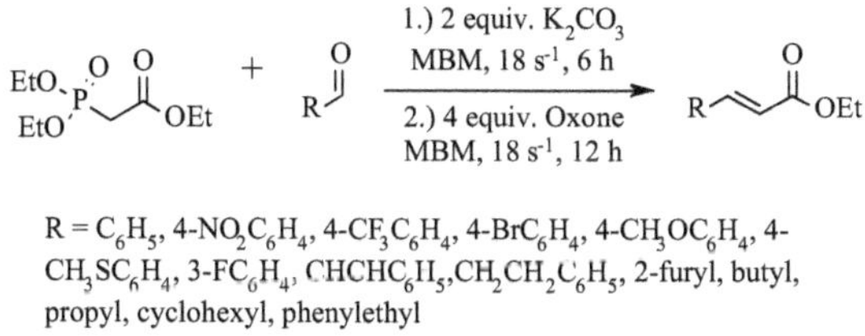

Scheme 2.29 Horner–Wadsworth–Emmons reaction catalyzed by K_2CO_3.

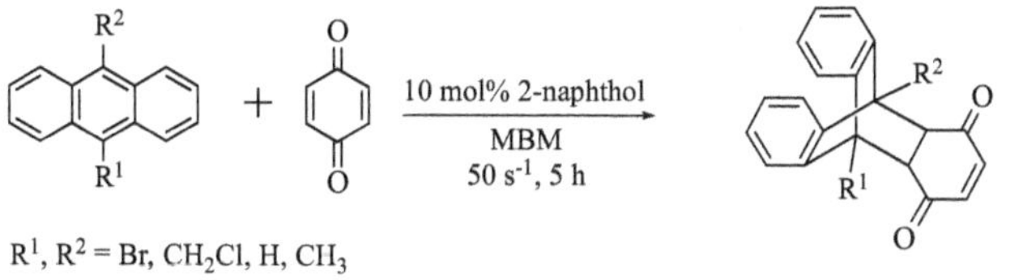

Scheme 2.30 Diels–Alder reaction of anthracene derivatives with 4-benzoquinone.

product in 61–95% yield. Therefore, substrates were milled in a planetary ball-mill at 550 min^{-1} for 7 h. Cs_2CO_3 was found to be the most suitable base. Several aromatic and aliphatic aldehydes reacted while ketones underwent no reaction. Stereochemistry was the same as in solution, the (*Z*)-isomer was preferred, in contrast to the results of Shearouse.[51]

2.6 Diels–Alder Reaction

The Diels–Alder reaction is an important synthetic tool and has wide applications in total synthesis of pharmaceutics and agrochemicals.[53,54]

Solid-state Diels–Alder-reaction of anthracene derivatives and 4-benzoquinone was successfully performed in a vibration micro ball mill.[55] However, only anthracene derivatives with electron-donating groups could be reacted and yields were low (26%). By adding 2-naphthol or BINOL in catalytic amounts, the yield could be increased to 75% and 85%, respectively, after 5 h milling (Scheme 2.30). With higher amounts of catalyst the yield could be raised to 99%. This solvent-free ball milling protocol was superior to reactions in solution where yields were lower with 56%. The catalytic effects were based on the formation of charge transfer complexes (CTCs) with strong hydrogen bonds and the formation of specific crystal network of the CTC that served as local reaction field for the reaction of diene and dienophile.

Zhang *et al.* reported the synthesis of *endo*-norbornenes by Diels–Alder reaction of cyclopentadiene with maleic anhydride or maleimide derivatives.[56] They achieved high yields (nine examples, 90–98%) after 30 min milling in a mixer ball mill at 30 Hz. During the reaction the *endo*-product was formed selectively. Advantages of this reaction protocol were shorter reaction times in comparison to solution based synthesis and higher yields. Furthermore the reaction needed only 4% excess of cyclopentadiene while in solution 3.6 equiv. were necessary. The authors suggested that the high efficiency of the solvent-free ball milling method was based on an enhanced second-order reaction rate resulting from high concentrations of reactants under solvent-free conditions.

The Diels–Alder reaction of graphite as diene with maleic anhydride and maleimide as dienophiles was reported by Seo and co-workers.[57] *In situ* generated active carbon species (carboradicals, carboanions and -cations), produced by mechanochemical breaking of graphitic carbon–carbon bonds, reacted with the dienophile to yield edge-selectively functionalized graphene nanoplatelets. This reaction was executed in a planetary ball mill at 500 min^{-1} for 48 h under argon atmosphere.

Wang *et al.* described the Diels–Alder reaction of fullerene (C_{60}) with 9-hydroxymethylanthracene or 9-methoxymethylanthracene in a mixer ball mill. After 30 min they achieved yields of 70%.[58] Prolonging the reaction time to 45 min led only to a slight increase (74%). In contrast, reaction in solution at room temperature afforded the product in 27% after 24 h.

2.7 Grignard Reaction

Organomagnesium compounds, especially Grignard reagents, have a huge importance in synthetic chemistry and have a wide range of application. As nucleophiles they can be used for the formation of carbon–carbon bonds by reaction with electrophiles like carbonyl groups.

Harrowfield *et al.* have described the synthesis of Grignard reagents by milling 1-chloro- or 1-bromonaphthalene with 4 equiv. Mg in a mixer ball mill under argon atmosphere (Scheme 2.31).[59] The Grignard reagents could be obtained in 95% yield, which is in same range as solvent-based methods. In a second reaction step the Grignard reaction took place by adding benzophenone or frozen acetophenone to the reaction mixture. Yields of 74–80% were realized after milling for further 20 min.

Waddell *et al.* observed that grinding methylmagnesium bromide in diethyl ether for 1 h with 4-bromobenzophenone yields 1-bromo-4-(1-phenylethenyl)benzene in 42% and 1-(4-bromophenyl)-1-phenylethanol in 58% yield (Scheme 2.32).[32] By milling the substrate without solvent for 17 h, 1-bromo-4-(1-phenylethenyl)benzene was formed as the major product. Therefore, the diethyl ether was evaporated from methylmagnesium

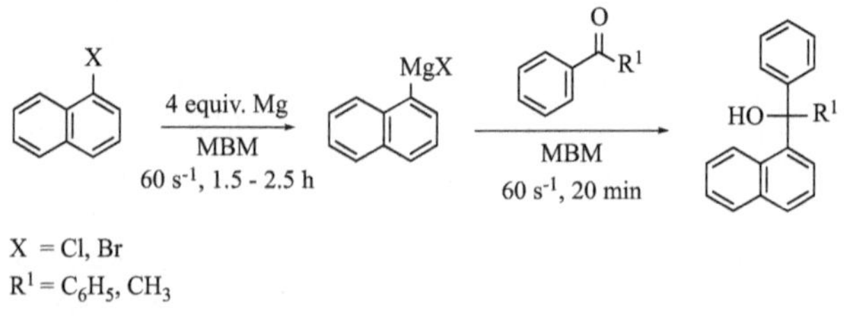

Scheme 2.31 Solvent-free Grignard reaction.

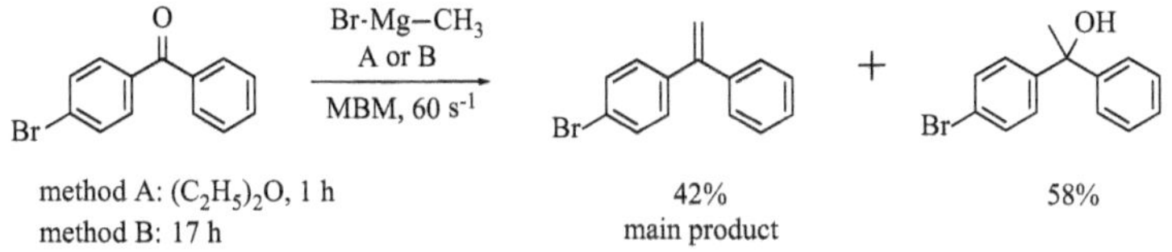

Schemc 2.32 Grignard reaction of methyl-Grignard and benzophenone.

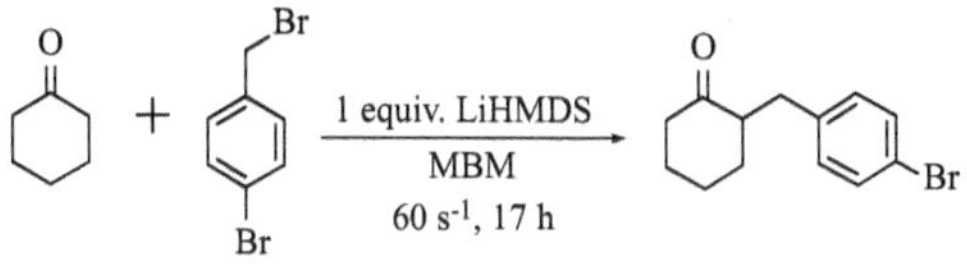

Scheme 2.33 Reaction of cyclohexanone and 4-bromobenzyl bromide yielding cyclohexanone enolate.

bromide before milling. The authors further observed that no precautions to avoid moisture in the air were necessary when reactions were performed in sealed vessels. They assumed that this was due to the small moisture content of the reactants or in the air before the vial was sealed. Therefore, chemistry under mechanochemical conditions was much less sensitive to ambient moisture and oxygen in the reaction environment than reaction under solvent conditions where the special requirements of a dry atmosphere were necessary.

2.8 McMurry Reactions

While investigating Grignard reactions in ball mills Harrowfield *et al.* reported the occurrence of McMurry reactions.[59] When benzophenone or acetophenone were milled together with 4 equiv. Mg in a mixer ball mill under argon atmosphere a broad product spectrum of McMurry products, pinacol products and reduced species was observed. Thereby, the alkene could be isolated in 40% yield for benzophenone and 15% for acetophenone.

2.9 Further Miscellaneous Reactions

Waddell *et al.* studied the reaction of cyclohexanone, LiHMDS and 4-bromobenzyl bromide in a mixer ball mill (Scheme 2.33).[60] The mono-substituted product was formed in 64% yield along with disubstituted products. In dependence on the base, the authors could show that for reactions with 2-methylcyclohexanone the thermodynamically controlled as well as the kinetic controlled products could be achieved. In comparison to reaction with mortar and pestle, yields in mixer ball mills were 4–5 times higher.[32] As well as 4-bromobenzyl bromide as electrophile the aliphatic halide 1-bromohexane could be reacted successfully.

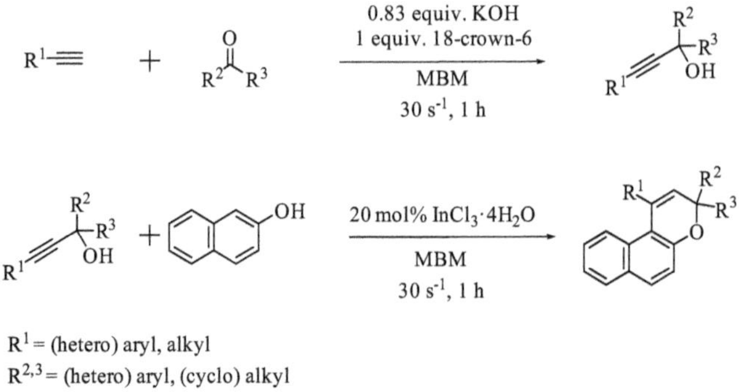

Scheme 2.34 Synthesis of propargyl alcohols by nucleophilic addition and their reaction to give naphthopyrans.

Alkynylation of carbonyl compounds is an important means of carbon–carbon coupling, delivering propargyl alcohols that are of great interest for synthetic chemistry. Dong *et al.* developed the solvent-free synthesis of propargyl alcohols by nucleophilic addition of terminal alkynes to carbonyl compounds in a ball mill (Scheme 2.34).[61] Reaction occurred in the presence of KOH and 18-crown-6. Moderate to high yields (54–97%, 18 examples) could be achieved by milling for 1 h at 30 Hz in a mixer ball mill. On applying other inorganic or organic bases no product could be obtained. The crown ether has to be added in stoichiometric amounts otherwise only moderate yields resulted. In a further reaction step propargyl alcohols were found to react with 2-naphthol, catalyzed by $InCl_3{\bullet}4H_2O$, to give the corresponding naphthopyrans (0–97%, 18 examples). In comparison to reactions in solution higher yields could be achieved in shorter reaction times.

2.10 Summary

Using ball mills as reactor for organic reactions is of considerable interest. Various carbon–carbon bond forming reactions have been successfully performed by ball-milling with high efficiency. Thereby, ball-milling offers some advantages over conventional reactions in solutions. Reactions in ball mills are often characterized by shorter reaction times, higher yields and in some cases even different products were formed compared to solution based methods. Furthermore, the avoidance of solvents during ball milling contributes to a greener synthesis by reducing waste and costs.

Abbreviations

Ac	acetyl
Boc	*tert*-butyloxycarbonyl
Et	ethyl
*t*Bu	*tert*-butyl
LAG	liquid assisted grinding

MBM mixer ball mill
MW microwave
PBM planetary ball mill
US ultrasound

References

1. S. L. James, C. J. Adams, C. Bolm, D. Braga, P. Collier, T. Friščić, F. Grepioni, K. D. M. Harris, G. Hyett, W. Jones, A. Krebs, J. Mack, L. Maini, A. G. Orpen, I. P. Parkin, W. C. Shearouse, J. W. Steed and D. C. Waddell, *Chem. Soc. Rev.*, 2012, **41**, 413–447.
2. G.-W. Wang, *Chem. Soc. Rev.*, 2013, **42**, 7668–7700.
3. K. Tanaka and F. Toda, *Chem. Rev.*, 2000, **100**, 1025–1074.
4. B. Rodríguez, A. Bruckmann, T. Rantanen and C. Bolm, *Adv. Synth. Catal.*, 2007, **349**, 2213–2233.
5. R. B. N. Baig and R. S. Varma, *Chem. Soc. Rev.*, 2012, **41**, 1559–1584.
6. A. Bruckmann, A. Krebs and C. Bolm, *Green Chem.*, 2008, **10**, 1131–1141.
7. K. Komatsu, G. W. Wang, Y. Murata, T. Tanaka, K. Fujiwara, K. Yamamoto and M. Saunders, *J. Org. Chem.*, 1998, **63**, 9358–9366.
8. D. A. Fulmer, W. C. Shearouse, S. T. Medonza and J. Mack, *Green Chem.*, 2009, **11**, 1821–1825.
9. R. Thorwirth, A. Stolle and B. Ondruschka, *Green Chem.*, 2010, **12**, 985–991.
10. A. Stolle and B. Ondruschka, *Pure Appl. Chem.*, 2011, **83**, 1343–1349.
11. S. F. Nielsen, D. Peters and O. Axelsson, *Synth. Commun.*, 2000, **30**, 3501–3509.
12. J. Gálvez, M. Gálvez-Llompart and R. Garcia-Domenech, *Green Chem.*, 2010, **12**, 1056–1061.
13. L. M. Klingensmith and N. E. Leadbeater, *Tetrahedron Lett.*, 2003, **44**, 765–768.
14. F. Schneider and B. Ondruschka, *ChemSusChem*, 2008, **1**, 622–625.
15. F. Bernhardt, R. Trotzki, T. Szuppa, A. Stolle and B. Ondruschka, *Beilstein J. Org. Chem.*, 2010, **6**, 7.
16. F. Schneider, T. Szuppa, A. Stolle, B. Ondruschka and H. Hopf, *Green Chem.*, 2009, **11**, 1894.
17. N. Linares, A. E. Sepulveda, M. C. Pacheco, J. R. Berenguer, E. Lalinde, C. Najera and J. Garcia-Martinez, *New J. Chem.*, 2011, **35**, 225–234.
18. G. Cravotto, D. Garella, S. Tagliapietra, A. Stolle, S. Schüssler, S. E. S. Leonhardt and B. Ondruschka, *New J. Chem.*, 2012, **36**, 1304–1307.
19. C. Schmöger, T. Szuppa, A. Tied, F. Schneider, A. Stolle and B. Ondruschka, *ChemSusChem*, 2008, **1**, 339–347.
20. A. R. Siamaki, Y. Lin, K. Woodberry, J. W. Connell and B. F. Gupton, *J. Mater. Chem. A*, 2013, **1**, 12909–12918.
21. E. Tullberg, D. Peters and T. Frejd, *J. Organomet. Chem.*, 2004, **689**, 3778–3781.

22. E. Tullberg, F. Schacher, D. Peters and T. Frejd, *Synthesis*, 2006, 1183–1189.
23. V. Declerck, E. Colacino, X. Bantreil, J. Martinez and F. Lamaty, *Chem. Commun.*, 2012, 11778–11780.
24. X. Zhu, J. Liu, T. Chen and W. Su, *Appl. Organomet. Chem.*, 2012, **26**, 145–147.
25. K. Shayesteh, J. Moghaddas, M. Haghighi and H. Eskandari, *Asian J. Chem.*, 2010, **22**, 2106–2116.
26. M. O. Rasmussen, O. Axelsson and D. Tanner, *Synth. Commun.*, 1997, **27**, 4027–4030.
27. R. Schmidt, R. Thorwirth, T. Szuppa, A. Stolle, B. Ondruschka and H. Hopf, *Chem. Eur. J.*, 2011, **17**, 8129–8138.
28. L. Chen, B. E. Lemma, J. S. Rich and J. Mack, *Green Chem.*, 2014, **16**, 1101–1103.
29. J. Yu, Z. Li, K. Jia, Z. Jiang, M. Liu and W. Su, *Tetrahedron Lett.*, 2013, **54**, 2006–2009.
30. J. Wang, B.-F. Sun, K. Cui and G.-Q. Lin, *Org. Lett.*, 2012, **14**, 6354–6357.
31. C. L. Raston and J. L. Scott, *Green Chem.*, 2000, **2**, 49–52.
32. D. C. Waddell, T. D. Clark and J. Mack, *Tetrahedron Lett.*, 2012, **53**, 4510–4513.
33. G. Kaupp, M. R. Naimi-Jamal and J. Schmeyers, *Chem. Eur. J.*, 2002, **8**, 594–600.
34. A. S. Heintz, J. E. Gonzales, M. J. Fink and B. S. Mitchell, *J. Mol. Catal. A: Chem.*, 2009, **304**, 117–120.
35. K. Tanaka, A. Asakura, T. Muraoka, P. Kalicki and Z. Urbanczyk-Lipkowska, *New J. Chem.*, 2013, **37**, 2851–2855.
36. S. Wada and H. Suzuki, *Tetrahedron Lett.*, 2003, **44**, 399–401.
37. R. Trotzki, M. M. Hoffmann and B. Ondruschka, *Green Chem.*, 2008, **10**, 767–772.
38. Z. Zhang, J. Gao, J.-J. Xia and G.-W. Wang, *Org. Biomol. Chem.*, 2005, **3**, 1617–1619.
39. G. Kaupp, M. R. Naimi-Jamal and J. Schmeyers, *Tetrahedron*, 2003, **59**, 3753–3760.
40. A. Stolle, R. Schmidt and K. Jacob, *Faraday Discuss.*, 2014, DOI: 10.1039/C1033FD00144J.
41. C. F. Burmeister, A. Stolle, R. Schmidt, K. Jacob, S. Breitung-Faes and A. Kwade, *Chem. Eng. Technol.*, 2014, **37**, 857–864.
42. Z. Zhang, Y. W. Dong, G. W. Wang and K. Komatsu, *Synlett*, 2004, 61–64.
43. G.-W. Wang and J. Gao, *Org. Lett.*, 2009, **11**, 2385–2388.
44. G. Kaupp, J. Schmeyers, A. Kuse and A. Atfeh, *Angew. Chem., Int. Ed.*, 1999, **38**, 2896–2899.
45. S. Mashkouri and M. R. Naimi-Jamal, *Molecules*, 2009, **14**, 474–479.
46. E. M. C. Gérard, H. Sahin, A. Encinas and S. Bräse, *Synlett*, 2008, 2702–2704.
47. J. Mack and M. Shumba, *Green Chem.*, 2007, **9**, 328–330.

48. K. M. Exner, K. Massonne, H. Laas, D. Glas and L. Szarvas, (BASF), *Pat. Appl.* WO2005058811 A1, 2005.
49. W. C. Shearouse and J. Mack, *Green Chem.*, 2012, **14**, 2771–2775.
50. V. P. Balema, J. W. Wiench, M. Pruski and V. K. Pecharsky, *J. Am. Chem. Soc.*, 2002, **124**, 6244–6245.
51. W. C. Shearouse, C. M. Korte and J. Mack, *Green Chem.*, 2011, **13**, 598–601.
52. A. Baron, J. Martinez and F. Lamaty, *Tetrahedron Lett.*, 2010, **51**, 6246–6249.
53. K. C. Nicolaou, S. A. Snyder, T. Montagnon and G. Vassilikogiannakis, *Angew. Chem., Int. Ed.*, 2002, **41**, 1668–1698.
54. R. H. Mueller, S. Wang, P. D. Pansegrau, J. Q. Jannotti, M. A. Poss, J. K. Thottathil, J. Singh, M. J. Humora, T. P. Kissick and B. Boyhan, *Org. Process Res. Dev.*, 1997, **1**, 14–19.
55. H. Watanabe and M. Senna, *Tetrahedron Lett.*, 2005, **46**, 6815–6818.
56. Z. Zhang, Z. W. Peng, M. F. Hao and J. G. Gao, *Synlett*, 2010, 2895–2898.
57. J.-M. Seo, I.-Y. Jeon and J.-B. Baek, *Chem. Sci.*, 2013, **4**, 4273–4277.
58. G. W. Wang, Z. X. Chen, Y. Murata and K. Komatsu, *Tetrahedron*, 2005, **61**, 4851–4856.
59. J. M. Harrowfield, R. J. Hart and C. R. Whitaker, *Aust. J. Chem.*, 2001, **54**, 423–425.
60. D. C. Waddell, I. Thiel, T. D. Clark, S. T. Marcum and J. Mack, *Green Chem.*, 2010, **12**, 209–211.
61. Y. W. Dong, G. W. Wang and L. Wang, *Tetrahedron*, 2008, **64**, 10148–10154.

CHAPTER 3

Oxidation and Reduction by Solid Oxidants and Reducing Agents using Ball-Milling

GIANCARLO CRAVOTTO* AND EMANUELA CALCIO GAUDINO

Dipartimento di Scienza e Tecnologia del Farmaco and Interdepartmental Centre "Nanostructured Interfaces and Surfaces" - NIS, University of Turin, Via P. Giuria 9, I-10125 Turin, Italy
*Email: giancarlo.cravotto@unito.it

3.1 Introduction

The use of mechanical forces to induce chemical reactivity is the oldest and simplest technique used by mankind and one that has existed since before the dawn of science. Since ancient times, in most civilizations mechanical treatments with pestle and mortar in food preparation and popular medicine have been well documented (Figure 3.1). The mechanical action in the presence of air may easily promote organic and inorganic oxidation reactions, typically the oxidation of natural products and the formation of metallic oxides.

Several transformations of plant materials may involve cell lyses and complex oxidative enzymatic reactions, with the generation of pleasant flavours or new chemical species suitable for food dressing or as natural remedies and cosmetics.

This chapter covers several oxidation and reduction reactions in the solid state phase by application of ball milling generally under solvent-free conditions. Based on a literature survey, the graphic depicted in Figure 3.2

RSC Green Chemistry No. 31
Ball Milling Towards Green Synthesis: Applications, Projects, Challenges
Edited by Brindaban Ranu and Achim Stolle

Published by the Royal Society of Chemistry, www.rsc.org

Figure 3.1 Medicaments preparation by mortar treatment in old China (author's photograph in public park Tainan city, Taiwan).

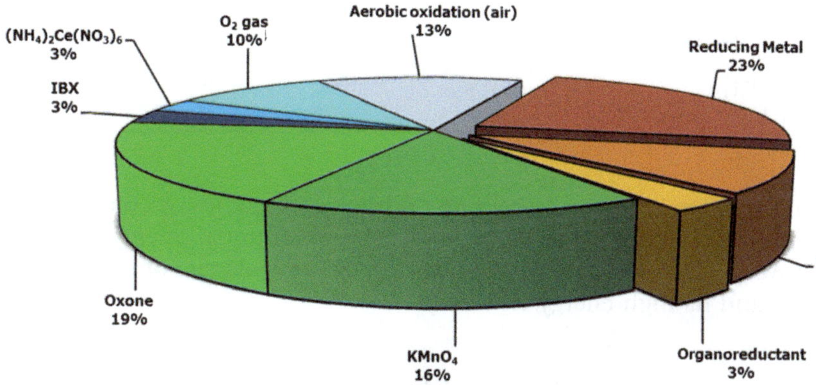

Figure 3.2 Solid oxidants and reductants used in mechanochemistry (the graph refers to a recent literature search based on 30 scientific articles).

reports the percentage use of the main solid reagents (or gas): oxidants in the green–blue portion and reducing agents in the brown area (right).

Tables 3.1 and 3.2 should be helpful to the reader to easily associate one specific reagent (solid oxidant or reducing agent) with the mechanochemical procedures described in the literature. The efficiency and scalability of this oxidation/reduction technique, the absence of solvents and the low energy consumption strongly encourage chemists to consider it in most cases as a method of choice. Clearly, technical parameters such as beaker dimension, size, material and number of milling balls as well as vibration/revolutions

Table 3.1 Solid oxidizing agents for mechanochemical protocols.

Oxidizing agents	References
Cu/CuO/ZnO	21
$KMnO_4$	1–4
$KMnO_4$ or $(NH_4)_2S_2O_8$	5
Oxone	4,6–8,10
Oxone + NaBr	9
$(NH_4)_2Ce(NO_3)_6$	11
3-Chloroperbenzoic acid	12
IBX, 2-iodoxybenzoic acid	13
O_2 gas	15*b*,16
NiO, Cr_2O_3, Pt/Al_2O_3, $Co_3O_4 + O_2/He$ 2 : 8 v/v% (10 MPa)	20
Aerobic oxidation (air) + Al_2O_3 (auxiliary)	14
Aerobic oxidation (air)	15*a*,17,18

Table 3.2 Solid reducing agents for mechanochemical protocols.

Reducing agent	References
Cu/CuO/ZnO	21
$Zn/ZnCl_2$	22
Bi or Pb	23
Mg	24,25
Mg or Al	26
Mg–butylamine	27
$NaBH_4$	28,29
$LiAlH_4$	30
1,4-Dihydropyridine	31

speed and milling time need accurate optimization studies. A comparison with conventional heating or with other non-conventional energy sources (microwaves, ultrasound, *etc.*) highlights the advantages of mechanochemistry and its high energy efficiency.

3.2 Mechanochemical Oxidation

The literature of the last two decades reports several oxidation processes with various solid oxidants. Nüchter *et al.* reported the oxidation of several types of organic compounds using an inorganic carrier supported potassium permanganate in a Pulverisette 7 mill at 400 rpm (6.7 Hz) for 10 min. The presence of water enhances the yields of almost all educts. Olefins were oxidized to carboxylic acids. The conversion of benzyl-type arenes was selective, providing ketones with good yields (Scheme 3.1).[1]

Scheme 3.2 shows a new method for the oxidative cleavage of β-pinene, yielding nopinone, using potassium permanganate as the oxidant in solvent-free conditions.[2] The reaction was performed in a conventional ball mill using a grinding auxiliary (Al_2O_3 or SiO_2) that was able to absorb liquid

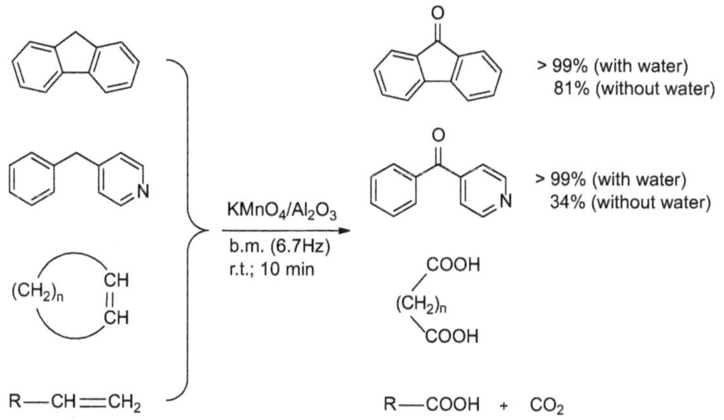

Scheme 3.1 Mechanochemical oxidation of model organic substances using $KMnO_4/Al_2O_3$.

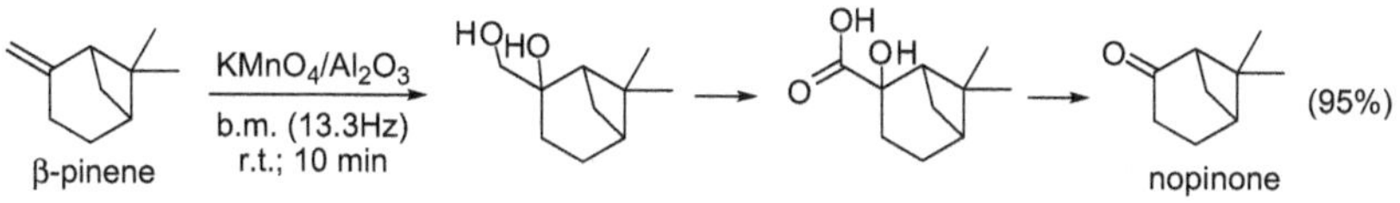

Scheme 3.2 Oxidation of β-pinene to nopinone with $KMnO_4/Al_2O_3$. Reaction condition: planetary ball mill (PULVERISETTE 7 classic line, Fritsch GmbH, Idar-Oberstein, Germany).

Table 3.3 Different milling materials (beaker and ball) and their effect on the yield of nopinone by solvent free oxidation of β-pinene.

Milling materials	Nopinone yield (%) Alumina	Quartz sand
Hardened steel	85	92
Zirconium oxide	82	93
Sintered corundum	95	96
Agate	84	93

reactants such as β-pinene onto its surface, making liquids accessible for mechanical impact. Various reaction parameters and technical variables were assessed to understand their influence on yield and selectivity of nopinone. Chemical parameters, such as oxidant, grinding auxiliaries and quantities, were also investigated. In addition, tuning parameters including milling time, rotation frequency and the number of milling balls were explored (Table 3.3). Using an excess of potassium permanganate high yields (95%) were obtained in short reaction times (10 min). However, due to the self-decomposition of $KMnO_4$ during the milling process, it was only necessary to remove MnO_2 and the grinding auxiliary *via* simple filtration.

Stolle have also reported the solvent-free dehydrogenation of γ-terpinene in a planetary ball mill to give *p*-cymene as the predominant reaction product (Scheme 3.3).[3] It was also shown that $KMnO_4$ can be replaced by other, less-toxic and environmentally-benign oxidants (Oxone®, $NaIO_4$, I_2). It was possible to fine-tune the transformation using variations in the oxidant-to-substrate ratio, the rotation frequency, the number of milling balls and the type of grinding auxiliary used. A quantitative yield was obtained after 5 min milling at 13.3 Hz, using $KMnO_4$ as the oxidant in the presence of alumina (Table 3.4).

A fast and solvent-free method for the direct oxidation of primary aromatic amines to the corresponding azo and azoxy homocoupling products using a planetary ball mill is shown in Scheme 3.4.[4] Several oxidants and milling auxiliaries were tested along with various substituted anilines. The selectivity of the solvent-free reaction could be switched by applying different combinations of oxidant and milling auxiliary; oxidations using $KMnO_4$ and Oxone furnished the azo and azoxy compounds, respectively. The solvent-free procedure in the ball mill is more efficient than in solution methods in terms of both chemical yield and energy consumption (Table 3.5). The lack of organic solvents and the fact that the reaction is easy, fast and energy-saving make this ball-milling method a real alternative to conventional reaction protocols.

Posudievsky *et al.* have described the soft mechanochemical synthesis of graphite oxide and subsequently graphene oxide, in the absence of concentrated acids. Graphite and oxidants, such as potassium permanganate and ammonium persulfate, were used as starting reagents ($KMnO_4$ or $(NH_4)_2S_2O_8$, molar ratio C/Ox 1 : 3 or 1 : 2 correspondingly). With the mechanochemical treatment (planetary ball mill Pulverisette 6 by Fritsch GmbH, agate grinding bowls: 3 h, 500 rpm) of the graphite–oxidant mixture, besides graphite oxidation a decrease in the size of its particles and a shift in the graphene layers, which form these particles, relative to each other due to a weakening of the bonding between them were also observed.[5]

Gao and Wang have found that Oxone (potassium peroxymonosulfate, $2KHSO_5 \cdot KHSO_4 \cdot K_2SO_4$) is an effective oxidant for the oxidative amidation of aldehydes using anilines to furnish amides in a one-pot process under mechanical milling conditions (Scheme 3.5). The solvent-free reaction was conducted in a mixer mill at 30 Hz for 90 min and amides were obtained in 38–78% yields. Two pathways (Scheme 3.6) were proposed for this oxidative amidation.

In pathway A, imines are rapidly formed from aldehydes and anilines and are then oxidized by Oxone to generate oxaziridines. The subsequent rearrangement of oxaziridines, *via* initial cleavage of the N–O bond, followed

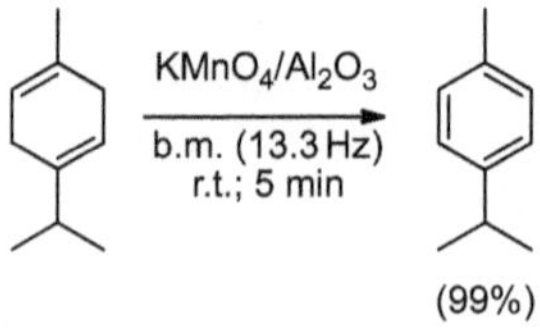

Scheme 3.3 Oxidative dehydrogenation of γ-terpinene.

Table 3.4 Solvent-free oxidative dehydrogenation of γ-terpinene to *p*-cymene in the presence of alumina or quartz sand as milling auxiliary and different oxidation agents.[a]

Oxidant agent	Yield (%) Alumina	Quartz sand
$KMnO_4$	99	53
p-Benzoquinone	69	37
I_2	93	52
Oxone	84	8
$NaMnO_4$	82	73
$(NaH_4)_4Ce(SO_4)_4 \cdot H_2O$	71	11
$NaIO_4$	67	41
$K_2Cr_2O_7$	12	7
$Na_2S_2O_8$	11	10
$(NH_4)_2S_2O_8$	7	7
Urea–H_2O_2-adduct (28%)	5	6

[a]Batch (per batch): 3.8 g auxiliary, 2 mmol γ-terpinene, 6 mmol oxidant, 0.1 g H_2O.

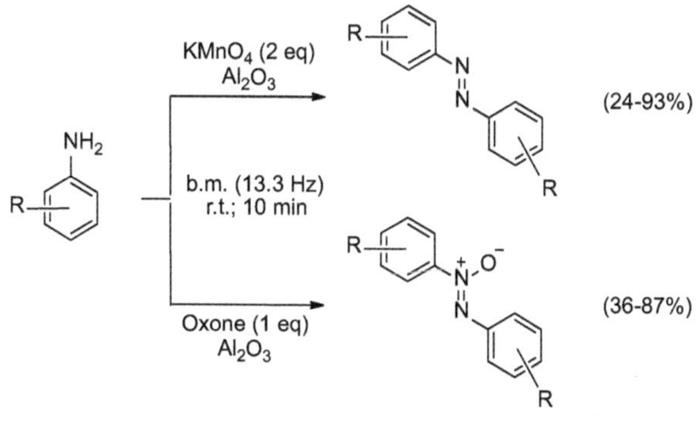

Scheme 3.4 Solvent-free chemoselective oxidation of anilines either using $KMnO_4$ to give the corresponding azo-benzenes or with Oxone to give the respective azoxy compounds. Reactions were performed in a Fritsch "Pulverisette 7 classic line" planetary ball mill using 45 mL grinding beakers (agate or ZrO_2) and milling balls (6–15 mm; agate or ZrO_2).

by the migration of the substituent (hydrogen), which is *trans* to the nitrogen lone pair, results in the formation of amides. In pathway B, carbinolamine intermediates are generated after the nucleophilic addition of anilines to aldehydes, followed by oxidation using Oxone to form the amide products. Despite the fact that the exact mechanism remains obscure and that both pathways could well be operating simultaneously, this novel solvent-free and metal catalyst free procedure makes the synthesis of amides more efficient, low cost, and eco-friendly as compared to the traditional liquid-phase reaction. Other oxidants, including $K_2S_2O_8$ and I_2, were also examined; however, no reaction was observed (entries 6–11, Table 3.6).[6]

In a drive to further deepen the work of Stolle's group on the oxidation of anilines (using acidic inorganic auxiliaries to direct the formation of

Table 3.5 Influence of energy input on the oxidation of anilines with $KMnO_4$ and Al_2O_3.[a]

Amine	Conversion of amine (%)/selectivity for the azo product (%)[b]					
	Solvent-free		Solution (30 mL; H_2O/acetonitrile, 1 : 1 v/v)			
	Planetary ball mill[c]	Vibration ball mill[d]	Conventional heating[e]	Multimode microwave[f]	Monomode microwave[g]	Ultrasound[h]
Aniline	93/95	71/89	83/88	90/91	58/91	88/94
m-Anisidine	70/90	40/91	53/89	22/91	27/90	73/96
o-Toluidine	94/89	81/97	89/88	39/87	63/88	82/87
m-Toluidine	95/91	87/2	84/72	54/81	84/72	67/87
p-Toluidine	>99/96	>99/95	96/90	91/93	96/90	97/97

[a]Standard batch size: amine (2 mmol), $KMnO_4$ (4 mmol), γ-Al_2O_3 (4 g; activity 90, neutral).
[b]Determined by GC-FD measurements of extracted products in relation to amine.
[c]Fritsch Pulverisette 7 classic line: two agate beaker (45 mL), six agate milling balls per beaker (15 mm), 800 rpm, 20 min.
[d]Retsch MM301: two agate beaker (50 mL), six agate milling balls per beaker (15 mm), 800 rpm, 20 min.
[e]Round-bottom flask (100 mL); 80 °C, 60 min.
[f]MLS ETHOS Plus: 80 °C (P_{max} = 300 W), 30 min.
[g]CEM Discover: 80 °C (P_{max} = 130 W), 30 min.
[h]Hielscher UP200 s: 33% of standard batch size; 100% P_{max}, 20 °C, 20 min.

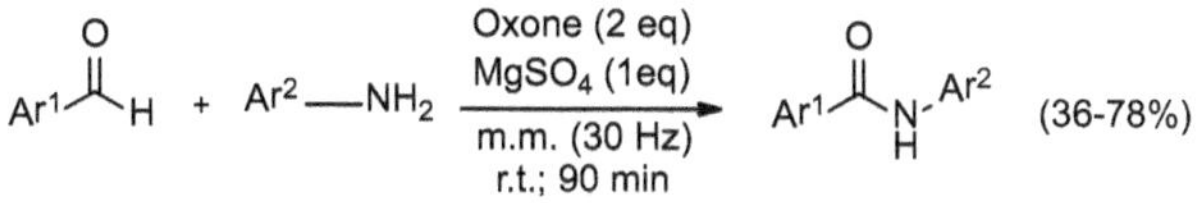

Scheme 3.5 Oxidative amidation of aldehydes using Oxone.

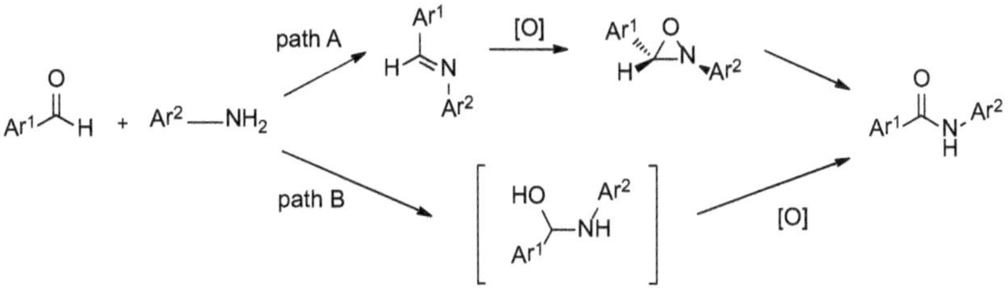

Scheme 3.6 Possible pathways for the direct oxidative amidation of aldehydes using anilines.

Table 3.6 Comparison of oxidants in direct oxidative amidation.

Entry[a]	Oxidant	Additives	Solvent	Yield (%)[b]
1	Oxone	—	—	61
2	Oxone	4Å MS	—	68
3	Oxone	$MgSO_4$	—	75
4	Oxone	$MgSO_4$	CH_3CN[c]	11
5	Oxone	$MgSO_4$	Toluene[c]	Trace
6	I_2	$MgSO_4$	—	0
7	$K_2O_2O_8$	$MgSO_4/ZnCl_2$	—	0
8	$K_2O_2O_8$	$MgSO_4$/CuCl	—	Trace
9	$K_2O_2O_8$	$MgSO_4$/CuBr	—	0
10	$K_2O_2O_8$	$MgSO_4$/TsOH	—	0
11	$K_2O_2O_8$	$MgSO_4/KHSO_4$	—	0

[a]3-Nitrobenzaldheyde (1.0 eq), *p*-toluidine (1.0 eq), $MgSO_4$ (1.0 eq), and Oxone (2.0 eq) in MM200 ball-mill.
[b]Isolated yields.
[c]3-Nitrobenzaldehyde (1.0 eq), *p*-toluidine (1.0 eq), $MgSO_4$ (1.0 eq) and Oxone (2.0 eq) in solvent.

azobenzenes and azoxybenzenes), Huskić *et al.* have more recently demonstrated that nitrosobenzene intermediates can be obtained in 26–85% yields from the mechanochemical Oxone oxidation of selected *p*-substituted anilines in the presence of $NaHCO_3$ at a frequency of 25 Hz for 20–30 min (Scheme 3.7, Table 3.7).[7] The use of $NaHCO_3$ to neutralize potassium hydrogen sulfate and other acidic species was very important in facilitating the formation of nitrosobenzenes. Besides shortening reaction times and avoiding the use of bulk solvents, the authors were able to exploit the high vapour pressure of nitrosobenzenes for the solvent-free separation of the products by sublimation, demonstrating an entirely solvent-free route to chemical synthesis and product isolation.

Cravotto *et al.* have reported the solventless selective oxidation of several organosulfur compounds to sulfones, under mechanical milling conditions

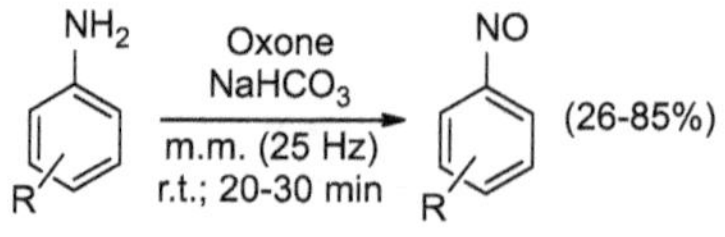

Scheme 3.7 Oxidation of selected substituted anilines in the presence of Oxone and $NaHCO_3$.

Table 3.7 Results of nitrosobenzene syntheses by mechanochemical Oxone oxidation of anilines. The only impurity found following sublimation was the corresponding nitrobenzene.

Substituent	*p*-I	*p*-Br[a]	*p*-Cl	*p*-NO_2	*p*-Me
Yield[b] (%)	85	80(70)	70	26	52
Purity[c] (%)	92	90(95)	80	90	69
Milling time (min)	20	20	20	30	20
$NaHCO_3$	Yes	Yes	Yes	No	Yes

[a]Numbers in brackets relate to solution synthesis.
[b]Yield is calculated only for the nitrosobenzene taking into account the mixture composition after sublimation.
[c]Purity is reported as the molar fraction of nitrosobenzene in a mixture with the corresponding nitrobenzene obtained after sublimation, determined by 1H NMR.

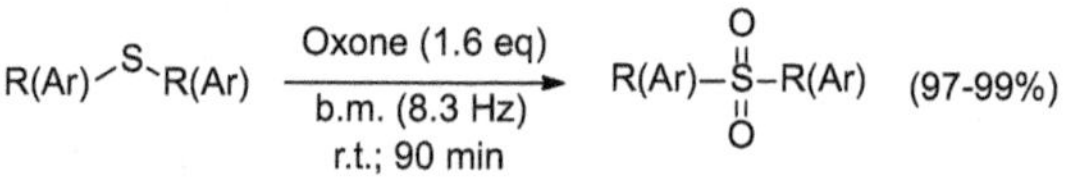

Scheme 3.8 Selective oxidation of several organosulfur compounds using a planetary ball mill PM100 Retsch GmbH, equipped with a single milling beaker (stainless steel, 125 mL).

using Oxone. Furthermore, they have applied this optimized procedure to the desulfurization of medium to high sulfur content paraffins (up to 3000 mg kg^{-1}). Substrates such as 3,6-dithiaoctane-1,8-diol (Lindlar catalyst poison), 4-(methylthio)benzaldehyde and a series of *S*-heterocycles including thiophene, benzothiophene, dibenzothiophene and substituted derivatives were all examined. All substrates exclusively afforded sulfones as the oxidation products in 97–99% yields, after 90 min milling in a planetary mill (Scheme 3.8). No traces of sulfoxides were detected, reflecting the high chemoselectivity of Oxone based mechanochemical oxidation.[8]

Stolle *et al.* have described a solvent-free method for the halogenation of aromatic compounds in a mixer ball mill using $KHSO_5$ (Oxone) and NaX (X = Cl, Br) as reagents (Scheme 3.9; Table 3.8).[9] High yields 71–95% have been achieved in 20 min treatment. The choice of the milling auxiliary, γ-alumina (basic or neutral), α-alumina (acidic) or silica gel, is an important factor in determining the performance of the reaction and influences both conversion and selectivity.

Results indicated that the oscillation frequency, milling time and the number of milling balls are important parameters that directly affect conversion. An assessment of liquid-assisted grinding (LAG) conditions revealed

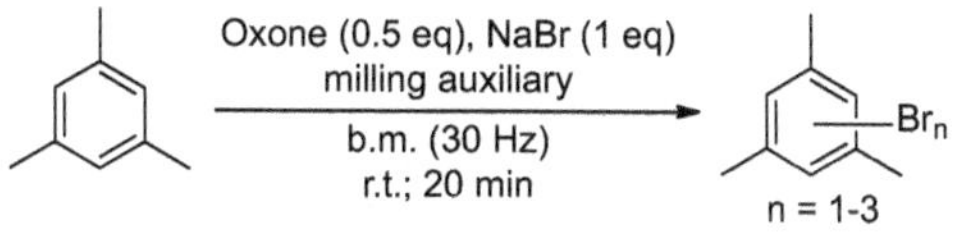

Scheme 3.9 Bromination of mesitylene under standard reaction conditions.

Table 3.8 Bromination of aromatic compounds using Oxone and NaBr.[a]

Aromatic compound	Conversion (%)	Product	Selectivity (%)[b]
Mesitylene	85	Bromomesitylene	96
Toluene	95	Bromotoluene	99 (*o*/*p* = 1 : 2)
o-Xylene	90	4-Bromo-1, 2-dimethylbenzene	97
m-Xylene	94	1-Bromo-2, 4-dimethylbenzene	98
p-Xylene	91	2-Bromo-1, 4-dimethylbenzene	94
Ethylbenzene	71	Bromoethylbenzene	90 (*o*/*p* = 1 : 3)
Anisole	88	Bromoanisole	97 (*o*/*p* = 1 : 3)
Nitrobenzene	No reaction	—	—
Chlorobenzene	No reaction	—	—
Naphthalene	78	α-Bromonaphthalene	94

[a]Reaction conditions: 1.5 mmol aromatic compound, 0.75 mmol Oxone, 1.5 mmol NaBr, 4 g quartz sand, milling beaker (YSZ, 35 mL) (Retsch GmbH), 10×10 mm milling balls (MSZ), $\nu_{osc} = 30$ Hz, $t = 20$ min.
[b]Ratio of *o*- and *p*-isomers determined by ratio of GC-MS signals.

that the addition of solvents has no advantageous effect in this case. Preferably, activated arenes are halogenated and bromination afforded higher product yields than chlorination. Reactions were most often regio- and chemoselective, since *p*-substitution was preferred, and the concurrent side-chain oxidation of alkylated arenes by $KHSO_5$ was not observed.[9]

Collom *et al.* have described the mechanochemical oxidation of methoxylated aromatic chemicals and have thus provided an example of selectivity that is very different to solution-based chemistry. Oxone was shown to react with 1,2,3-trimethoxybenzene to yield predominantly 2,6-dimethoxybenzoquinone in the solid state or 2,3,4-trimethoxyphenol in solution (Scheme 3.10). The oxidation of 1,2,3-trimethoxybenzene in aqueous solution was screened using varying amounts of Oxone (Table 3.9). The principal oxidized product was 2,3,4-trimethoxyphenol, which was obtained with a selectivity of up to 77% at 17% conversion, with Oxone (0.5 eq). The mechanochemical oxidation of the same compound was performed using four dram glass vials with two stainless steel milling balls (~ 3.5 g ea) along with Oxone inside a rock tumbler (7 days) and the starting material was selectively converted into 2,6-dimethoxybenzoquinone with no evidence of any other organic products. This result is in sharp contrast with the solution based reaction, which gave 2,3,4-trimethoxyphenol as the major product, with high selectivity, but at low conversion.[10]

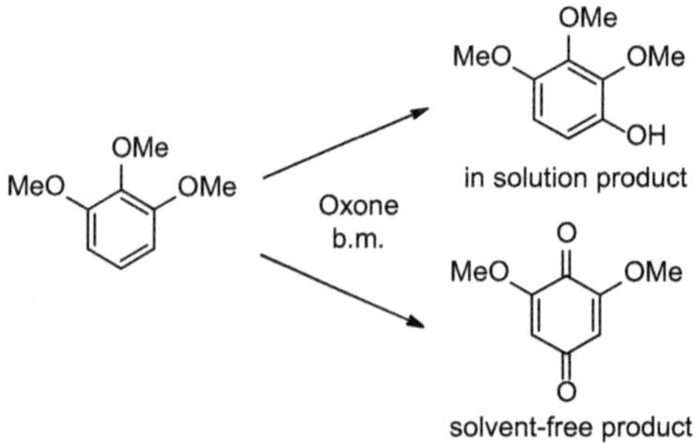

Scheme 3.10 Mechanochemical oxidation of methoxylated aromatic compounds with Oxone.

Table 3.9 Solution based oxidation of 1,2,3-trimethoxybenzene using Oxone.[a]

Entry	Oxone (eq)	Co-solvent	Conversion (%)[b]	Product (2,3,4-Trimethoxy-phenol)	(2,6-Dimethoxy-benzoquinone)
1	0.5	d_3-AcCN	17	77	8
2	1	d_3-AcCN	26	69	8
3[c]	2	d_3-AcCN	55(53)	49(54)	9(10)
4	4	d_3-AcCN	81	24	9
5	2	d_3-MeOH	67	34	3
6	2	d_6-Acetone	55	31	2

[a]Reaction conditions: 50 μmol substrate, 1 mL of 10% (v/v) co-solvent in D_2O (9 mL), room temp, 5 h.
[b]Conversion as percentage of starting material consumed.
[c] Numbers in parenthesis were obtained when the reaction was degassed and run under inert (N_2) atmosphere.

The solid-state oxidation of cinnamic, 4-methoxy- and 3,4-dimethoxy-cinnamic acids using CAN–MX (CAN = ceric ammonium nitrate ($(NH_4)_2Ce(NO_3)_6$); M = Li, Na; X = Br, Cl) systems in a vibratory mill has been described by Nikishin and co-workers. The oxidation led to the β-halostyrenes in 69–91% yields after for 4 h of grinding at a frequency of 12 Hz. A similar procedure, in the absence of a metal halide, resulted in the cleavage of the C=C bond to afford the corresponding benzaldehydes in 20–60% yields (Scheme 3.11).[11]

It is also possible to oxidize a tertiary amine under ball-milling conditions. Pankrushina *et al.* found that the nitroxide derivative was formed from the mechanochemical oxidation of lappaconitine using 3-chloroperbenzoic acid (3-Cl-PBA) in a SPEX-8000 vibratory mill using a solid absorptive material (aluminium oxide) to loosen the reaction mixture (Scheme 3.12).[12] In fact, the mechanochemical solid-state oxidation and de-ethylation of lappaconitine leads to the formation of lappaconitine *N*(20)-oxide II and *N*(20)-diethyl-lappaconitine III, respectively, in good isolated yields (57% and 42% respectively). This mechanochemical approach was also applied to the

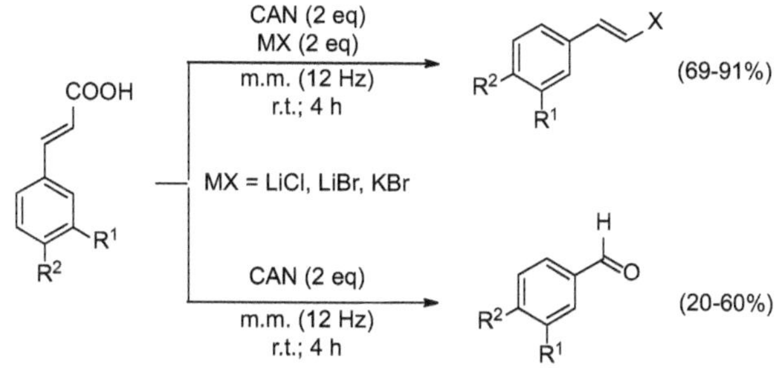

Scheme 3.11 Solid-state ball mill oxidation of cinnamic, 4-methoxy- and 3,4-dimethoxycinnamic acid.

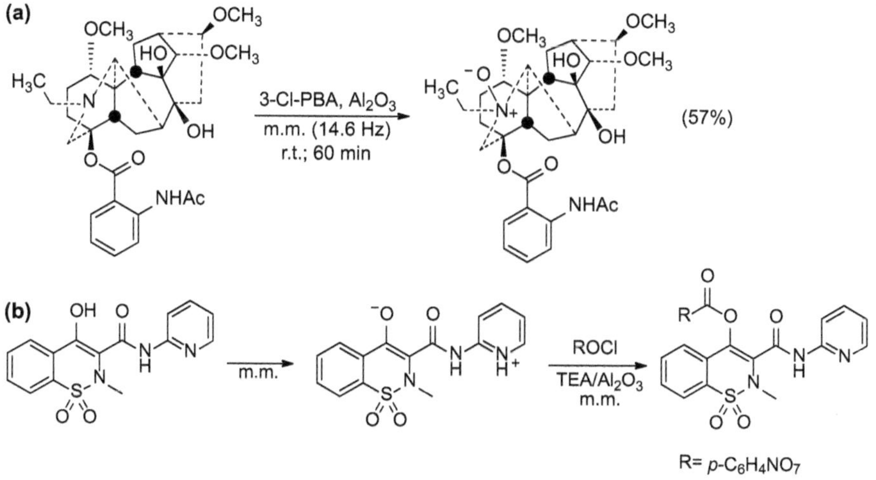

Scheme 3.12 General scheme of mechanochemical reactions of lappaconitine (a) and piroxicam-4-nitrobenzoate mechanochemical synthesis (b). Reaction condition: Vibratory mill SPEX-8000 (CertiPrep Inc., USA) with 60 mL steel vials 3 mm diameter grinding balls.

acylation of the anti-inflammatory drug piroxicam to improve its therapeutic properties. Consequently, piroxicam 4-nitrobenzoate was obtained in solvent-free reaction conditions, giving a moderate yield as a result of the mechanical activation of the reaction between piroxicam and 4-nitrobenzoyl chloride in the presence of triethylamine (TEA) impregnated aluminium oxide.[12]

Mal *et al.* have investigated the potential use of IBX (2-iodoxybenzoic acid, an oxidant whose extensive use is impeded by its explosiveness at high temperature and poor solubility in common organic solvents except DMSO) in solvent free, milling conditions at room temperature in the following reactions: the oxidation of primary/secondary alcohols to corresponding carbonyl compounds, amine to imine; the conversion of olefins to α-bromo/iodoketones, sulfide to sulfoxide; dithiane deprotection and the synthesis of benzimidazoles from primary alcohols.[13] This methodology has been

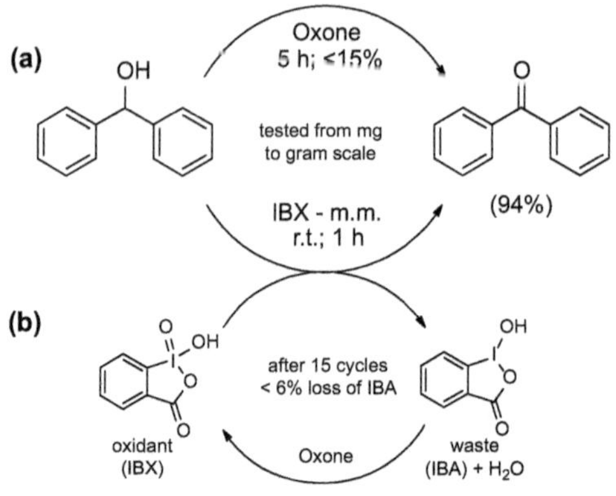

Scheme 3.13 Mechanochemical oxidation of benzhydrol to benzophenone with IBX (a); recycling of IBA waste with Oxone by mechanical milling (b).

Table 3.10 Comparative statement of economic benefit that may be obtained from milling methodology compared to a traditional method.

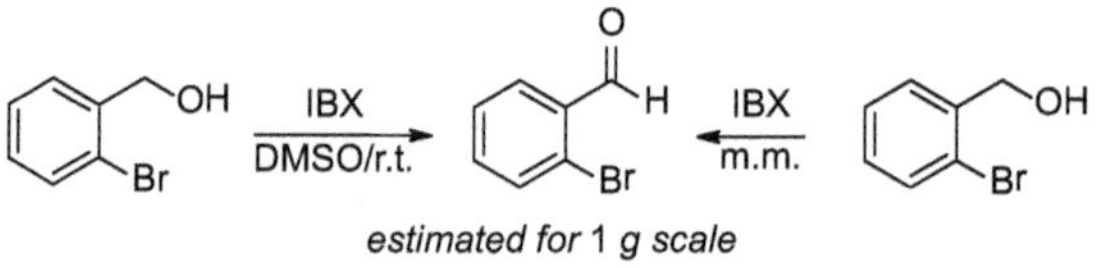

	Method	
Parameter	Traditional	Milling[a]
Solvents	DMSO (reaction), 50 mL CH_2Cl_2 (extraction), 300 mL	—
Purification	Chromatography: petroleum ether (1.2 L)/ethyl acetate (400 mL)	Filtration: CH_2Cl_2 (120 mL)
Time	2 h	45 min
Yield (%)	82	87
Total cost ($)	108	56.4

[a]Reaction conditions: one grinding ball (15 mm diameter, ZrO_2) in a ball milling 10 mL ZrO_2 jar.

advantageously used for long-range working windows, *e.g.* 10 mg to 2.5 g. Furthermore, the waste IBA (2-iodosobenzoic acid) produced from the reactions was oxidized *in situ* to IBX in the following step using Oxone® and thus reused in multiple cycles while conserving its efficiency (only 6% loss after 15 cycles) (Scheme 3.13). The oxidized products of primary, secondary and aliphatic alcohols were obtained in very good to excellent yields (79–99%) in shorter times (45–120 min). This methodology does not require any aqueous work-up, avoids chromatographic purification and was found to be highly cost effective for use in the pharmaceutical/chemical industries (Table 3.10).

Ranu and Chatterjee reported the aerobic oxidation of thiols to disulfides under ball-milling conditions over neutral aluminium oxide (grinding

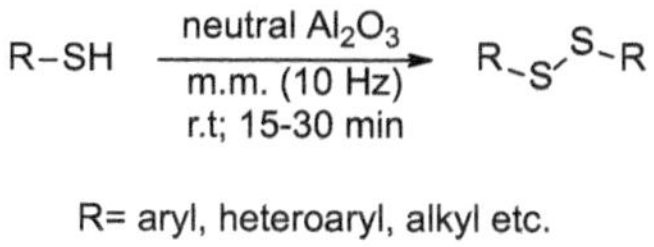

Scheme 3.14 Aerobic oxidation of thiol to disulfide. Reaction conditions: thiophenol (110 mg, 1 mmol) adsorbed on neutral Al_2O_3 (3 g) was ball-milled in a 25 mL stainless steel vessel at 600 rpm using six balls ($d = 10$ mm) for 15 min (PM 100, Retsch GmbH, Germany).

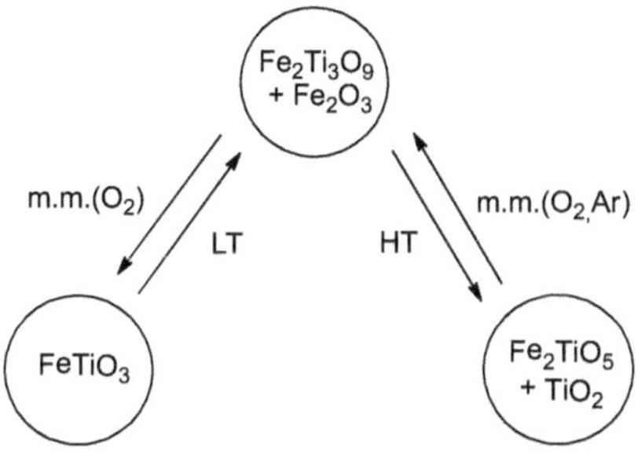

Scheme 3.15 Chemical reactions observed during both milling and annealing processes (HT = high-temperature annealing, LT = low-temperature annealing).

auxiliary) in the absence of catalysts, co-oxidants, solvents and bases (Scheme 3.14). The reactions were generally very clean and fast (15–30 min). The products (a wide variety of functionalized diaryl, diheteroaryl and dialkyl disulfides) were obtained in high purity and excellent yields after a simple work-up. The grinding auxiliary, Al_2O_3, is recovered, dried and recycled. This procedure is thus highly cost effective.[14]

Oxidation reactions of ilmenite ($FeTiO_3$) have been investigated under thermal annealing and ball milling conditions, while the use of a static vacuum (10–2 kPa) (1997) and O_2 atmosphere (1998) using X-ray diffraction and thermal analyses have been investigated by Chen and co-workers.[15] Thermal oxidation reactions consist of low-temperature (<1000 °C) reactions that form intermediate phases ($Fe_2Ti_3O_9$ and Fe_2O_3) and high-temperature (>1000 °C) that form the final thermally stable phases (Fe_2TiO_5 and TiO_2). Only the low-temperature reaction occurred during ball milling (conduced in a vertical planetary ball mill using hardened steel balls with a diameter of 25.4 mm and a stainless steel cell) at room temperature; the high-temperature reactions were not observed even after prolonged milling. It was also found that the thermally stable Fe_2TiO_5 phase transforms into the metastable $Fe_2Ti_3O_9$ phase during milling in oxygen (Scheme 3.15).

Oxidation kinetics and the mechanisms of ilmenite being milled in air and oxygen atmospheres were subsequently investigated by Chun and Bin.[16] The effects of mill spin rate, milling balls/mass ratio and oxygen pressure on the oxidation were investigated. The results showed that the two former parameters significantly affected the oxidation process while the influence of

$$ZrAl_3 \xrightarrow[T>500^\circ C]{O_2} 3/2\ Al_2O_3\ (amorphous)$$

$$+\ ZrO_{2\ 2\text{-}x} \xrightarrow[T>800^\circ C]{O_2} 3/2\ (\gamma, \delta, \theta,)\ \text{-}\ Al_2O_3 + t\ \text{-}\ ZrO_2$$

Scheme 3.16 Milling oxidation conduced in a Spex CertiPrep Freezer/Mill 6800.

oxygen pressure was found to be significant only at a low initial O_2 pressure range. O_2 adsorption (macroadsorption) was evident at high O_2 pressures (≥50 kPa), but disappeared with decreasing O_2 pressure. The oxidation was mechanochemically limited at the initial O_2 pressure range of 152.0 to 33.4 kPa. Upon a further decrease in O_2 pressure to ≤27.4 kPa, oxidation was found to be controlled by O_2 adsorption onto the active sites of ilmenite particle surfaces (micro-adsorption). Mechanochemical oxidation (conduced in a vertical planetary ball mill (QM-1SP2, Nanjing, China)) occurred mainly at active sites that present *c*-axis lattice strain.

Geßwein and Binder have dealt with the heterogeneous oxidation of $ZrAl_3$ particles in a gaseous oxidant (oxygen-containing atmosphere).[17] The oxidation kinetics of ball milled $ZrAl_3$ powder was investigated by thermogravimetry at temperatures of up to 1100 °C. The non-selective oxidation of $ZrAl_3$ results in the formation of α-Al_2O_3 and tetragonal and monoclinic ZrO_2. According to X-ray, TG/DSC and kinetic results, the following simplified oxidation scheme for $ZrAl_3$ powder is proposed (Scheme 3.16).

Ultra-high temperature ceramics (UHTCs) have recently become the focus of widespread attention due to their critical applications in the thermal protection systems of hypersonic aerospace vehicles. Of the UHTCs available, zirconium diboride (ZrB_2) is a particularly highly covalent refractory material. Oxide impurity in ZrB_2 powder promotes coarsening, resulting in lower sinterability. Because of this effect on sinterability, Ortiz *et al.* (2012) have carried out an in-depth study into ZrB_2 oxidation in high-energy ball milling conditions. They reported that high-energy grinding in air introduces twice as much oxygen into the ZrB_2 powders as the more conventional attrition milling. Furthermore, this oxygen does not form solid-solutions with ZrB_2, but amorphous oxides (*i.e.* ZrO_2 and B_2O_3) that preferentially locate onto the surface of the ultrafine agglomerates that result from the cold-welding of the primary nanoparticles that form during the ball milling.[18]

Aside from the case of CO_2 hydrogenation,[19] a few observations of catalytic reactivity induced by the application of friction under high vacuum conditions are described in the literature. Schüth *et al.* have hypothesized that it may also be possible to activate a solid catalyst *in situ* to make it more efficient in catalyzing the reaction between gas-phase molecules.[20] They have reported a case of catalytic CO oxidation under *in situ* ball-milling conditions (reaction conduced in a planetary ball mill (Fritsch Pulverisette 6) where the reaction rate was increased by three orders of magnitude by the milling. In an initial exploratory study to assess the potential of *in situ* milling in heterogeneous catalysis, a set of batch experiments using various metal oxides, such as catalysts (NiO, Cr_2O_3, Co_3O_4, Fe_2O_3, $Mg(OH)_2$ and 1%Pt/g-Al_2O_3), were carried out. A planetary ball mill (Fritsch Pulverisette 6) was filled with the solid (1 g) and three balls (12 g, $d = 10$ mm, hardened steel), and pressurized to 10 MPa with a mixture of reaction gases (CO/O_2/He = 1 : 20 : 79 v/v%). The gas/solid mixture was milled for 2–4 h at room temperature and the gas phase was analyzed at the end of the experiment. CO conversion was observed for NiO (22% after 2 h), Cr_2O_3 (10%), Pt/Al_2O_3 (48%) and Co_3O_4 (7.3% after 4 h). The use of Fe_2O_3 and $Mg(OH)_2$ did not lead to any conversion after 2 h.

Selected active oxidants were then studied in more detail in continuous experiments to provide closer insight into the process. For this purpose, a commercial shaker mill (Retsch MM200) was modified to allow for the continuous operation of heterogeneously catalyzed gas-phase reactions, which led to a dramatic increase in the reaction rate. Without milling, no CO conversion into CO_2 was observed at room temperature over Cr_2O_3. The nature of the highly active sites is not clear, although defects induced by the high mechanical-energy input probably contribute strongly. Although it is very difficult to obtain insight into the processes that occur during milling, a recent study has introduced a method for *in situ* XRD (X-ray diffraction) experiments during ball milling that may be able to shed light onto the catalytic reactions during milling. Ball milling is possible on a production scale (in the cement industry, mills with volumes of 100 m^3 are in use) and if improvements in catalytic effectiveness outweigh the more complex process and potentially high running costs, which are due to milling energy requirements, it may become suitable for industrial applications.[20]

3.3 Mechanochemical Reduction

Castricum *et al.* have found that the creation of various types of defects when zinc oxide (ZnO) is milled (in a vibratory ball mill under vacuum or in the presence of oxygen) results in an increased amorphous fraction, as well as higher surface area.[21] Mechanochemical reactions occur when copper and copper oxides are milled together with zinc oxide; the oxidation of copper and copper oxides takes place in the presence of oxygen, whereas reduction takes place in vacuum. These reactions are promoted by the presence of ZnO. The formation of a Cu_2O-like intermediate is suggested, which is not observed when ZnO is not present.

The high-intensity milling of ZnO results in an increased fraction of amorphous ZnO and in a higher surface area, probably due to surface roughening or increased dispersion. The milling of mixtures of Cu, Cu_2O or CuO and ZnO results in the oxidation of Cu precursors when milling occurs in synthetic air, and reduction when milling occurs under vacuum. These mechanochemical reactions are promoted by the presence of ZnO. The formation of a Cu_2O-like intermediate, which is closely attached to ZnO, has been observed as well as several copper oxide species that reduce at different

Scheme 3.17 Mechanochemical reaction of aromatic ketones and aldehydes in the presence of $Zn/ZnCl_2$.

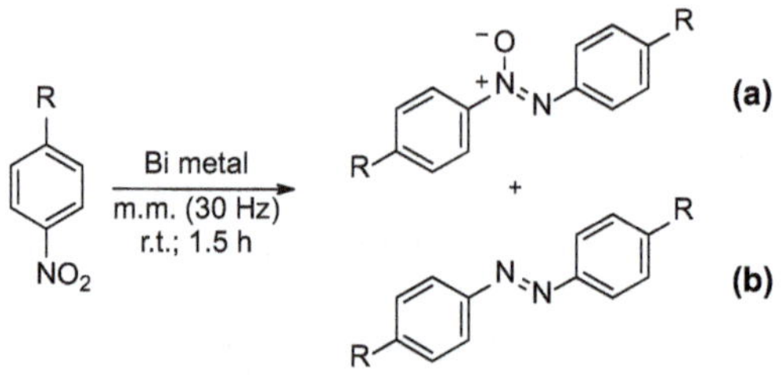

Scheme 3.18 Solvent free reduction of nitroarenes to azoxy and azoarenes using bismuth shots; milling apparatus used: Retsch mixer mill, MM200 (Retsch GmbH, Haan, Germany).

temperatures. Milled samples possess much higher surface areas than both unmilled mixtures and mixtures of separately milled components.[21]

Wang and Liu have reported that the mechanochemical reaction of aromatic ketones and aldehydes in the presence of Zn-$ZnCl_2$ in a high-speed vibration mill (Retsch MM200 mixer mill and Retsch RM100 mortar grinder) gave the pinacol coupling product and reduction product in varying yields (Scheme 3.17). Of the three possible activation methods, the gentler Retsch RM100 mortar grinder technique afforded a much higher yield and improved pinacol coupling product selectivity. However, notably, the pinacol coupling products were formed predominantly or exclusively in all cases.[22]

A solvent-free, one-step reduction of nitroarenes to azoxy and azoarenes using bismuth shots or lead shots under milling conditions was first described by Suzuki and co-workers.[23] The reaction was highly chemoselective, and gave good yields in short reaction times (Scheme 3.18). Although the reaction with Bi shot gave azoxyarenes as major products in aerobic conditions, azoarenes were predominantly formed under the protection of a nitrogen atmosphere (Table 3.11).

Interestingly, the addition of a drop of an inert organic solvent such as hexane or benzene to the nitroarene bismuth mixture prior to milling led to the complete formation of azoxyarenes. Replacement of Bi with Pb selectively provided azoarenes instead. Scheme 3.19 shows the proposed pathway for the deoxygenative dimerization of nitroarenes on the activated bismuth surface.

Nitroarenes were adsorbed and deoxygenated on the newborn bismuth surface to form nitrosoarenes as the initial products, which underwent dimerization to afford azoxyarenes. In the presence of oxygen or an additive, part of the active metal surface is destroyed or occupied by these molecular

Table 3.11 Reaction of *p*-nitroanisole and bismuth shots under different conditions.

Condition	Product ratio[a] (a/b, Scheme 3.18)	Conversion (%)
Under air	78/22	~100
Under air[b]	100/0	7
Under N_2	17/83	~100
Under O_2	100/0	~100
1 Drop of hexane added	100/0	91
1 Drop of benzene added	100/0	~100

[a]Determined by ^{1}H NMR.
[b]Powdery bismuth (~100 mesh) was used.

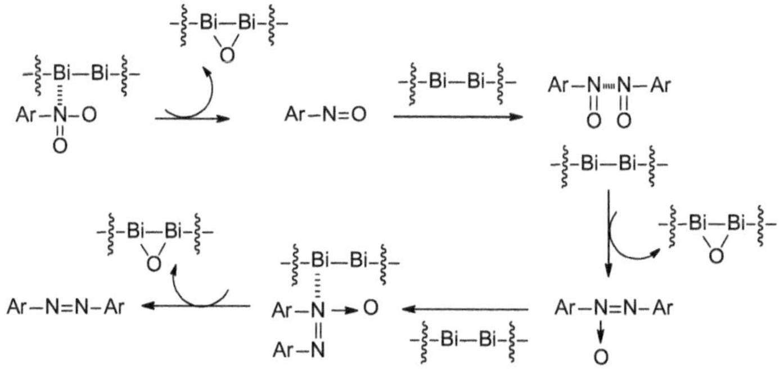

Scheme 3.19 Proposed pathway for the deoxygenative dimerization of nitroarenes.

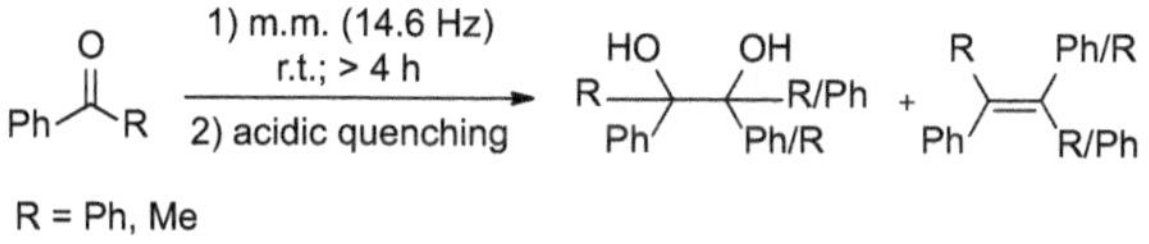

Scheme 3.20 Mechanochemical reactions of benzophenone and acetophenone with Mg.

species and further reduction is suppressed. Otherwise, the reaction proceeded further to produce azoarenes (Scheme 3.19).[23] The direct mechanochemical reactions of benzophenone and acetophenone with Mg (4 eq.) have been explored by Harrowfield *et al.*, who found that the reaction included coupling products to give pinacol, alkene and alkane species, as well as simple reduced species.[24] After acid quenching, benzophenone provided 1,1,2,2-tetraphenylethanediol, tetraphenylethene and 1,1,2,2-tetraphenylethane as the main products, along with small quantities of diphenylmethanol and diphenylmethane. Acetophenone principally gave *cis*- and *trans*-2,3-diphenylbut-2-ene plus *meso*- and *rac*-2,3-diphenylbutane-2,3-diol, as well as some 1-phenylethanol (Scheme 3.20). Moreover, the replacement of Mg with other metals (Al, Fe, Ti and Zn) led to much slower reactions and only McMurray coupling products were readily detected when Ti and Zn were used.

El-Eskandarany *et al.* have reported a unique process for obtaining carbon-free metallic iron *via* a single-step reduction and described how a single-step displacement reaction occurred during the mechanical milling (using a Fritsch P6 ball mill) of a mixture of haematite (Fe_2O_3) and magnesium powders at room temperature under moderate rotation speed (2.2 s^{-1}). During the early milling stage (0–12 h), the Fe_2O_3 particles were embedded in the soft Mg matrix (reducing agent) particles to form coarse reactant material composite powders as a result of the repeated impact and shear forces generated by the balls. After 24 h of milling, a mechanical solid-state reduction takes place between the fresh Fe_2O_3 and Mg. Complete solid-state reduction is successfully achieved after 120 h. The end product, obtained after 150 h, is a nanocrystalline mixture of metallic Fe and MgO powders with average grain size of <20 nm in diameter.[25] On the other hand, metallic Fe particles and non-magnetic oxide were prepared by reducing $MgFe_2O_4$ with Al or Mg in a ball mill, in work by Takacs *et al.* (2002) (Scheme 3.21). The reaction took place as a fast self-propagating process at some point between 0.5 and 1 h of milling. It was shown, using XRD and Mössbauer spectroscopy, that the Fe phase contains a small percentage of dissolved Al, that it shows high defect concentration and that the surface tension of the ultrafine (about 15 nm) grains results in local compressive strains. Magnetization is 25% less than expected for pure ferromagnetic Fe.[26]

A kinetic study using dichloro- and monochlorobenzene (DCB, MCB) as model pollutants and an excess of magnesium and *n*-butylamine as reagents in a laboratory centrifugal ball mill at 25 °C was carried out by Birke *et al.* to investigate the reaction mechanisms for mechanochemical dehalogenation reactions using magnesium and slightly acidic hydrogen donors. The focus of the study was on polyhalogenated aromatics, such as PCBs or hexachlorobenzene, which have chlorine–aromatic ring carbon bonds, which display higher bond strengths than chlorine–aliphatic carbon bonds. An analysis of the findings strongly suggested that one-pot, consecutive Grignard–Zerewitinoff-like reactions occur; firstly, the Grignard intermediates are formed from DCB or MCB and then, in a stepwise manner, their protonation to monochlorobenzene or benzene, respectively, occurs *via* the amine. Furthermore, a rationale was derived to explain the observation of complete reductive dechlorination to benzene, which was formed in approximately 100% yield (mole/mole, based on DCB or MCB).[27]

Mack *et al.* have reported a practical method for the solvent-free reduction of aldehydes and ketones *via* high speed ball milling (using a Spex CertiPrep mixer/mill 8000 M apparatus) in an open atmosphere. Using $NaBH_4$ (1 eq.),

$$MgFe_2O_4 + 2\,Al \longrightarrow MgAl_2O_4 + 2\,Fe$$

$$MgFe_2O_4 + 3\,MgAl \longrightarrow 4MgO + 2\,Fe$$

Scheme 3.21 $MgFe_2O_4$ ball mill reduction in presence of Al or Mg.

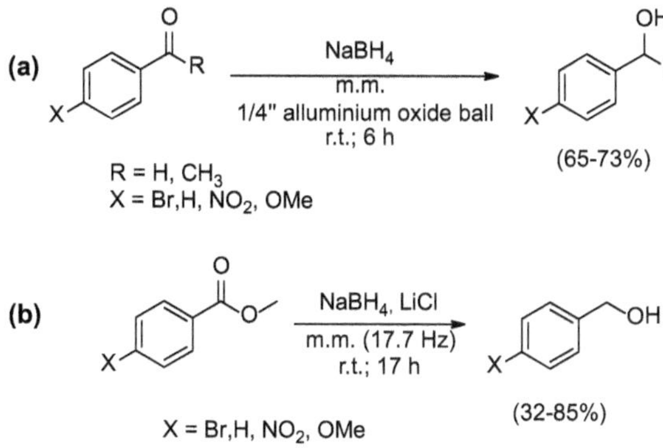

Scheme 3.22 Reduction of *p*-substituted aryl aldehydes and ketones (a) and esters (b) by HSBM.

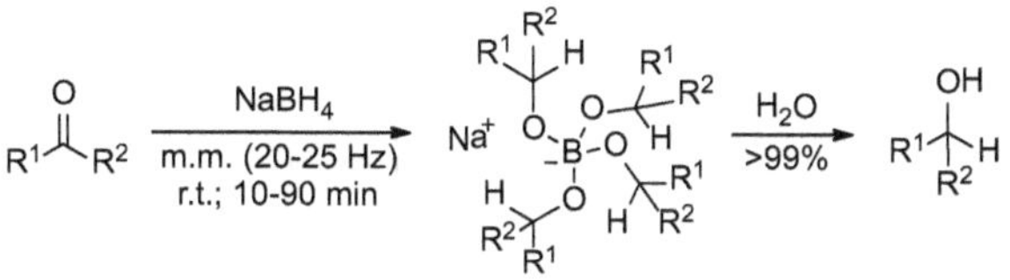

Scheme 3.23 Reduction of carbonyl compounds by $NaBH_4$ under solvent-free ball milling (4 : 1 stoichiometric ratio) using a Retsch MM 200 swing mill. (The yield of solid or oily alcohols was always 99% after hydrolysis of sodium tetraalkoxyborates and workup.)

both aldehydes and ketones were successfully reduced in 65–73% yields after milling for 6 h (Scheme 3.22). The reduction was also carried out with 1/4 eq. of $NaBH_4$, although much longer reaction times were required to achieve comparable yields. These conditions were further extended to the reduction of esters, amides, carboxylic acids and nitriles *via* the addition of lithium chloride, which presumably generated the more powerful and hazardous reducing agent $LiBH_4$.[28]

Naimi-Jamal *et al.* have proved that three solvent-free mechanochemical procedures (ball milling, vibrational ball milling and kneading ball milling) of different aldehydes and ketones with $NaBH_4$ under temperature control give, in fast and reproducible reductions, quantitative yields of thermally stable sodium tetraalkoxyborates (Scheme 3.23).[29]

Nevertheless, the easily isolated solids are extremely sensitive to hydrolysis, which leads to quantitative yields of the corresponding alcohols. The rapid syntheses are regiospecific and stereoselective. Conjugated aldehydes and ketones provide quantitative yields of allylic alcohols that are free from the saturated alcohol contamination that would occur in the solution reaction. Moreover, differing stoichiometric ratios will give different products; benzil is quantitatively reduced to benzoin (4:1 ratio) or dihydrobenzoin (2:1 ratio) (Scheme 3.24).[29]

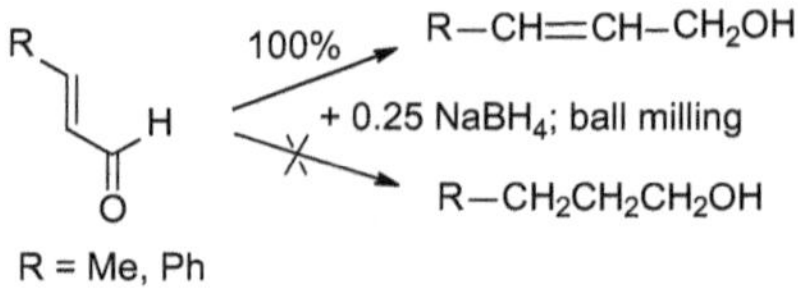

Scheme 3.24 Regiospecific reduction of α,β-unsaturated aldehydes by solvent-free stoichiometric ball milling with $NaBH_4$.

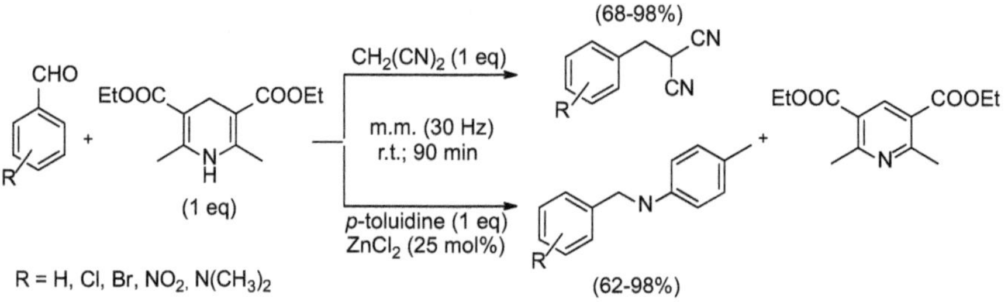

Scheme 3.25 Direct reductive benzylization of malononitrile and 4-methylaniline (*p*-toluidine) using a 1,4-dihydropyridine (HEH).

Blanchard *et al.* have analyzed the reduction reaction of VCl_3 and $LiAlD_4$ at varying ball-mill energies (from 400 to 720 rpm). VCl_3 was reduced by the formation of Li–V–Cl metastable phases, LiCl and free Al and V or Al–V phases, from low to high ball-milling energies. Measurements of the hydrogen released were made using Sievert's apparatus as well as X-ray and neutron diffraction experiments that helped the authors to demonstrate that kinetics enhancement reaches a limit with increased ball-mill energy. They reported that even under mild conditions, at or close to room temperature, the two first steps of the decomposition of VCl_3-enhanced $LiAlD_4$ occurs during the first weeks after milling at temperatures in the range 20–50 °C.[30]

The direct reductive benzylization of malononitrile and 4-methylaniline using aromatic aldehydes was achieved by Wang *et al.* using Hantzsch 1,4-dihydropyridine (HEH) as the only organo-reductant, under solvent-free one-pot mechanochemical conditions (Scheme 3.25). The benzylidenemalononitriles and imines formed *in situ* were efficiently reduced to give benzylated malononitriles and anilines in 62–98% yields at 30 Hz after 90 min milling (Retsch MM200 mixer mill, Retsch GmbH, Haan, Germany). Furthermore, the HEH organo-reductant demonstrates complete selectivity, as no reduction of any other functional group such as aldehydes, nitro or cyano groups was observed.[31]

3.4 Conclusion

This chapter provides a comprehensive overview of solvent-free mechanochemical procedures for oxidation/reduction steps in the synthesis of fine chemicals or the preparation of new material. The advantages over classic liquid-phase methods in terms of higher product yield, better selectivity, shorter reaction time and simpler work-up procedure have been highlighted. Beyond their fundamental importance, the results reported in this chapter may have substantial practical implications because ball milling oxidation/reduction is in most cases scalable up to production scale and therefore suitable for industrial applications.

References

1. M. Nüchter, B. Ondruschka and R. Trotzki, *J. Prakt. Chem.*, 2000, **342**, 720.
2. T. Szuppa, A. Stolle, B. Ondruschka and W. Hopfe, *ChemSusChem*, 2010, **3**, 1181.
3. T. Szuppa, A. Stolle, B. Ondruschka and W. Hopfe, *Green Chem.*, 2010, **12**, 1288.
4. R. Thorwirth, F. Bernhardt, A. Stolle, B. Ondruschka and J. Asghari, *Chem. Eur. J.*, 2010, **16**, 13236.
5. O. Y. Posudievsky, O. A. Khazieieva, V. G. Koshechko and V. D. Pokhodenko, *J. Mater. Chem.*, 2012, **22**, 12465.
6. J. Gao and G.-W. Wang, *J. Org. Chem.*, 2008, **73**, 2955.
7. I. Huskić, I. Halasz, T. Friščić and H. Vančik, *Green Chem.*, 2012, **14**, 1597.
8. G. Cravotto, D. Garella, D. Carnaroglio, E. Calcio Gaudino and O. Rosati, *Chem. Commun.*, 2012, **48**, 11632.
9. R. Schmidt, A. Stolle and B. Ondruschka, *Green Chem.*, 2012, **14**, 1673.
10. S. L. Collom, P. T. Anastas, E. S. Beach, R. H. Crabtree, N. Hazari and T. J. Sommer, *Tetrahedron Lett*, 2013, **54**, 2344.
11. G. I. Nikishin, L. L. Sokova, V. D. Makhaev and N. I. Kapustina, *Russ. Chem. Bull.*, 2008, **57**, 118.
12. N. Pankrushina, I. Nikitina, E. Chernjak, C. Myz, T. Shakhtshneider and V. Boldyrev, *Mater. Manuf. Processes.*, 2008, **23**, 561.
13. T. K. Achar, S. Maiti and P. Mal, *RSC Adv.*, 2014, **4**, 12834.
14. T. Chatterjee and B. C. Ranu, *RSC Adv.*, 2013, **3**, 10680.
15. (a) Y. Chen, *J. Alloy. Compd*, 1997, **257**, 156; (b) Y. Chen, *J. Alloy. Compd*, 1998, **266**, 150.
16. L. Chun and L. Bin, *J. Alloy. Compd*, 2008, **459**, 354.
17. H. Geßwein and J. R. Binder, *Thermochim. Acta*, 2006, **444**, 6.
18. A. L. Ortiz, V. Zamora and F. Rodríguez-Rojas, *Ceram. Int*, 2012, **38**, 2857.
19. S. Mori, W.-C. Xu, T. Ishidzuki, N. Ogasawara, J. Imai and K. Kobayashi, *Appl. Catal., A*, 1996, **137**, 255.
20. S. Immohr, M. Felderhoff, C. Weidenthaler and F. Schüth, *Angew. Chem., Int. Ed.*, 2013, **52**, 12688.
21. H. L. Castricum, H. Bakker and E. K. Poels, *Mater. Sci. Eng. A*, 2001, **304**, 418.
22. G. W. Wang and L. Liu, *Chin. Chem. Lett.*, 2004, **15**, 587.

23. S. Wada, M. Urano and H. Suzuki, *J. Org. Chem.*, 2002, **67**, 8254.
24. J. M. Harrowfield, R. J. Hart and C. R. Whitaker, *Aust. J. Chem.*, 2001, **54**, 423.
25. M. S. El-Eskandarany, H. N. El-Bahnasawy, H. A. Ahmed and N. A. Eissa, *J. Alloy. Compd*, 2001, **314**, 286.
26. L. Takacs, V. K. Garg, V. Soika and A. C. Oliveira, *Hyperfine Interact.*, 2002, **139/140**, 345.
27. V. Birke, C. Schütt and W. K. L. Ruck, Small particle size magnesium in one-pot Grignard–Zerewitinoff-like reactions under mechanochemical conditions: on the kinetics of reductive dechlorination of persistent organic pollutants (POPs), in *Environmental Applications of Nanoscale and Microscale Reactive Metal Particles*, ed. C. L. Geiger and K. M. Carvalho-Knighton, American Chemical Society, Washington DC, 2009, ACS Symposium Series, vol. 1027, ch. 3, pp. 39–54.
28. J. Mack, D. Fulmer, S. Stofel and N. Santos, *Green Chem.*, 2007, **9**, 1041.
29. M. R. Naimi-Jamal, J. Mokhtari, M. G. Dekamin and G. Kaupp, *Eur. J. Org. Chem.*, 2009, 3567.
30. D. Blanchard, A. I. Lem, S. Øvergaard, H. W. Brinks and B. C. Hauback, *J. Alloy. Compd*, 2008, **458**, 467.
31. Z. Zhang, J. Gao, J.-J. Xia and G.-W. Wang, *Org. Biomol. Chem.*, 2005, **3**, 1617.

CHAPTER 4

Asymmetric Organocatalytic Reactions under Ball Milling

ELIZABETH MACHUCA AND EUSEBIO JUARISTI*

Departamento de Química, Centro de Investigación y de Estudios Avanzados del Instituto Politécnico Nacional, Apartado Postal 14-740, 07000-México, D.F. Mexico
*Email: juaristi@relaq.mx

4.1 Introduction

Nowadays, following the convincing arguments presented by the pioneers who advanced the principles of green chemistry,[1] the development of chemical processes that are friendlier to the environment is increasingly pursued.

Several strategies are presently being explored to improve working conditions in chemistry laboratories, both academic and industrial, in order to address salient environmental concerns.[2] In particular, the high-speed ball milling (HSBM) technique is being successfully applied in various areas, such as synthetic organic chemistry,[3] structural modification of fullerenes,[4] co-crystallization,[5] degradation of biomass,[6] and peptide synthesis,[7] among others.[8] Indeed, in many instances, HSBM has been shown to afford more efficient processes relative to the corresponding reactions in solution.

Some advantages of the application of HSBM are (1) the possibility to carry out reactions in the absence of solvent, with an immediate reduction of costs and handling procedures, (2) the frequently observed diminished production of secondary products and potentially toxic wastes, (3) the

RSC Green Chemistry No. 31
Ball Milling Towards Green Synthesis: Applications, Projects, Challenges
Edited by Brindaban Ranu and Achim Stolle

Published by the Royal Society of Chemistry, www.rsc.org

reduction of heating time, which implies energy savings, and (4) the frequent observation of reduced reaction times.

In this context, an important goal in this area is the application of HSBM in asymmetric organocatalysis,[9] *i.e.*, the aim here is to make organocatalysis even more sustainable.[10] Indeed, the high concentration of reactants achieved under solvent-free conditions, the increased mixing efficiency, and lower activation energies are all factors that improve the efficiency of the reaction by decreasing the reaction time and simultaneously improving the enantioselectivity of the reaction.

4.2 Asymmetric Organocatalysis and the Ball Milling Technique

4.2.1 Asymmetric Aldol Reaction

Since the emergence of the organocatalysis, aldol reactions have been widely used as test reaction.[9] One reason for the great interest in the study of asymmetric aldol reactions lies in the feasibility of enantioselective preparation of β-hydroxyketones through C–C bond formation, with the concomitant generation of one or more stereogenic centers. Indeed, a great diversity of molecules presenting the β-hydroxycarbonyl moiety are present in bioactive molecules that are of great importance in the pharmaceutical industry.[11,12]

Aldol reactions are carried out employing various strategies, including metal catalysis,[13] enzymatic catalysis,[14] and more recently by means of small organic molecules in the absence of metals – that is by means of organocatalysts.[9] In this regard, the success of (*S*)-proline (**1**) as organocatalyst in asymmetric intra- or intermolecular aldol reactions[15] has led to the study of the reaction under solvent-free conditions.[16] In particular, Bolm and co-workers reported the first successful asymmetric aldol reaction organocatalyzed by (*S*)-proline (**1**) under HSBM, demonstrating the advantages of this technique when compared with other reaction conditions.[17] With this study, Bolm *et al.* made a revolutionary contribution to organocatalysis: the reaction was performed under solvent-free reaction conditions. Various aldehydes and ketones were evaluated and the anticipated aldol products were obtained in good to excellent yields (42–99%), high diastereoselectivity (up to 99 : 1) and moderate to excellent enantiomeric excess (45–99% ee) (Scheme 4.1).[17,18]

Today, various chiral organocatalysts are used to carry out asymmetric aldol reactions. Indeed, the most successful organocatalysts in solution conditions have been evaluated under solvent-free conditions employing the HSBM technique.

While solvents help homogenize reaction parameters such as temperature, the use of HSBM presents important advantages such as the fact that reagents are generally used in stoichiometric amounts, reaction times are short compared with conventional solution reactions with traditional stirring, and the reactions are much cleaner. It is worth mentioning that

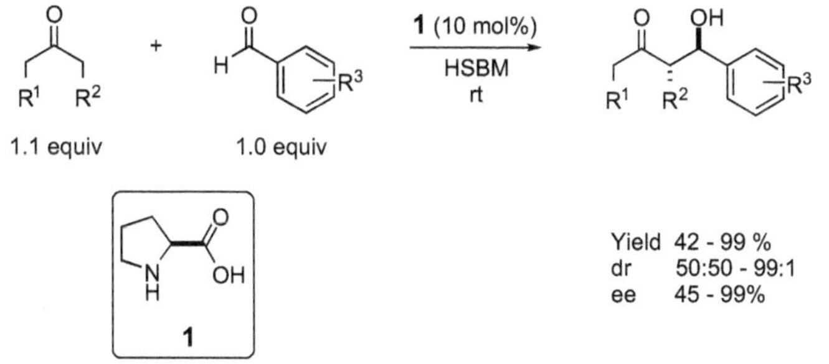

Scheme 4.1 First example of an aldol reaction organocatalyzed by (*S*)-proline (**1**) in a ball-mill under solvent-free conditions.[17,18]

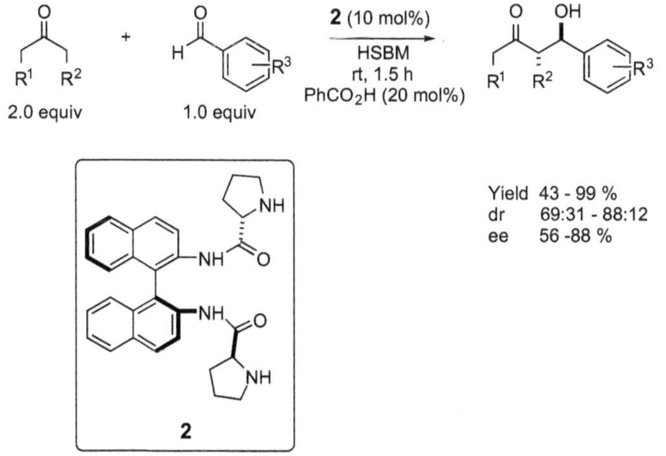

Scheme 4.2 Asymmetric aldol reaction organocatalyzed by (*S*)-Binam-L-prolinamide (**2**) in a ball-mill.[19]

temperature control can be achieved by including pauses ("resting" periods), as well as constant temperature containers.[18]

(*S*)-Binam-L-prolinamide (**2**), another successful organocatalyst employed in asymmetric aldol reactions in solution was also tested by Nájera and co-workers under the ball milling technique.[19] The results show that the reaction proceeds better under HSBM relative to the use of THF/H_2O solvent with conventional stirring. In particular, using 5–10 mol% catalyst in reactions between aldehyde and ketone (1 : 2 ratio), the desired product was obtained in good to excellent yields (43–100%), good diastereoselectivity (up to 88 : 12) in favor of the *anti* isomer, and moderate to good enantioselectivity 56–88% ee (Scheme 4.2).[19]

The use of peptidic organocatalysts has also been evaluated. In particular, the efficacy of peptide organocatalysts (*S*,*S*)-Pro-Phe and (*S*,*S*)-Pro-Trp (**3** and **4**, Scheme 4.3) was examined under the HSBM technique by Hernández and Juaristi.[20,21] These organocatalysts afforded the corresponding aldol

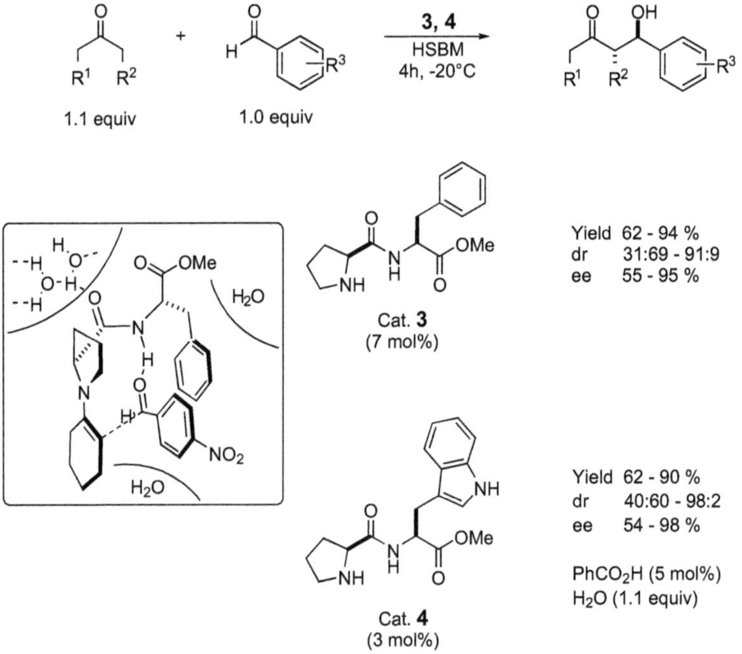

Scheme 4.3 Asymmetric aldol reaction organocatalyzed by (*S*,*S*)-dipeptides **3** and **4** in a ball-mill, with benzoic acid and water as additives.[20,21]

product in 62–94% yield, with high diastereoselectivity (up to 91 : 9) and enantioselectivity ranging from 55% to 95% ee. With the use of additives such as benzoic acid and water, it was possible to improve by more than 10 percentage units the enantioselectivity. Furthermore, the presence of water molecules apparently induces the formation of chiral micelles in the reaction medium, which results in a more rigid transition state and higher stereoselectivities. Finally, the Brønsted acid effect may be explained in terms of the protonated organocatalyst making the amide NH more acidic, and efficient anchoring of the aldehyde substrate in the transition state (Scheme 4.3).[20,21]

Taking into consideration a possible stabilization of the transition state by hydrogen bonding and π–π stacking interactions,[22] it was anticipated that the presence of thioamide segments instead of amide segments in the organocatalyst would give rise to more acidic N–H protons and therefore a stronger hydrogen-bonding interaction between the organocatalyst and the substrate, *i.e.*, a more rigid transition state.[23] In the event, peptide thioamides **5** and **6** led to the formation of aldol products with increased diastereoselectivity, up to 98 : 2 *anti* : *syn* ratio. Nevertheless, the observed enantioselectivity was essentially unchanged.

Thioamides **5** and **6** were also examined in aldol reactions with isatins as substrates.[24] The products of this reaction are of significant interest in view of the biological activity exhibited by tertiary alcohols derived from oxindoles.[25] In the event, the anticipated products were obtained in good yield (54–68%) and high enantioselectivity, up to 86% ee (Scheme 4.4).[24]

4.2.2 Asymmetric Michael Addition

Several chiral organocatalysts have been used successfully in asymmetric Michael reactions in solution. The first organocatalytic asymmetric Michael reactions were reported by Barbas and co-workers just over a decade ago. The reaction proceeded with good yield and high diastereoselectivity, although the observed enantioselectivity was only moderate.[26] This pioneering work motivated the design of chiral organocatalysts that provide better stereocontrol in the reaction.[27,28]

The first example of Michael reactions carried out under solvent-free conditions employing the HSBM technique was reported by Wang *et al.* in 2004. The catalyst was the achiral, inorganic base K_2CO_3. The reaction took place in less than one hour, affording good yields of the anticipated racemic adducts with good *anti*:*syn* diastereoselectivity.[29] As could have been anticipated, interest then arose in the use of chiral organocatalysts that can provide enantioselectivity. Indeed, several chiral organocatalysts that had been used previously in the asymmetric Michael reaction in solution were then examined under solvent-free reaction conditions.[27,28,30]

Michael additions organocatalyzed by thioureas have been particularly successful.[31] Studies by Houk and co-workers showed that the relevance of hydrogen bonding in Michael reactions decreases with increasing solvent

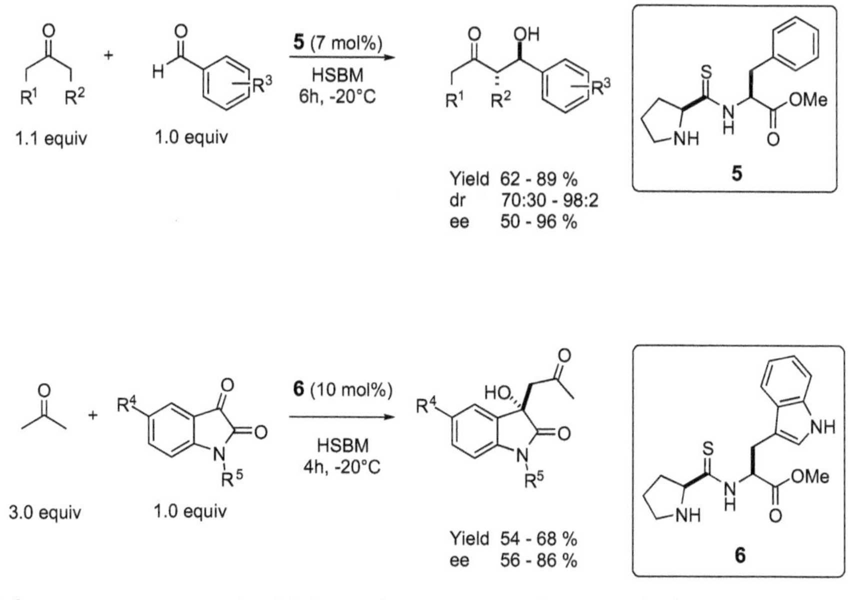

Scheme 4.4 Asymmetric aldol reaction organocatalyzed by (*S*,*S*)-thiodipeptides **5** and **6** in a ball-mill.[24]

polarity.[32] As a consequence of a less rigid transition state, both the reaction efficiency and the stereoselectivity diminish significantly in polar media.[33]

In this context, the use of the HSBM technique can be highly convenient, since the formation of the desired hydrogen bonding is more efficient under solvent-free conditions. Thus, the development of asymmetric Michael reactions under the HSBM technique is a very promising area of study.

The first study in this topic was conducted by Šebesta *et al.* who examined the asymmetric Michael reaction between aldehydes and β-nitrostyrene in solvent-free conditions under HSBM with catalysis by chiral pyrrolidine derivatives.[34] In particular, α,α-diphenylprolinol trimethylsilyl ether (7) gave the desired Michael products in yields in the range 44–97%, and with diastereoselectivities that reached 95 : 5 diastereomeric ratios, and good enantioselectivities (62–94% ee) (Scheme 4.5). A salient observation is that by employing HSBM the yields of the reaction can be quantitative after just 1 h reaction time, which corresponds to a reduction of more than twenty times in the reaction time required under the best conditions in solution employing organocatalyst *O*-lauroyl-*trans*-4-hydroxyproline (Scheme 4.5).[34]

Recently, Xu and co-workers reported the use of chiral squaramide derivatives as hydrogen bonding donor catalysts for the Michael addition reaction of 1,3-dicarbonyl compounds with nitroolefins under solvent-free conditions using a planetary ball-mill. The importance of carrying out the reaction in the absence of solvents was evident as only 0.5 mol% of the organocatalyst was required to successfully perform the asymmetric reaction between 2,4-pentanedione and various aromatic β-nitrostyrenes.[35] Indeed, after only 5–25 min of reaction, the desired adducts were obtained in good yields (63–95%), good diastereoselectivities (up to 84 : 16 dr) and excellent enantioselectivity (91–99% ee). It was concluded that HSBM did improve the performance of the organocatalyst relative to the reaction in solution (Scheme 4.6).[35]

Interestingly, application of grinding with a mortar and pestle in the highly stereoselective Michael addition of trisubstituted β-ketoesters to nitroalkene derivatives was recently reported by Chimni's group.[36]

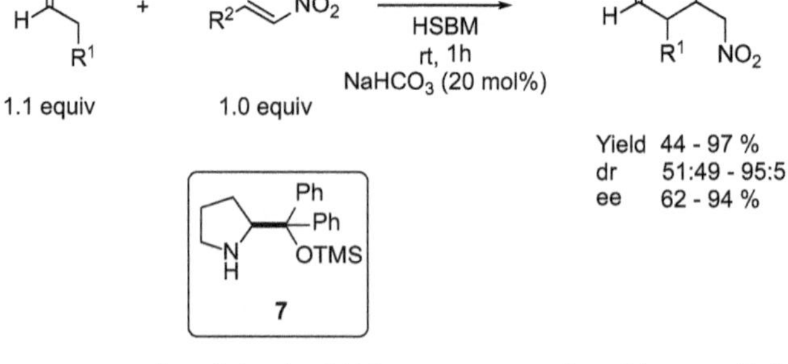

Scheme 4.5 Asymmetric Michael addition organocatalyzed by α,α-diphenylprolinol trimethylsilyl ether (7) in a ball-mill.[34]

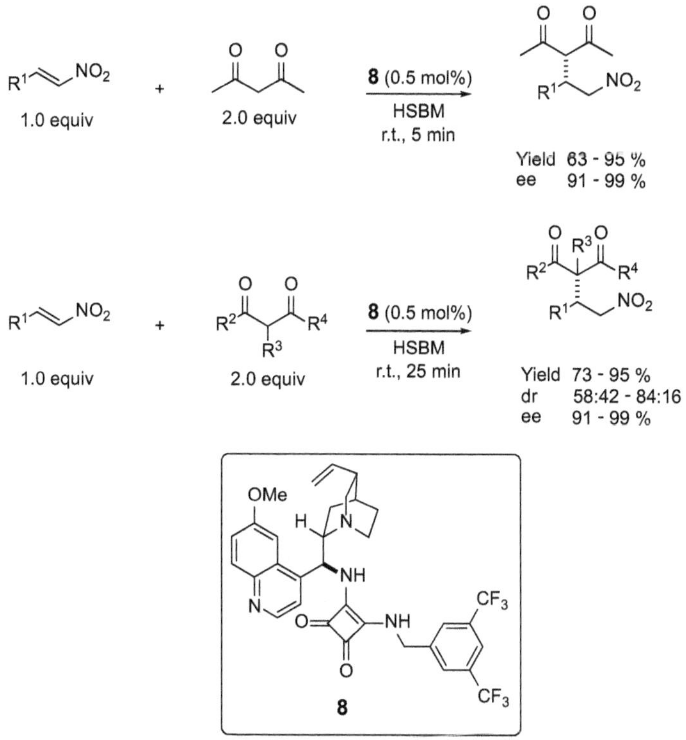

Scheme 4.6 Asymmetric Michael additions organocatalyzed by squaramide **8** in a ball-mill.[35]

As already mentioned, the use of chiral thioureas as organocatalysts in asymmetric Michael additions has been highly successful. In an additional example, the reaction of α-nitrocyclohexanone with various nitrostyrenes, in the presence of 2.5 mol% of catalyst **9**, resulted in efficient Michael adduct formation, in high yields (80–95%), with high diastereoselectivity (94 : 6 to 98 : 2 diastereomeric ratios), and excellent enantiomeric ratios (96 : 4 to 99 : 1) (Scheme 4.7). These results reveal that the efficiency of the stereoinduction by chiral organocatalysts through hydrogen bonding is greatly increased under the solvent-free reaction conditions characteristic of HSBM.[37]

Unlike the previously mentioned examples, the approach developed by Friščić when using hydrogen-bonding organocatalysts in HSBM-assisted addition reactions makes use of dichloromethane in liquid assisted grinding (LAG). As the name indicates, in this technique a liquid additive helps improve the grinding efficiency. Even when a liquid is used in LAG, the reaction is still considered a solvent-free reaction, since the amount of solvent present is much less than that required to dissolve the reagents.[38]

Very recently, 15 structurally different hydrogen-bonding chiral organocatalysts were evaluated in Michael additions to β-nitrostyrene as substrate under solvent-free HSBM conditions and compared with the corresponding

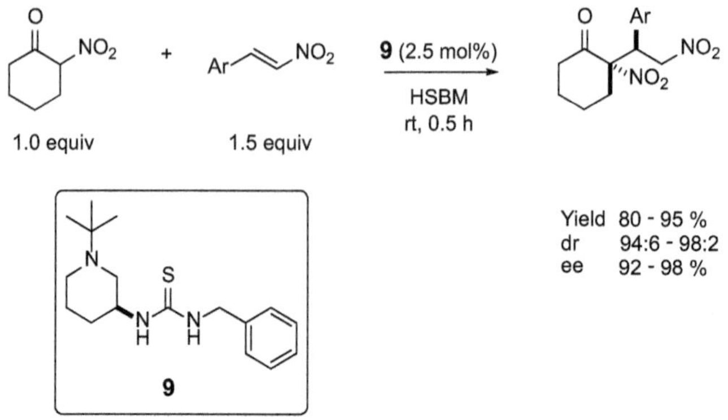

Scheme 4.7 Asymmetric Michael addition organocatalyzed by thiourea **9** in a ball-mill.[37]

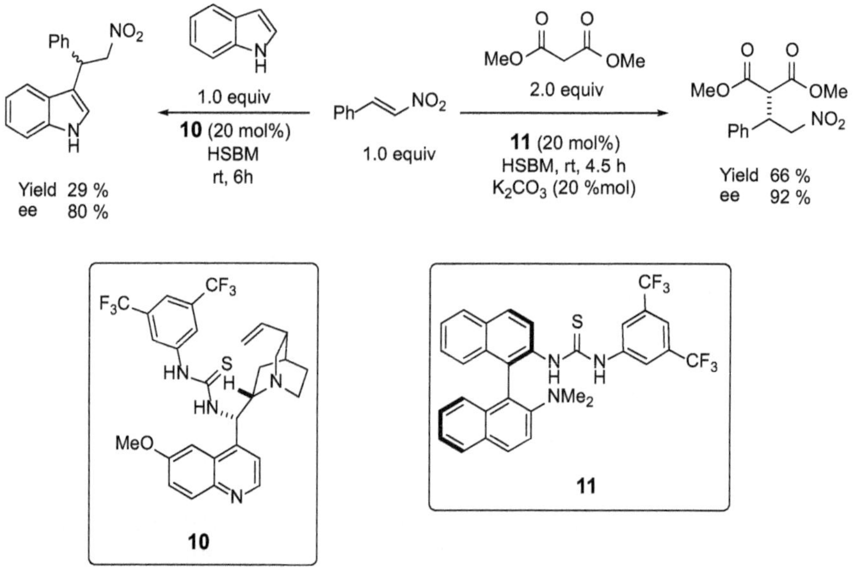

Scheme 4.8 Asymmetric Michael addition organocatalyzed by cinchona and L-Binam-thioureas **10** and **11** in a ball-mill.[39]

experiments in solution. In particular, evaluation was reported of several chiral thiourea-containing cinchonas, binaphthyls, and squaramides and bis-thioureas as organocatalysts under solvent-free conditions using LAG to obtain enantioenriched Michael adducts (Scheme 4.8).[39] Reaction yields were significantly improved relative to the reactions carried out in solution. Evidently, HSBM can be an excellent strategy to carry out reactions activated by hydrogen bonding.[40] The lower efficiency observed in solution was tentatively ascribed to heating of the containers that takes place under traditional heating conditions and that disrupts the otherwise hydrogen-bonded transition states.[39]

4.2.3 Asymmetric Morita–Baylis–Hillman (MBH) Reaction

The Morita–Baylis–Hillman (MBH) reaction constitutes a useful method for the synthesis of β-hydroxycarbonyl compounds with an α-alkylidene group.[41] The use of small organic molecules as catalysts to perform asymmetric MBH transformations has received increasing attention over the past decade.[9] In this context, Friščić's group recently reported the *in situ* synthesis of thiourea organocatalyst **12**, which was then evaluated in the asymmetric MBH reaction (Scheme 4.9).[42]

4.2.4 Asymmetric Alkylation of Imines

Since the development of the enantioselective alkylation of O'Donnell's Schiff bases,[43] this methodology has been widely exploited for the preparation of chiral amino acids.[44] Recently, Lamaty and co-workers used several cinchona organocatalysts with proven efficiency in amine alkylation processes in solution and prepared various chiral amino acids by enantioselective alkylation of alkylidene imines under the HSBM technique. Over a period of 1 h under solvent-free reaction conditions, it was possible to obtain the desired compounds with excellent yields (in the range 91–97%) and with up to 75% ee (Scheme 4.10).[45]

4.2.5 Desymmetrization of Aromatic Diamines and *meso* Anhydrides by Mechanochemistry

The desymmetrization of prochiral precursors such as *meso* diols, diamines, and anhydrides into non-symmetrical and/or chiral building blocks is an important goal in synthetic organic chemistry.[46] In this context, Friščić and

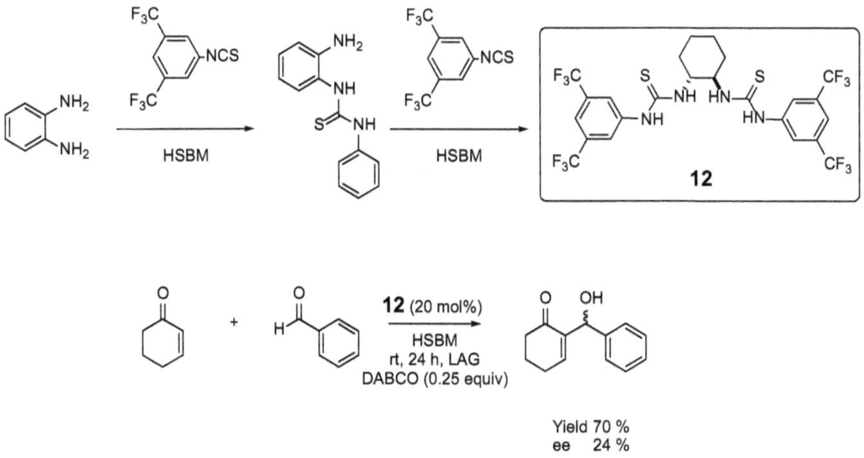

Scheme 4.9 Asymmetric Morita–Baylis–Hillman addition reaction organocatalyzed by bis(thiourea) **12**, synthesized *in situ* in a ball-mill.[42]

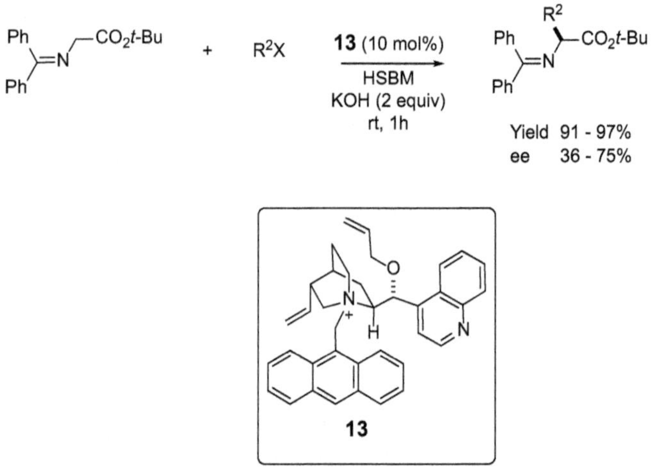

Scheme 4.10 Asymmetric alkylation of alkylidene imines organocatalyzed by cinchona derivative **13**.[45]

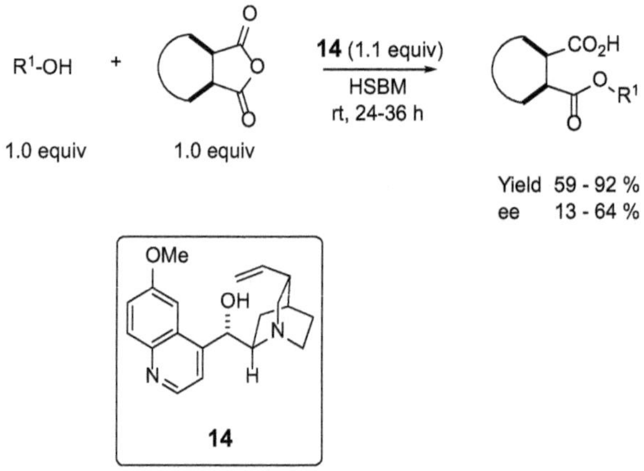

Scheme 4.11 Asymmetric ring opening of *meso* anhydrides organocatalyzed by quinidine (**14**).[48]

co-workers[47] recently reported that *ortho-* and *para*-phenylenediamines were desymmetrized and quantitatively transformed into mono- and bis-(thio)ureas through a one-pot mechanochemical "click" reaction sequence.

To carry out the opening of cyclic anhydrides with alcohols in enantioselective fashion, Bolm *et al.* made use of quinidine organocatalyst **14** under HSBM conditions. Yields ranging from 59% to 92% and enantiomeric excesses of 13–64% were obtained (Scheme 4.11).[48]

4.3 Closing Remarks

Ball-milling constitutes a powerful strategy for the development of environmentally benign chemical transformations. The application of HSBM in asymmetric synthesis is relatively recent, but the accumulated observations already provide evidence that this methodology generally affords higher stereoselectivity, higher yields, and cleaner reaction products. Furthermore, the technique can be scaled-up,[49] and it is already being applied in materials science and large-scale enterprises such as the paint and pharmaceutical industries.

Nevertheless, the field is still open to experimental modifications, both by the experimental practitioner and by instrument developers, in order to achieve better control of the reaction conditions, as well as to reach a better understanding of the reaction mechanism in the absence of solvent and under purely mechanical activation. Particularly relevant are the very recent advances in real-time and *in situ* monitoring of mechanochemical milling reactions.[50] Of course, the rapid development of asymmetric organocatalysis in solution offers limitless possibilities for application and study under solvent-free, ball-milling conditions. Finally, additional innovations in the field are possible by means of the combination of metallic and organocatalysts.[51]

Acknowledgements

We are indebted to José Gregorio Hernández for a careful revision of the manuscript prior to submission.

References

1. (a) P. T. Anastas and J. C. Warner, *Green Chemistry: Theory and Practice*; Oxford University Press, New York, 1998; (b) I. Horvath and P. T. Anastas, *Chem. Rev.*, 2007, **107**, 2167; (c) N. Eghbali and P. T. Anastas, *Chem. Soc. Rev.*, 2010, **39**, 301; (d) R. A. Sheldon, *Chem. Soc. Rev.*, 2012, **41**, 1437; (e) C. Jiménez-González, D. J. Constable and C. S. Ponder, *Chem. Soc. Rev.*, 2012, **41**, 1485; (f) P. G. Jessop, *Green Chem.*, 2011, **13**, 1391.
2. (a) R. S. Varma, *Green Chem.*, 2008, **10**, 1129; (b) A. Stolle, T. Szuppa, S. E. S. Leonhardt and B. Ondruschka, *Chem. Soc. Rev.*, 2011, **40**, 2317; (c) R. S. Varma, *Green Chem.*, 2014, **16**, 2027.
3. (a) J. Mack and M. Shumba, *Green Chem.*, 2007, **9**, 328; (b) J. Mack, D. Fulmer, S. Stofel and N. Santos, *Green Chem.*, 2007, **9**, 1041; (c) T. Szuppa, A. Stolle, B. Ondruschka and W. Hopfe, *ChemSusChem*, 2010, **3**, 1181; (d) M. O'Brien, R. Denton and S. V. Ley, *Synthesis*, 2011, 1157; (e) N. Mukherjee, T. Chatterjee and B. C. Ranu, *J. Org. Chem.*, 2013, **78**, 11110.
4. G. W. Wang, K. Komatsu, Y. Murata and M. Shiro, *Nature*, 1997, **387**, 583.
5. (a) M. D. Eddleston, M. Arhangelskis, T. Friščić and W. Jones, *Chem. Commun.*, 2012, **48**, 11340; (b) G. Springuel and T. Leyssens, *Cryst. Growth Des.*, 2012, **12**, 3374; (c) P. Thorey, P. Bombicz, I. M. Szilagyi, P. Molnar, G. Bansaghi, E. Szekely, B. Simandi, L. Parkanyi, G. Pokol and J. Madarasz, *Thermochim. Acta*, 2010, **497**, 129; (d) J. C. Espinosa-Lara,

D. Guzman-Villanueva, J. I. Arenas-García, D. Herrera-Ruiz, J. Rivera-Islas, P. Roman-Bravo, H. Morales-Rojas and H. Hoepfl, *Cryst. Growth Des.*, 2013, **13**, 169; (e) I. Halasz, A. Puškarić, S. A. J. Kimber, P. J. Beldon, A. M. Belenguer, F. Adams, V. Honkimäki, R. E. Dinnebier, B. Patel, W. Jones, V. Štrukil and T. Friščić, *Angew. Chem., Int. Ed.*, 2013, **52**, 11538.
6. (a) T. Kleine, J. Buendia and C. Bolm, *Green Chem.*, 2013, **15**, 160; (b) Y. Liao, Q. Liu, T. Wang, J. Long, L. Ma and Q. Zhang, *Green Chem.*, 2014, **16**, 3305; (c) H. Zhang, T. R. Carlson, R. Xiao and G. W. Huber, *Green Chem.*, 2012, **14**, 98; (d) A. Klapiszewski, M. Wysokowski, I. Majchrzak, T. Szatkowski, M. Nowacka, K. Siwińska-Stefańska, K. Szwarc-Rzepka, P. Bartczak, H. Ehrlich and T. Jesionowski, *J. Nanomater.*, 2013, Article ID 425726 (DOI: http://dx.doi.org/10.1155/2013/425726).
7. (a) V. de Clerck, P. Nun, J. Martinez and F. Lamaty, *Angew. Chem., Int. Ed.*, 2009, **48**, 9318; (b) J. G. Hernández and E. Juaristi, *J. Org. Chem.*, 2010, **75**, 7107; (c) T.-X. Métro, J. Bonnamour, T. Reidon, J. Sarpoulet, J. Martinez and F. Lamaty, *Chem. Commun.*, 2012, **48**, 11781; (d) D. Margetić, *Chem. Commun.*, 2012, **48**, 12100; (e) J. Bonnamour, T.-X. Métro, J. Martinez and F. Lamaty, *Green Chem.*, 2013, **15**, 1116.
8. For additional applications of HSBM, see: (a) S. L. James, C. J. Adams, C. Bolm, D. Braga, P. Collier, T. Friščić, F. Grepioni, K. D. M. Harris, G. Hyett, W. Jones, A. Krebs, J. Mack, L. Maini, A. G. Orpen, I. P. Parkin, W. C. Shearouse, J. W. Steed and D. C. Waddell, *Chem. Soc. Rev.*, 2012, **41**, 413; (b) C.-C. Hsu, N.-C. Chen, C.-C. Lai, Y.-H. Liu, S.-M. Peng and S.-H. Chiu, *Angew. Chem., Int. Ed.*, 2008, **47**, 7475; (c) J. G. Hernández, N. A. J. Macdonald, C. Mottillo, I. S. Butler and T. Friščić, *Green Chem.*, 2014, **16**, 1087; (d) M. Ferguson, N. Giri, X. Huang, D. Apperley and S. L. James, *Green Chem.*, 2014, **16**, 1374.
9. See, for example: (a) P. I. Dalko and L. Moisan, *Angew. Chem., Int. Ed.*, 2004, **43**, 5138; (b) J. Seayad and B. List, *Org. Biomol. Chem.*, 2005, **3**, 719; (c) D. Enders, C. Grondal and M. R. M. Huettl, *Angew. Chem., Int. Ed.*, 2007, **46**, 1570; (d) D. W. C. MacMillan, *Nature*, 2008, **455**, 304; (e) A. Dondoni and A. Massi, *Angew. Chem., Int. Ed.*, 2008, **47**, 4638; (f) P. Melchiorre, M. Marigo, A. Carlone and G. Bartoli, *Angew. Chem., Int. Ed.*, 2008, **47**, 6138; (g) X. Yu and W. Wang, *Chem. Asian J.*, 2008, **3**, 516; (h) S. Bertelsen and K. A. Jørgensen, *Chem. Soc. Rev.*, 2009, **38**, 2178; (i) P. M. Pihko, I. Majander and A. Erkkila, *Top. Curr. Chem.*, 2010, **291**, 29; (j) A. Moyano and R. Ríos, *Chem. Rev.*, 2011, **111**, 4703; (k) H. Pellissier, *Recent Developments in Asymmetric Organocatalysis*, RSC Publishing, Cambridge, 2012; (l) R. Ríos Torres (ed.), *Enantioselective Catalysis: Bond Formation Methodologies and Activation Modes*, John Wiley & Sons, Inc., Hoboken, 2013.
10. (a) J. G. Hernández and E. Juaristi, *Chem. Commun.*, 2012, **48**, 5396; (b) P. Chauhan and S. S. Chimni, *Beilstein J. Org. Chem.*, 2012, **8**, 2132.
11. (a) R. Mahrwald, *Modern Aldol Reactions*, Wiley-VCH Verlag GmbH, Weinheim, 2004, Vols. 1-2; (b) R. Mahrwald (ed.), *Modern Methods in*

Stereoselective Aldol Reactions, Wiley-VCH Verlag GmbH, Weinheim, 2013.
12. (a) B. M. Trost and C. S. Brindle, *Chem. Soc. Rev.*, 2010, **39**, 1600; (b) V. Bisai, A. Bisai and V. K. Singh, *Tetrahedron*, 2012, **68**, 4541.
13. L. Bernardi, M. Fochi, M. C. Franchini and A. Ricci, *Org. Biomol. Chem.*, 2012, **10**, 2911.
14. See, for example: (a) H. Gröger, E. M. Vogland and M. Shibasaki, *Chem. Eur. J.*, 1998, **4**, 1137; (b) S. G. Nelson, *Tetrahedron: Asymmetry*, 1998, **9**, 357; (c) E. M. Carreira, in *Comprehensive Asymmetric Catalysis*, ed. E. N. Jacobsen, A. Pfalz and H. Yamamoto, Springer, Heidelberg, 1999, vol. 3, pp. 997–1065; (d) R. Mahrwald, *Chem. Rev.*, 1999, **99**, 1095; (e) T. D. Machajewski and C.-H. Wong, *Angew. Chem., Int. Ed.*, 2000, **39**, 1352; (f) B. Alcaide and P. Almendros, *Eur. J. Org. Chem.*, 2002, 1595; (g) C. Palomo, M. Oiarbide and J. M. García, *Chem. Soc. Rev.*, 2004, **33**, 65; (h) R. Mestres, *Green Chem.*, 2004, **6**, 583; (i) J. L. Vicario, D. Badía, L. Carrillo, E. Reyes and J. Etxbarria, *Curr. Org. Chem.*, 2005, **9**, 219; (j) B. Schetter and R. Mahrwald, *Angew. Chem., Int. Ed.*, 2006, **45**, 7506.
15. B. List, R. A. Lerner and C. F. Barbas, *J. Am. Chem. Soc.*, 2000, **122**, 2395.
16. For salient examples of solvent-free asymmetric organocatalytic aldol reactions, see: (a) A. Berkessel, K. Roland and J. M. Neudörfl, *Org. Lett.*, 2006, **8**, 4195; (b) A. Carlone, M. Marigo, C. North, A. Landa and K. A. Jørgensen, *Chem. Commun.*, 2006, 4928; (c) Y. Hayashi, S. Aratake, T. Itoh, T. Okano, T. Sumiya and M. Shoji, *Chem. Commun.*, 2007, 957.
17. B. Rodríguez, T. Rantanen and C. Bolm, *Angew. Chem., Int. Ed.*, 2006, **45**, 6924.
18. B. Rodríguez, A. Bruckmann and C. Bolm, *Chem. Eur. J.*, 2007, **13**, 4710.
19. (a) G. Guillena, M. C. Hita, C. Nájera and S. F. Viózquez, *Tetrahedron: Asymmetry*, 2007, **18**, 2300; (b) G. Guillena, M. C. Hita, C. Nájera and S. F. Viózquez, *J. Org. Chem.*, 2008, **73**, 5933; (c) D. Almasi, D. A. Alonso and C. Nájera, *Adv. Synth. Catal.*, 2008, **350**, 2467.
20. J. G. Hernández and E. Juaristi, *J. Org. Chem.*, 2011, **76**, 1464.
21. J. G. Hernández and E. Juaristi, *Tetrahedron*, 2011, **67**, 6953.
22. Regarding the importance of π–π stacking interactions, see: (a) A. L. Fuentes de Arriba, L. Simón, C. Raposo, V. Alcázar, F. Sanz, F. M. Múñiz and J. R. Morán, *Org. Biomol. Chem.*, 2010, **8**, 2979; (b) Z. Jiang, H. Yang, X. Han, J. Luo, M. W. Wong and Y. Lu, *Org. Biomol. Chem.*, 2010, **8**, 1368; (c) M. R. Vishnumaya and V. K. Singh, *J. Org. Chem.*, 2009, **74**, 4289; (d) M. Raj and V. K. Singh, *Chem. Commun.*, 2009, 6687; (e) C. Wang, Y. Jiang, X.-X. Zhang, Y. Huang, B.-G. Li and G.-L. Zhang, *Tetrahedron Lett.*, 2007, **48**, 4281; (f) Z. Jiang, Z. Liang, X. Wua and Y. Lu, *Chem. Commun.*, 2006, 2801.
23. See, for example: (a) A. G. Doyle and E. N. Jacobsen, *Chem. Rev.*, 2007, **107**, 5713; (b) Z. Zhang and P. R. Schreiner, *Chem. Soc. Rev.*, 2009, **38**, 1187; (c) D. Gryko, M. Chromiński and D. J. Pielacińska, *Symmetry*, 2011, **3**, 265; (d) B. Wang, X.-W. Liu, L.-Y. Liu, W.-X. Chang and J. Li, *Eur. J. Org. Chem.*, 2010, 5951; (e) B. Wang, G.-H. Chen, L.-Y. Liu, W.-X. Chang

and J. Li, *Adv. Synth. Catal.*, 2009, **351**, 2441; (f) D. Gryko, M. Zimnicka and R. Lipiński, *J. Org. Chem.*, 2007, **72**, 964; (g) D. Gryko and R. Lipiński, *Adv. Synth. Catal.*, 2005, **347**, 1948; (h) P. Dziedzic, W. Zou, J. Háfren and A. Córdova, *Org. Biomol. Chem.*, 2006, **4**, 38; (i) D. Gryko and R. Lipiński, *Eur. J. Org. Chem.*, 2006, 3864.

24. J. G. Hernández, V. García-López and E. Juaristi, *Tetrahedron*, 2012, **68**, 92.
25. See, for example: (a) M. Somei and F. Yamada, *Nat. Prod. Rep.*, 2003, **20**, 216; (b) H.-P. Zhang, H. Shigemori, M. Ishibashi, T. Kosaka, G. R. Pettit, Y. Kamano and J. Kobayashi, *Tetrahedron*, 1994, **50**, 10201; (c) Y. Kamano, H.-P. Zhang, Y. Ichihara, H. Kizu, K. Komiyama and G. R. Pettit, *Tetrahedron Lett.*, 1995, **36**, 2783.
26. (a) K. Sakthivel, W. Notz, T. Bui and C. F. Barbas, *J. Am. Chem. Soc.*, 2001, **123**, 5260; (b) J. M. Betancort, K. Sakthivel, R. Thayumanavan and C. F. Barbas, *Tetrahedron Lett.*, 2001, **42**, 4441; (c) B. List, P. Pojarliev and H. J. Martin, *Org. Lett.*, 2001, **3**, 2423.
27. (a) C. Palomo, S. Vera, A. Mielgo and E. Gómez-Bengoa, *Angew. Chem., Int. Ed.*, 2006, **45**, 5984; (b) M. P. Lalonde, Y. Chen and E. N. Jacobsen, *Angew. Chem., Int. Ed.*, 2006, **45**, 6366; (c) E. Alza, S. Sayalero, P. Kasaplar, D. Almasi and M. A. Pericás, *Chem. Eur. J.*, 2011, **17**, 11585; (d) E. Alza and M. Pericás, *Adv. Synth. Catal.*, 2009, **351**, 3051.
28. Y. Hayashi, H. Gotoh, T. Hayashi and M. Shoji, *Angew. Chem., Int. Ed.*, 2005, **44**, 4212.
29. (a) Z. Zhang, Y.-W. Dong, G.-W. Wang and K. Komatsu, *Synlett*, 2004, 61; (b) Z. Zhang, Y.-W. Dong, G.-W. Wang and K. Komatsu, *Chem. Lett.*, 2004, **33**, 168; see also: (c) G. Choudhary and R. K. Peddinti, *Green Chem.*, 2011, **13**, 276.
30. (a) N. Mase, R. Thayumanavan, F. Tanaka and C. F. Barbas, *Org. Lett.*, 2004, **6**, 2527; (b) J. M. Betancort and C. F. Barbas, *Org. Lett.*, 2001, **3**, 3737; (c) J. P. Malerich, K. Hagihara and V. H. Rawal, *J. Am. Chem. Soc.*, 2008, **130**, 14416; (d) H. Y. Bae, S. Some, J. S. Oh, Y. S. Lee and C. E. Song, *Chem. Commun.*, 2011, **47**, 9621.
31. For reviews on bifunctional hydrogen-bonding organocatalysis, see: (a) S. Narayanaperumal, D. G. Rivera, R. C. Silva and M. W. Paixao, *ChemCatChem*, 2013, **5**, 2756; (b) Y. Takemoto, *Org. Biomol. Chem.*, 2005, **3**, 4299; (c) M. S. Taylor and E. N. Jacobsen, *Angew. Chem., Int. Ed.*, 2006, **45**, 1520; (d) S. J. Connon, *Chem. Commun.*, 2008, 2499; (e) S. J. Connon, *Synlett*, 2009, 354; (f) L.-Q. Lu, X.-L. An, J.-R. Chen and W.-J. Xiao, *Synlett*, 2012, 490.
32. J. Chen, M. A. McAllister, J. K. Lee and K. N. Houk, *J. Org. Chem.*, 1998, **63**, 4611.
33. (a) Y. Huang, A. K. Unni, A. N. Thadani and V. H. Rawal, *Nature*, 2003, **424**, 146; (b) T. Okino, Y. Hoashi and Y. Takemoto, *J. Am. Chem. Soc.*, 2003, **125**, 12672; (c) J. Wang, H. Li, W. Duan, L. Zu and W. Wang, *Org. Lett.*, 2005, 7, 4713.
34. E. Veverková, V. Poláčková, L. Liptáková, E. Kázmerová, M. Mečiarová, Š. Toma and R. Šebesta, *ChemCatChem*, 2012, **4**, 1013.

35. Y.-F. Wang, R.-X. Chen, K. Wang, B.-B. Zhang, Z.-B. Li and D.-Q. Xu, *Green Chem.*, 2012, **14**, 893.
36. P. Chauhan and S. S. Chimni, *Asian J. Org. Chem.*, 2012, **1**, 138.
37. (a) M. Jörres, I. Schiffers, I. Atodiresei and C. Bolm, *Org. Lett.*, 2012, **14**, 4518; (b) M. Jörres, S. Mersmann, G. Raabe and C. Bolm, *Green Chem.*, 2013, **15**, 612.
38. T. Friščić, *Chem. Soc. Rev.*, 2012, **41**, 3493.
39. M. Hestericová and R. Šebesta, *Tetrahedron*, 2014, **70**, 901.
40. In general, reactions activated by more acidic catalysts require shorter reaction times and afford higher enantioselectivities, see: (a) X. Li, H. Deng, B. Zhang, J. Li, L. Zhang, S. Luo and J. P. Cheng, *Chem. Eur. J.*, 2010, **16**, 450; (b) see, also: G. Jakab, C. Tancon, Z. Zhang, K. M. Lippert and P. R. Schreiner, *Org. Lett.*, 2012, **14**, 1724.
41. For recent reviews of the Morita–Baylis–Hillman (MBH) reaction, see: (a) D. Basavaiah, B. S. Reddy and S. S. Badsara, *Chem. Rev.*, 2010, **110**, 5447; (b) G. Masson, C. Housseman and J. Zhu, *Angew. Chem., Int. Ed.*, 2007, **46**, 4614; (c) D. Basavaiah, K. V. Rao and R. J. Reddy, *Chem. Soc. Rev.*, 2007, **36**, 1581; (d) D. Basavaiah, A. J. Rao and T. Satyanarayana, *Chem. Rev.*, 2003, **103**, 811.
42. V. Štrukil, M. D. Igrc, M. Eckert-Maksić and T. Friščić, *Chem. Eur. J.*, 2012, **18**, 8464.
43. (a) M. J. O´Donnell, J. M. Boniece and S. E. Earp, *Tetrahedron Lett.*, 1978, **30**, 2641; (b) M. J. O´Donnell, W. D. Bennett and S. Wu, *J. Am. Chem. Soc.*, 1989, **111**, 2353.
44. (a) D. Ferraris, *Tetrahedron*, 2007, **63**, 9581; (b) T. Hashimoto and K. Maruoka, *Chem. Rev.*, 2007, **107**, 5656; (c) M. J. O'Donnell, *Acc. Chem. Res.*, 2004, **37**, 506; (d) B. Lygo and B. I. Andrews, *Acc. Chem. Res.*, 2004, **37**, 518; (e) B. Lygo, B. I. Andrews, J. Crosby and J. A. Peterson, *Tetrahedron Lett.*, 2002, **43**, 8015; (f) K. Maruoka, *Org. Process Res. Dev.*, 2008, **12**, 679.
45. P. Nun, V. Pérez, M. Calmès, J. Martinez and F. Lamaty, *Chem. Eur. J.*, 2012, **18**, 3773.
46. A. C. Spivey and B. I. Andrews, *Angew. Chem., Int. Ed.*, 2001, **40**, 3131.
47. V. Štrukil, D. Margetić, M. D. Igrc, M. Eckert-Maksic and T. Friščić, *Chem. Commun.*, 2012, **48**, 9705.
48. T. Rantanen, I. Schiffers and C. Bolm, *Org. Process Res. Dev.*, 2007, **11**, 592.
49. A. Stolle, R. Schmidt and K. Jacob, *Faraday Discuss.*, 2014, DOI: 10.1039/c3fd00144j.
50. (a) D. Gracin, V. Štrukil, T. Friščić, I. Halasz and K. Užarević, *Angew. Chem., Int. Ed.*, 2014, **53**, 6193; (b) T. Friščić, I. Halasz, P. J. Beldon, A. M. Belenguer, F. Adams, S. A. J. Kimber, V. Honkimäki and R. E. Dinnebier, *Nat. Chem.*, 2013, **5**, 66.
51. K. Tanaka, A. Asakura, T. Muraoka, P. Kalicki and Z. Urbanczyk-Lipkowska, *New J. Chem.*, 2013, **37**, 2851.

CHAPTER 5

Cross Dehydrogenative Coupling Reactions by Ball Milling

JINGBO YU, ZHIJIANG JIANG AND WEIKE SU*

Collaborative Innovation Center of Yangtze River Delta Region Green Pharmaceuticals, College of Pharmaceutical Sciences, Zhejiang University of Technology, Hangzhou 310014, China
*Email: pharmlab@zjut.edu.cn

5.1 Introduction

On the basis of the conception of atom economy, the most fundamental way to eliminate waste formation is to develop new chemical reactivity that can greatly shorten the synthetic routes. The cross-dehydrogenative coupling (CDC) reaction of two different C–H bonds without prefunctionalization of starting materials, pioneered by Professor Chao-Jun Li, provided a highly attractive synthetic strategy as an atom-economic and environmentally benign method for C–C bond formation (Scheme 5.1). Hitherto, the CDC reaction has been successfully applied in dehydrogenative coupling reactions between various hybrid forms of C–H bond, for example C(sp^3)–H/C(sp)–H coupling, C(sp^3)–H/C(sp^2)–H coupling, C(sp^3)–H/C(sp^3)–H coupling and C(sp^2)–H/C(sp^2)–H coupling, *etc.*

As a powerful candidate for a next generation green chemistry technique, the CDC reaction has received much attention. To date, various catalytic systems have been developed for this important reaction.

RSC Green Chemistry No. 31
Ball Milling Towards Green Synthesis: Applications, Projects, Challenges
Edited by Brindaban Ranu and Achim Stolle

Published by the Royal Society of Chemistry, www.rsc.org

Scheme 5.1 Cross-dehydrogenative coupling (CDC) reaction.

Copper-catalyzed CDC reactions reported by Li and others use ^{t}BuOOH,[1] O_2,[2] NBS,[3] and diethyl azodicarboxylate (DEAD)[4] as efficient oxidants. Other transition metal catalyzed oxidation systems such as rhodium–T-HYDRO (70% in water),[5] iron–(^{t}BuO)$_2$/^{t}BuOOH,[6] vanadium–^{t}BuOOH,[7] and ruthenium–O_2[8] systems also showed high activities in CDC reactions. In addition, the reaction could proceed in the absence of oxidant by using platinum as catalyst.[9] Although good results could be achieved in all the above cases, these reactions still possess some limitations: (1) The use of unrecoverable metal catalysts. Only one example of adopting recyclable silica supported iron terpyridine complex for CDC reactions of tertiary amines with various carbon nucleophiles was reported, recently by Liu.[10] However, a relatively long reaction time and high temperature were required. (2) The use of explosive oxidants or volatile organic solvents along with long reaction time or high reaction temperature. (3) Although several solvent-free CDC reactions have been reported, relative excess of oxidants or reaction substrates were taken as the role of solvent.

The concept of green chemistry has inspired chemists to investigate the possibility for greener CDC reaction, which is characterized by solvent-free and mild conditions. The ball mill protocol, which is normally carried out without, or with minimal use of, solvents, could potentially offer a greener alternative to their solution-based counterparts, where a high concentration of materials in the reaction vials is presented.[11]

Different C–C coupling reactions under ball milling and their application to the preparation of various useful molecules have been illustrated in Chapter 2. The main goal of the present chapter is to introduce the combination of ball milling technique with CDC reactions, which lead to a more efficient and environment-friendly method for C–C bond formation.

5.2 Cross Dehydrogenative Coupling Reactions under Ball Milling Conditions Applied to the Synthesis of Functionalized Tetrahydroisoquinolines

The synthesis of nitrogen-containing compounds, especially tetrahydroisoquinoline derivatives which are widely present in nature, has attracted much attention in industrial and academic research because of their biological and pharmaceutical properties.[12]

A series of reactions of tetrahydroisoquinolines and pronucleophiles under high-energy vibrational ball milling was first reported by our group,[13] based on direct $C(sp^3)$–H bond activation adjacent to a nitrogen atom of tetrahydroisoquinoline.

Three types of pronucleophiles were coupled with tetrahydroisoquinoline in a stainless steel vial by using 2,3-dichloro-5,6-dicyano-1,4-benzoquinone (DDQ) as oxidant and silica gel as grinding auxiliary: (1) active methylene ($C(sp^3)$–$C(sp^3)$), (2) terminal alkyne ($C(sp^3)$–C(sp)), and (3) indole ($C(sp^3)$–aryl–$C(sp^2)$). A series of tetrahydroisoquinoline derivatives were prepared under ball milling conditions. In addition, milling frequency, milling time and grinding auxiliaries were studied and the reaction results were compared with the traditional solution-based reactions.

5.2.1 $C(sp^3)$–$C(sp^3)$ Coupling Reaction

The work began with CDC reactions between tetrahydroisoquinolines and nitroalkanes under ball milling conditions without additional metal catalyst. Several 2-substituted tetrahydroisoquinolines and 6,7-dimethoxytetrahydroisoquinolines were tested, affording desired products with moderate to good yields within 30 min (Table 5.1).

Comparative experiments were performed under stirring conditions by using nitroalkanes as solvent – the corresponding reaction times were much longer and most of the products yields were lower than that obtained under milling condition. (Table 5.1, see in parentheses).

This method has been extended to the synthesis of 1-aminomethyl-tetrahydroquinoline, a key intermediate of the anti-schistosomiasis praziquantel,[14] by reduction and deprotection of the CDC product of **3d** (Scheme 5.2). In the classic E. Merck's synthetic process, this intermediate was synthesized from isoquinoline using an excessive amount of highly toxic cyanide.

As a viable participant, malononitrile was reacted with 2-phenyltetrahydroisoquinoline, affording comparable yields under above solvent and metal-free condition. When using *N*-acryloyl-6,7-dimethoxytetrahydroisoquinoline (**1f**) as the pro-electrophile, DDQ should be added in portions and an extra pause during the milling is needed to avoid side reactions induced by a high concentration of the reactive iminiumion (Scheme 5.3).

Diethyl malonate was also available as pronucleophile to react with **1f** under intermittent milling conditions (Scheme 5.4). However, $Cu(OAc)_2$ or $Cu(OTf)_2$ should be used to promote the reaction, due to the relative lower reactivity of malonates. Subsequent transformation by intramolecular Michael addition, hydrolysis and decarboxylation gave **3m**″ as a member of the tetrahydrobenzo[*a*]quinolizidine system that can be found in various alkaloids.[15]

5.2.2 $C(sp^3)$–C(sp) Coupling Reaction

$C(sp^3)$–H bond and C(sp)–H bond coupling reactions of tetrahydroisoquinolines and alkynes were performed under ball milling conditions. Owing to the expected lower activity of terminal alkynes as pronucleophiles, various catalytic amount of copper catalysts such as Cu(0),

Table 5.1 CDC reaction of tetrahydroisoquinolines with nitroalkanes by ball milling.[a]

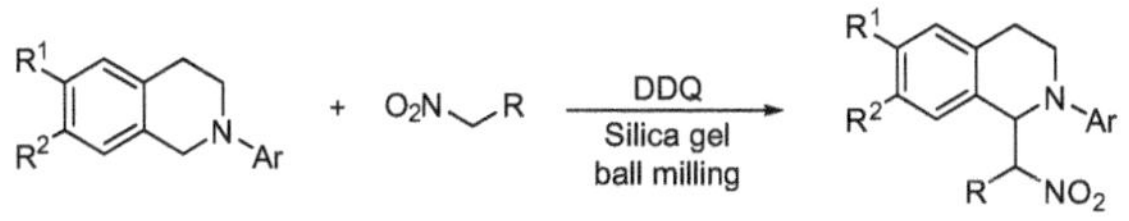

1 **2** **3**

1a: $R^1 = R^2 = H$, Ar = Ph
1b: $R^1 = R^2 = H$, Ar = 4-$MeOC_6H_4$
1c: $R^1 = R^2 = H$, Ar = 4-MeC_6H_4
1d: $R^1 = R^2 = OMe$, Ar = Ph
1e: $R^1 = R^2 = OMe$, Ar = 4-MeC_6H_4

2a: R = H
2b: R = Me
2c: R = Et

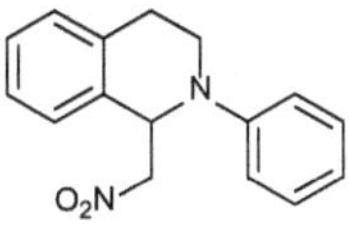

3a
85%, 10 min,
(89%, 120 min)[b]

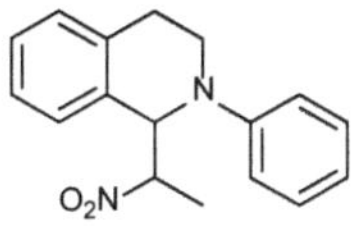

3b
75%, 15 min
(78%, 180 min)[b]

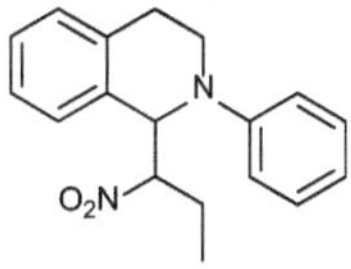

3c
76%, 15 min
(70%, 180 min)[b]

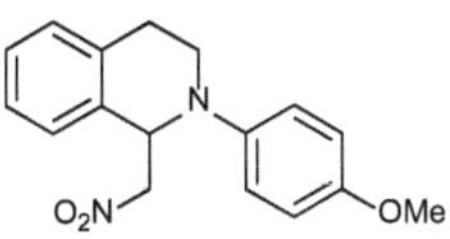

3d
80%, 20 min
(75%, 120 min)[b]

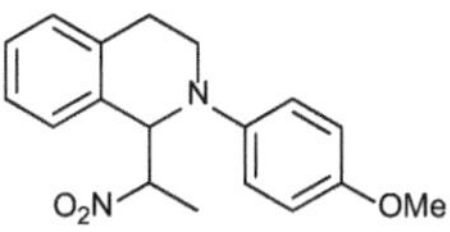

3e
70% 30 min
(65%, 210 min)[b]

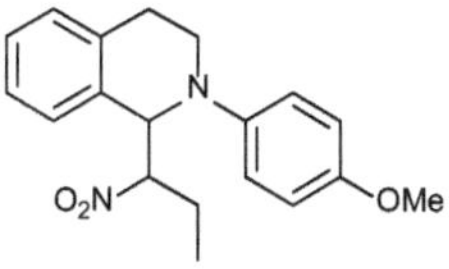

3f
70% 30 min
(60%, 210 min)[b]

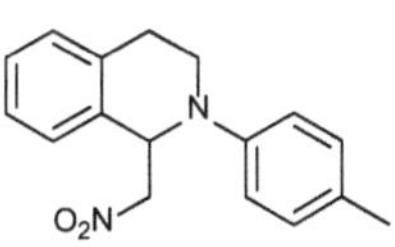

3g
75%, 30 min
(70%, 270 min)[b]

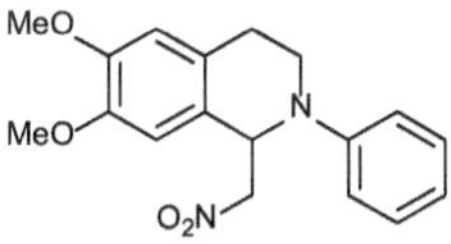

3h
65%, 15 min
(68%, 210 min)[b]

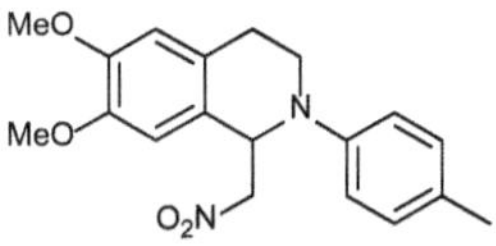

3i
60%, 30 min
(30%, 210 min)[b]

[a]Reactions were carried out with **1** (1.0 mmol), **2** (2 mmol), DDQ (1.0 mmol), silica gel (0.5 g) and two stainless steel balls ($d = 8.0$ mm) at 30 Hz, 10–30 min. Isolated yields based on **1**.
[b]Comparative experiments were carried out with **1** (1 mmol), **2** (10 mL), DDQ (1 mmol) under stirring conditions. Isolated yields based on **1**.

Cu(I) and Cu(II) were used to promote the reactions, and good results were obtained in all cases.[13]

Various α-alkynylamine derivatives were generated in good yields with various kinds of substituted aromatic alkynes, aliphatic alkynes and propiolate as substrates (Table 5.2). When *N*-(4-methoxyphenyl)tetrahydroisoquinoline (**1b**) or 6,7-dimethoxy-*N*-aryl-tetrahydro-isoquinoline **1d**/**1e** was

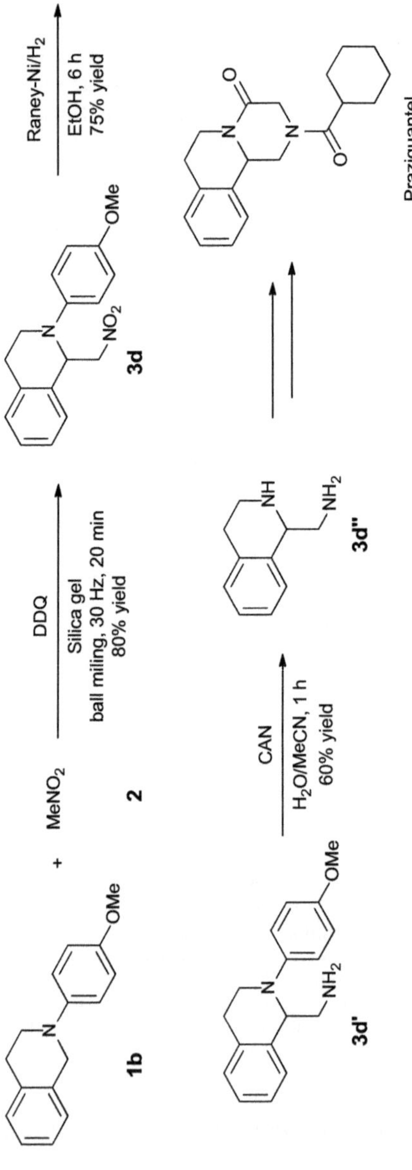

Scheme 5.2 Synthesis of 1-aminomethyl-tetrahydroquinoline (**3d′′**).

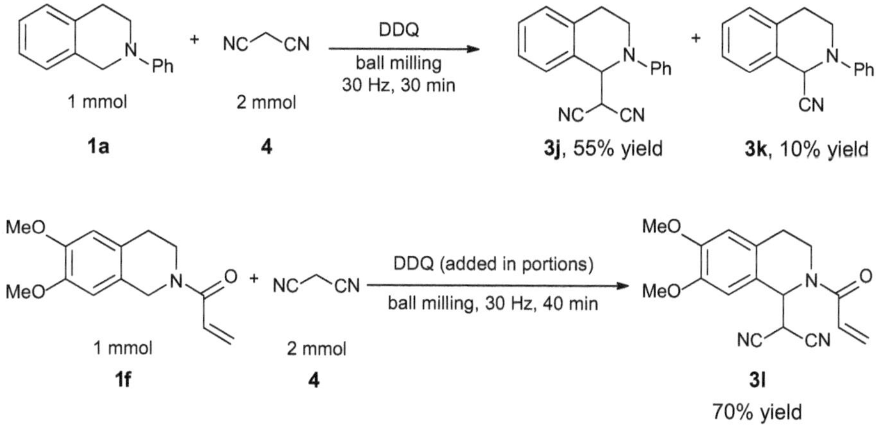

Scheme 5.3 CDC reaction of tetrahydroisoquinolines with malononitrile by ball milling.

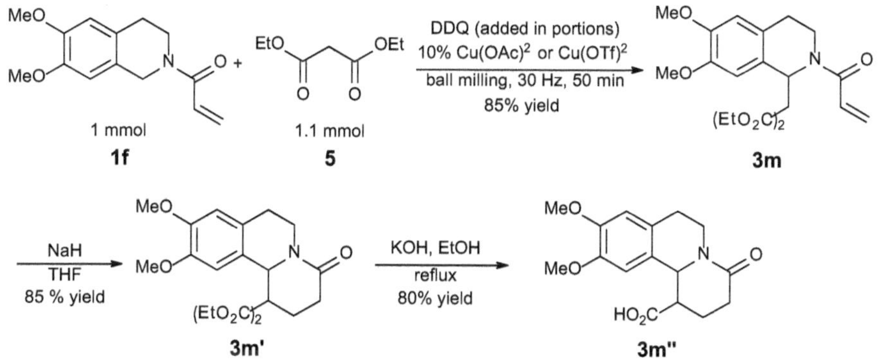

Scheme 5.4 Synthesis of tetrahydrobenzo[*a*]quinolizidine through CDC reaction.

used as pro-electrophile, a slightly lower yield was afforded, probably due to their lower electrophilicity.

To simplify the recovery of catalysts, copper balls were used instead of the original stainless steel balls as both catalyst and milling balls – the reaction proceeded smoothly to give similar results to those promoted with copper salts. After the reaction, the catalyst could be recovered simply by removing the copper ball from the reaction media and reused in the next reaction without any further treatment. The ball bearings were weighed after being used for more than 20 times; no significant change was noticed in their mass.

5.2.3 C(sp^3)–Aryl-C(sp^2) Coupling Reaction

C(sp^3)–H bond and aryl-C(sp^2)–H bond CDC reactions were investigated under ball milling conditions by taking tetrahydroquinolines and indoles as

Table 5.2 CDC reaction of tetrahydroisoquinolines with alkynes by ball milling.[a]

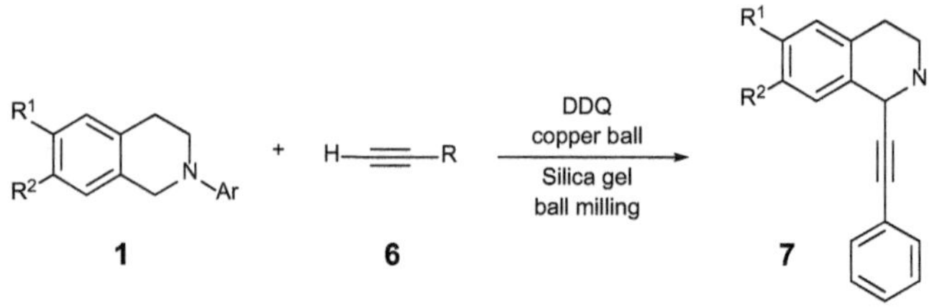

1a: $R^1 = R^2 = H$, Ar = Ph
1b: $R^1 = R^2 = H$, Ar = 4-MeOC_6H_4
1c: $R^1 = R^2 = H$, Ar = 4-MeC_6H_4
1d: $R^1 = R^2 =$ OMe, Ar = Ph
1e: $R^1 = R^2 =$ OMe, Ar = 4-MeC_6H_4

6a: R = Ph
6b: R = 4-MeC_6H_4
6c: R = 4-FC_6H_4
6d: R = 2-Py
6e: R = Pr
6f: R = CO_2Me

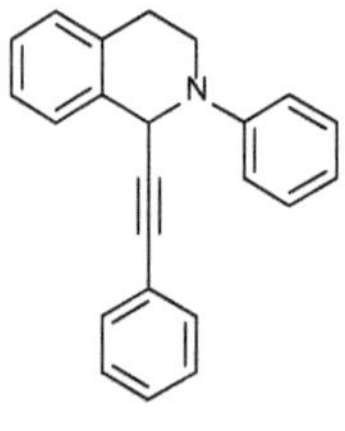

7a
78%, 20 min

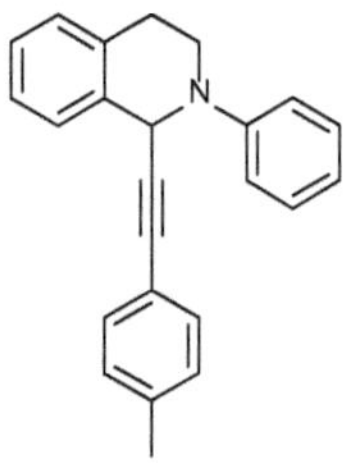

7b
74%, 20 min

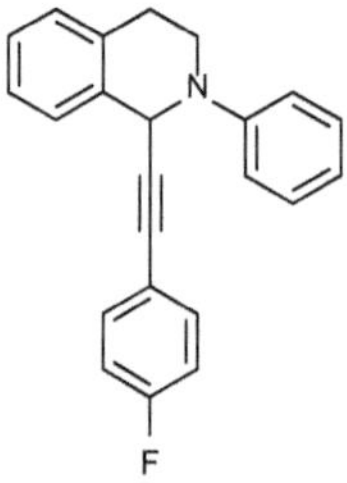

7c
87%, 20min

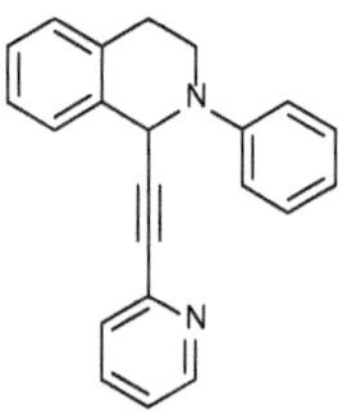

7d
69%, 40 min

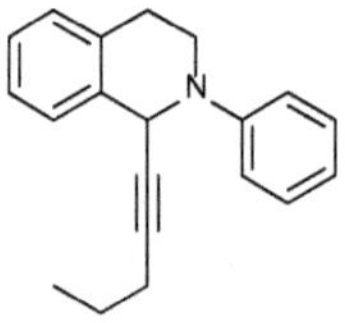

7e
67%, 30 min

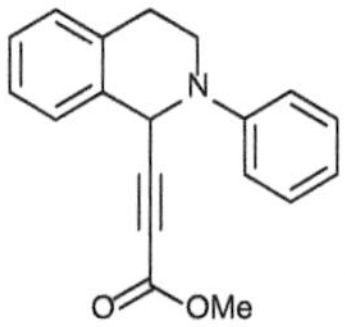

7f
73%, 30 min

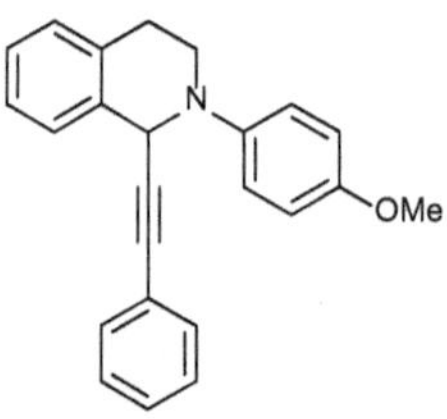

7g
75%, 30 min

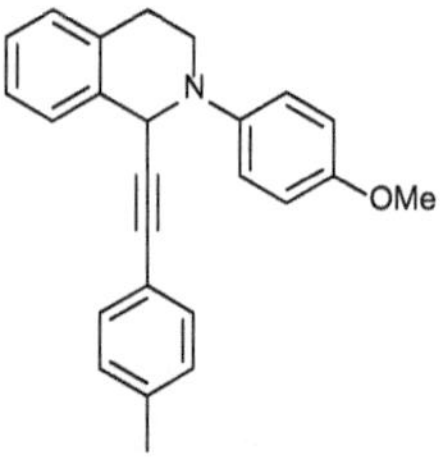

7h
70%, 30 min

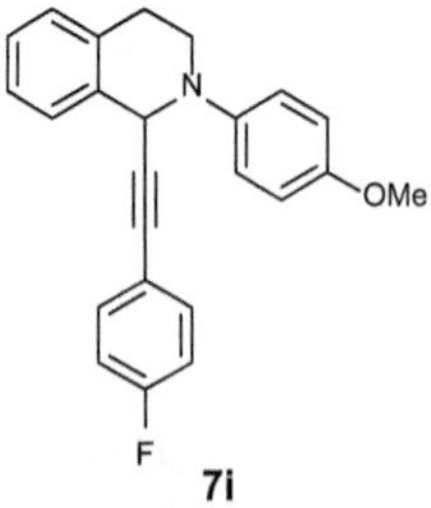

7i
84%, 20 min

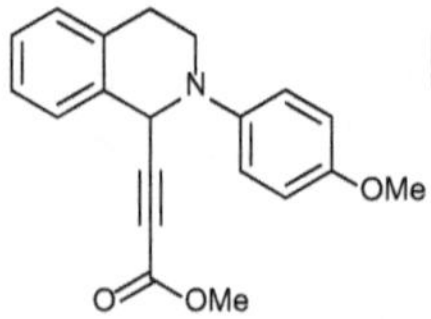

7j
72%, 30 min

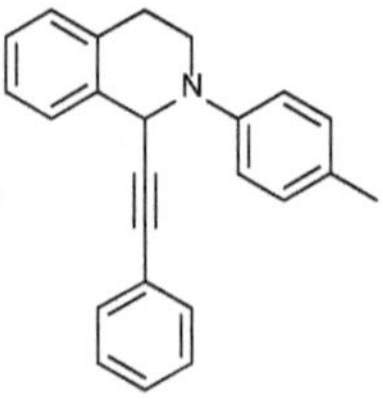

7k
79%, 20 min

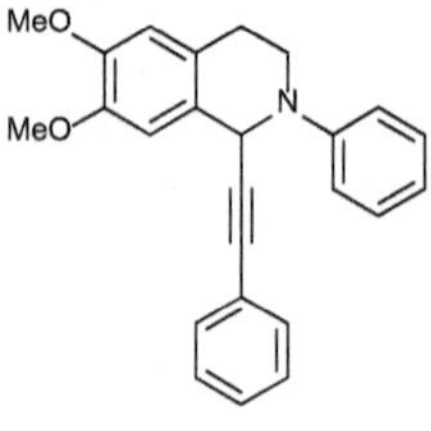

7l
75%, 40 min

Table 5.2 (*Continued*)

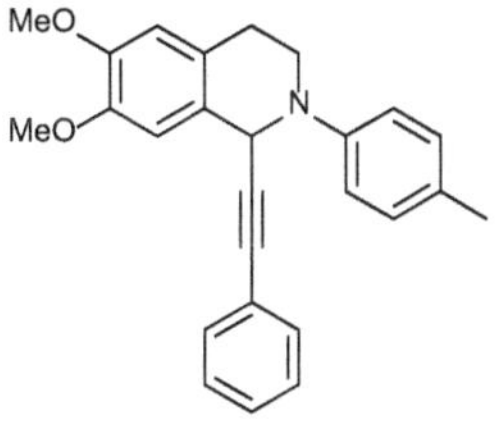

7m
65%, 50 min

[a]Reactions were carried out with **1** (1.0 mmol), **6** (1.1 mmol), DDQ (1.0 mmol), silica gel (0.5 g) and two copper balls ($d = 8.0$ mm) at 30 Hz, 20–50 min. Isolated yields based on **1**.

substrates. As with terminal alkynes, the coupling between indoles and tetrahydroquinolines also needs the participation of metal catalyst. Fortunately, the reaction could also be efficiently promoted by copper balls and a slight excess of the tetrahydroisoquinoline was conducive to the conversion. Various indoles were tested to afford desired products in moderate to good yields (Table 5.3).

Under the milling conditions, the electronic character of the substituents on different positions of indole did not show a significant influence on the yield. In contrast, using 6,7-dimethoxy-2-phenyl-tetrahydroisoquinoline (**1d**) as pro-electrophile leads to a relatively lower yield.

Compared with traditional solvent-free reaction under heating conditions, the mechanical activated solvent-free conditions proved to be advantageous due to a short reaction time and recoverable catalyst (Table 5.3; see in parentheses).

5.2.4 Milling Parameters

5.2.4.1 Milling Frequency (ν) and Milling Time (t)

The milling frequency ν was proven to be one of the most important parameters to be considered as it influences yield, selectivity or similar properties during ball milling. Besides ν, the consideration of milling time t is almost as important. Commonly, an increase of t leads to higher yields in ball milling promoted organic reactions.[16] However, in some cases a decrease in yield and enantioselectivity was attributed to the prolonged time of reaction. Even though some studies have paid attention to the combined effects of ν and t, this issue is still of specific interest due to its unpredictability.

Evaluation of the process parameters (t & ν) in the case of C(sp^3)–C(sp^3) coupling reaction was conducted by using tetrahydroisoquinoline **1a** and nitromethane **2a** as substrates. The results showed that the main influence on the yields come from the milling frequency ν, while milling time t has little effect (Figure 5.1). When the reaction was performed at lower frequencies, such as 20 Hz, a few substrates were still present in the crude product, probably due to the diminished amount of energy per impact. Furthermore, ball milling with pauses also resulted in lower conversion.

Table 5.3 CDC reaction of tetrahydroisoquinolines with indoles by ball milling.[a]

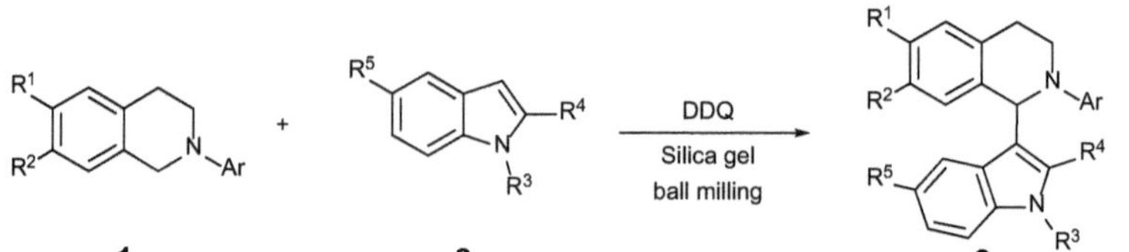

1a: $R^1 = R^2$ = H, Ar = Ph
1b: $R^1 = R^2$ = H, Ar = 4-$MeOC_6H_4$
1d: $R^1 = R^2$ = OMe, Ar = Ph

8a: R^3 = H, R^4 = H, R^5 = H
8b: $R^3 = CH_3$, R^4 = H, R^5 = H
8c: R^3 = H, R^4 = H, R^5 = OMe
8d: R^3 = H, R^4 = H, R^5 = Br
8e: R^3 = H, $R^4 = CH_3$, R^5 = H
8f: R^3 = H, R^4 = H, $R^5 = NO_2$

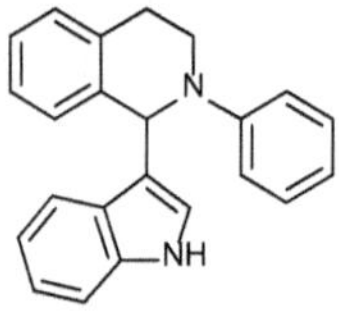

9a
77%, 30 min
(79%, overnight)[b]

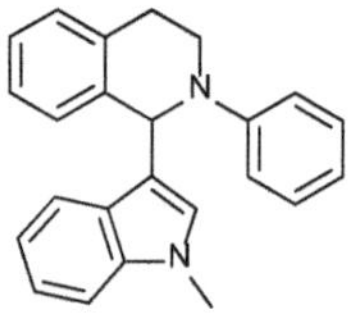

9b
80%, 40 min
(44%, overnight)[b]

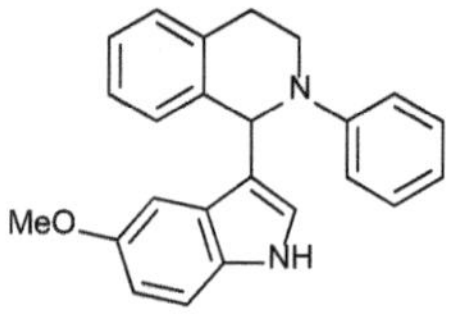

9c
70%, 20 min
(57%, overnight)[b]

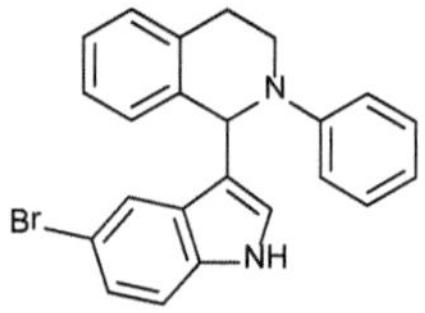

9d
70%, 30 min

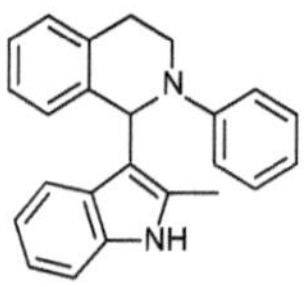

9e
75%, 40 min
(61%, overnight)[b]

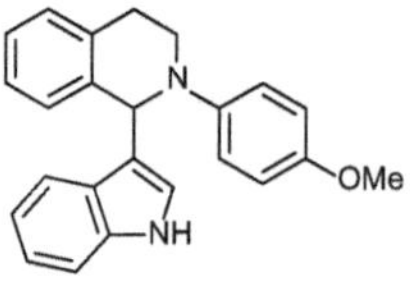

9f
67%, 30 min

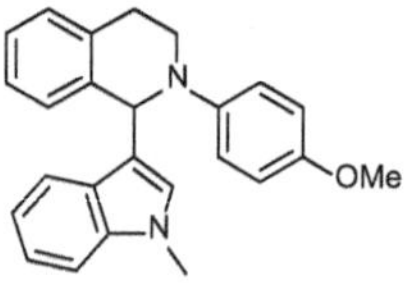

9g
80%, 40 min
(71%, overnight)[b]

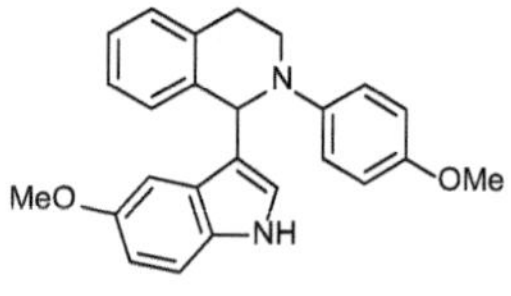

9h
72%, 40 min
(65%, overnight)[b]

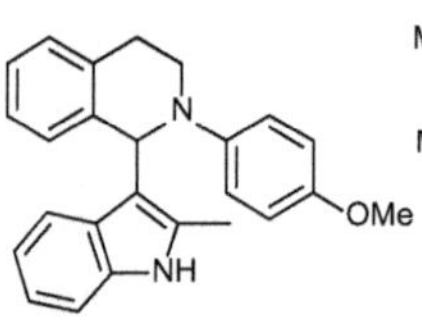

9i
70%, 40 min

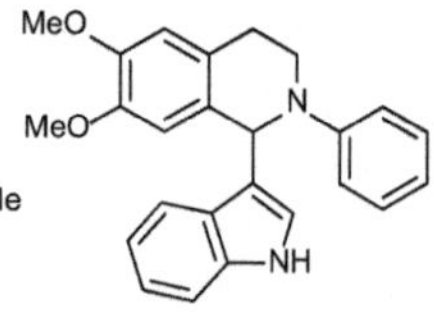

9j
65%, 50 min

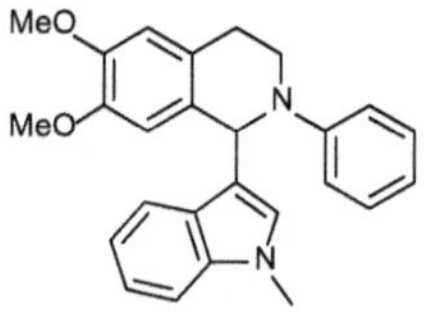

9k
67%, 45 min

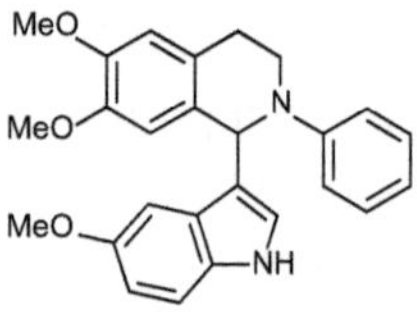

9l
63%, 50 min

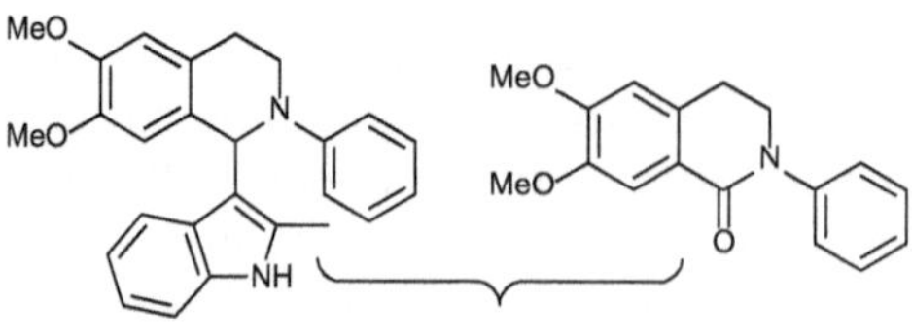

9m + 9m' (55% + 15%)
70%, 40 min

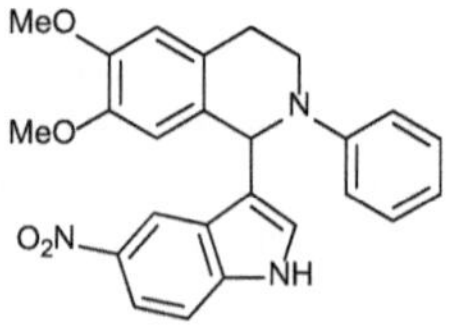

9n
61%, 50 min

[a]Reactions were carried out with **1** (1.0 mmol), **8** (0.8 mmol), DDQ (1.0 mmol), silica gel (0.5 g) and two copper balls (d = 8.0 mm) at 30 Hz, 20–50 min. Isolated yields based on **8**.
[b]The reaction time and isolated yields from ref. 1d.

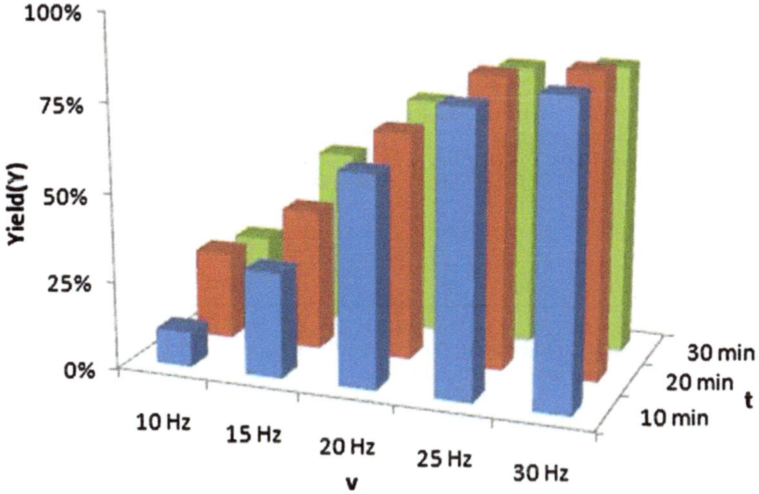

Figure 5.1 Influence of frequency (ν) and milling time (t) on the yield of **3a**. Reactions were carried out with **1a** (1.0 mmol), **2a** (2.0 mmol), DDQ (1.0 mmol), silica gel (0.5 g) and two stainless steel balls ($d = 8.0$ mm) (see Table 5.1). Isolated yields based on **1a**.
Reproduced from ref. 13, Copyright (2011), with permission from American Chemical Society.

5.2.4.2 *Grinding Auxiliary*

The grinding auxiliary can have a huge impact on the ball milling reaction, especially for grinding liquid substances. It is necessary to adsorb the substrates with a suitable grinding auxiliary, or the reaction may result in a mud-like mixture. Such mixtures would have different mechanical properties, which prevent inelastic interaction between balls and substances and result in inefficient energy transfer. Various grinding auxiliaries were examined in respect of the above three types of CDC reactions. Silica gel was found to be the most effective. It might act as both grinding aid and absorbent in the reaction. In the absence of grinding auxiliary, the formation of mud-like mixtures would lead to a poor yield (Table 5.4, entry 5). However, adding a relatively great amount of grinding auxiliary also decreased the yield dramatically owing to the dilution of reactant concentrations (Table 5.4, entry 2).

5.2.4.3 *Scalability*

The issue of scalability in mechanosynthesis has not yet been broadly addressed. Indeed, a common perception is that there are still difficulties in scaling up such mechanical-power promoted reactions.

Nonetheless, efforts have been made to examine the scalability of these CDC reactions under ball milling conditions. The results show that this method was amenable to relatively larger scale preparation, which provides

Table 5.4 Influence of grinding auxiliary on CDC reactions.

			sp^3–sp^3 Coupling[a]		sp^3–sp Coupling[b]		sp^3–sp^2 Coupling[c]	
Entry	Grinding auxiliary	Weight (g)	Time (min)	Yield (%)	Time (min)	Yield (%)	Time (min)	Yield (%)
1	Silica gel	0.5	10	85	20	78	30	77
2	Silica gel	2	10	60	20	60	30	55
3	NaCl	2	10	78	40	70	40	70
4	γ-Al_2O_3 (neutral)	0.5	10	70	40	58	40	57
5	–	–	30	58	40	35	40	30

[a]Reactions were carried out with **1a** (1.0 mmol), **2a** (2.0 mmol), DDQ (1.0 mmol) and two stainless steel balls (d = 8.0 mm) at 30 Hz, 10–30 min. Isolated yields based on **1a**.
[b]Reactions were carried out with **1a** (1.0 mmol), **6a** (1.1 mmol), DDQ (1.0 mmol) and two copper balls (d = 8.0 mm) at 30 Hz, 20–40 min. Isolated yields based on **1a**.
[c]Reactions were carried out with **1a** (1.0 mmol), **8a** (0.8 mmol), DDQ (0.8 mmol) and two copper balls (d = 8.0 mm) milling at 30 Hz, 30–40 min. Isolated yields based on **8a**.

the potential for amplification. For instance, the coupling of *N*-phenyltetrahydroisoquinoline (**1a**) with nitromethane and phenylacetylene were performed at 10 mmol scale resulted in the corresponding 1-(nitromethyl)-2-phenyl-1,2,3,4-tetrahydroisoquinoline (**3a**) (80%) and 1-(phenylethynyl)-2-phenyl-1,2,3,4-tetrahydroisoquinoline (**5a**) (75%) within 20 and 30 min, respectively. The reaction between indole (8.0 mmol) and *N*-phenyltetrahydroisoquinoline (**1a**) also proceeded mildly at 10 mmol scale in the presence of DDQ (8.0 mmol) affording 1-(1*H*-indol-3-yl)-2-phenyl-1,2,3,4-tetrahydroisoquinoline (**9a**) (70%) after 40 min.

5.3 Asymmetric Cross-dehydrogenative Coupling Reaction by Ball Milling

Asymmetric synthesis reactions are one of the most powerful tools to build up chiral compounds, which are important in pharmaceutical and biological science. However, most asymmetric reactions usually proceed at low temperature with long reaction times.

Typically, ball milling reactions take less time than similar ones in solution, thus providing a possibility to achieve high stereoselectivity in shorter time. Hitherto, several cases of organocatalyzed asymmetric reactions have been performed under ball milling conditions and shown their unique characteristics of high speed and high stereoselectivity.[11a,16a,17] However, there are few reports of metal-catalyzed asymmetric reactions under ball milling conditions.

In 2013, our group developed an asymmetric CDC reaction between tetrahydroisoquinolines and terminal alkynes catalyzed by copper balls. A series of 1-alkynyl tetrahydroquinolines were prepared efficiently under solvent-free ball milling conditions, and the effects on enantioselectivity were investigated in detail.

5.3.1 Asymmetric Alkynylation of Prochiral $C(sp^3)$–H Bonds

Traditional strategies for building chiral C1-substituted tetrahydroisoquinolines include (1) catalytic asymmetric transfer hydrogenation of C1-substituted dihydroisoquinolines or acyclic imines or iminium intermediates,[18] such as Pictet–Spengler reaction, Bischler–Napieralski reaction and Pomeranz–Frisch cyclization, (2) catalytic asymmetric addition reactions to the C=N bonds of isoquinoline scaffolds,[19] and (3) catalytic asymmetric intramolecular allylic-amination reactions.[20] Although these methods give target products with desired enantioselectivity, harsh conditions and complex operations limit their applications. Quite recently, Li *et al.* first reported their asymmetric CDC reaction between tetrahydroisoquinolines and terminal alkynes, providing an efficient method to build up chiral C1-alkyne tetrahydroisoquinolines.[21] In 2012, an organocatalysis/copper-catalyzed asymmetric CDC reaction of tertiary amines with olefins was reported by Wang *et al.* to access diverse optically active C1-alkene tetrahydroisoquinolines.[22] At the same time, Chi *et al.* described an enantioselective CDC reaction of aldehydes and tertiary amines, which allowed optically active β-amino alcohols that contain tetrahydroisoquinoline units to be synthesized.[23]

Inspired by Li's work, a faster and more efficient synthetic method was investigated to determine the possibility of finishing the asymmetric CDC reaction under ball milling conditions.[24] By adding Pybox-1 to the reaction mixture discussed in Section 5.2.2, the desired product with acceptable enantioselectivity was obtained in 40 min. Several copper catalysts and chiral ligands (Scheme 5.5) were tested; the system of copper balls and Pybox-1 was the optimum one.

A series of α-alkynylamine derivatives were prepared in good yields with moderate to good enantioselectivity in a short time (Table 5.5). Strong substitution effects were observed. Electro-donating groups on tetrahydroisoquinolines gave relatively higher enantioselectivities but lower

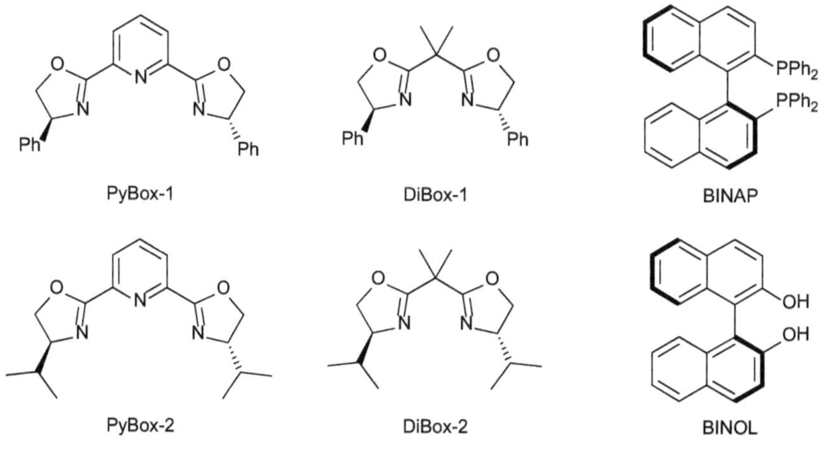

Scheme 5.5 Chiral ligands tested for the reaction.
Reproduced from ref. 24, Copyright (2013), with permission from Elsevier.

Table 5.5 Asymmetric CDC reaction of tetrahydroisoquinolines with alkynes by ball milling.[a,b]

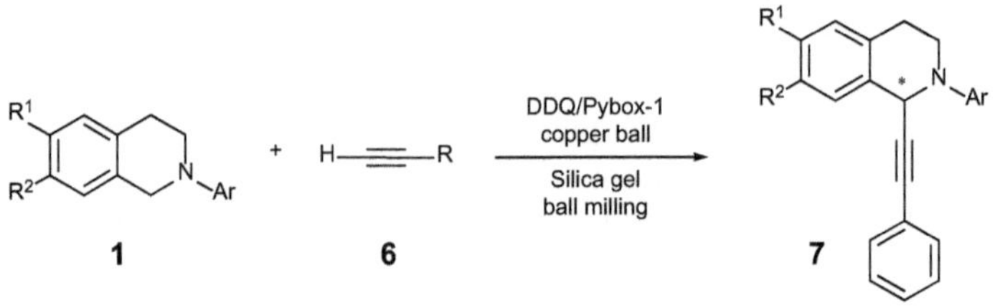

1a: $R^1 = R^2 = H$, Ar = Ph
1b: $R^1 = R^2 = H$, Ar = 4-$MeOC_6H_4$
1c: $R^1 = R^2 = H$, Ar = 4-MeC_6H_4
1f: $R^1 = R^2 = H$, Ar = 2-$MeOC_6H_4$

6a: R = Ph; **6b**: R = 4-MeC_6H_4
6c: R = 4-FC_6H_4; **6d**: R = 2-Py
6e: R = Pr; **6f**: R = CO_2Me
6g: R = CO_2Et

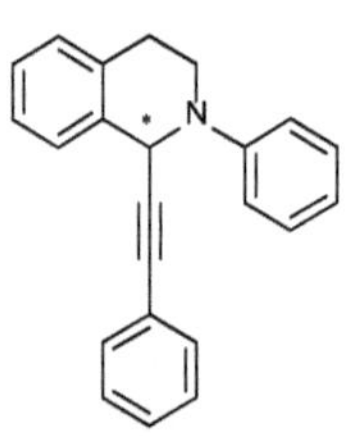

7a
74%, 43% *ee*,
50 min

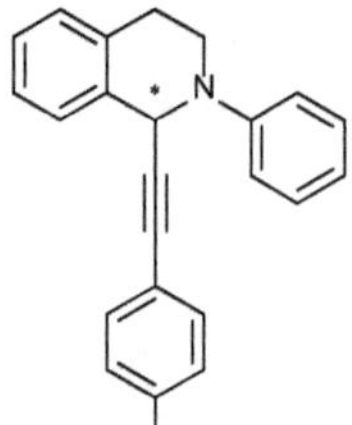

7b
75%, 32% *ee*,
60 min

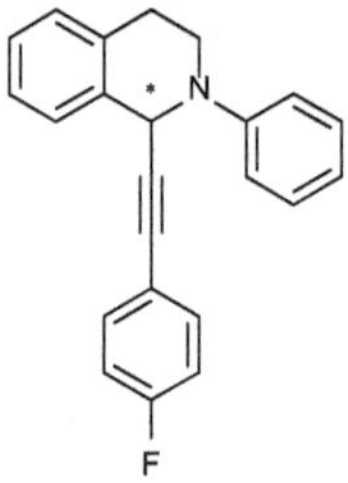

7c
77%, 34% *ee*,
50 min

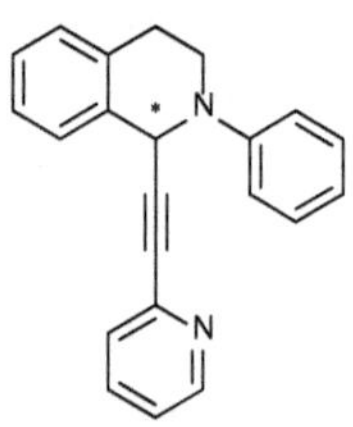

7d
60%, 44% *ee*,
90 min

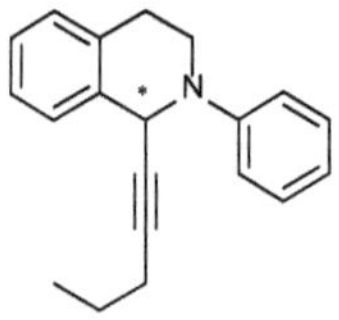

7e
61%, 10% *ee*,
30 min

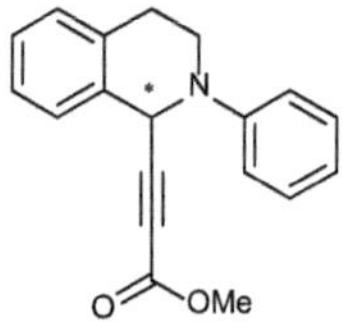

7f
70%, 51% *ee*,
40 min

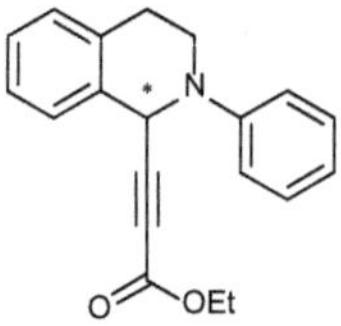

7n
67%, 56% *ee*,
50 min

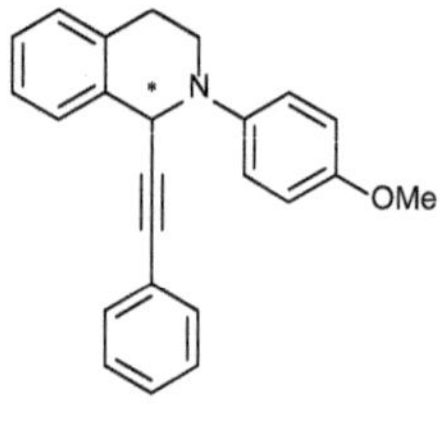

7g
69%, 74% *ee*,
40 min

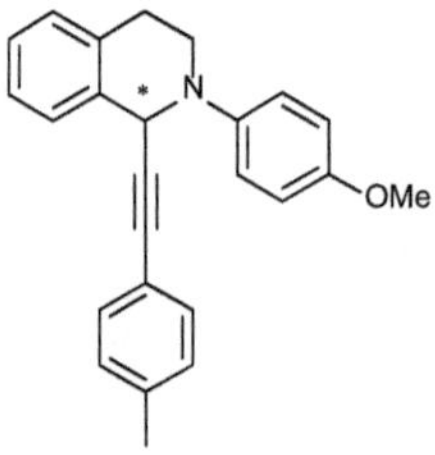

7h
68%, 46% *ee*,
40 min

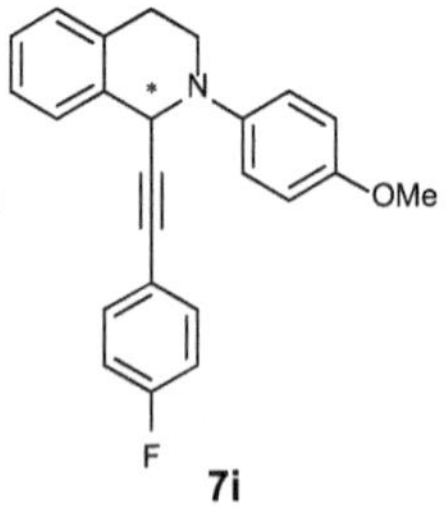

7i
75%, 55% *ee*,
30 min

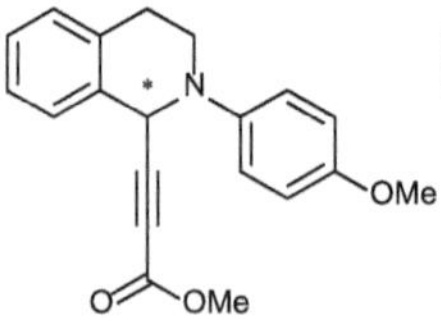

7j
60%, 55% *ee*,
40 min

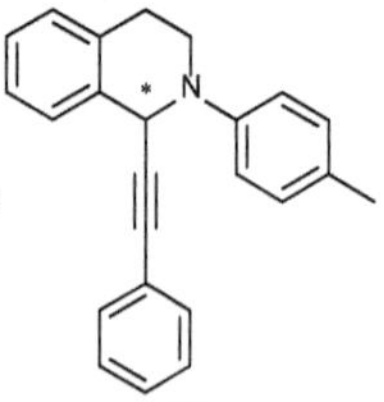

7k
75%, 38% *ee*,
30 min

Table 5.5 (*Continued*)

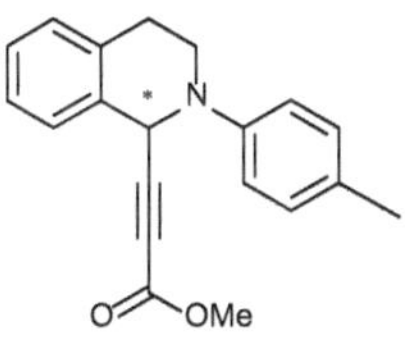	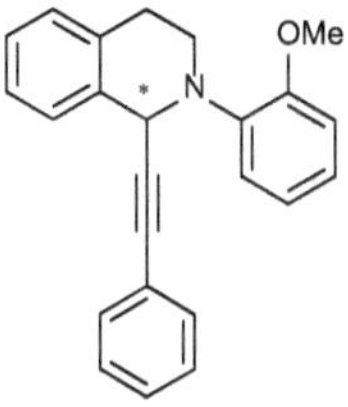	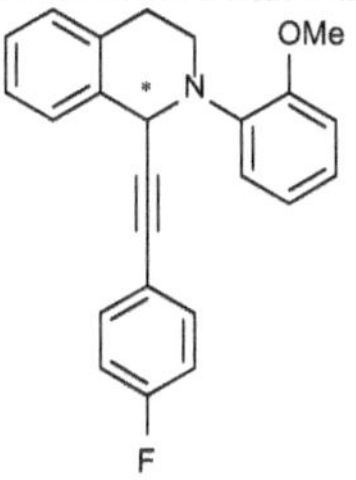	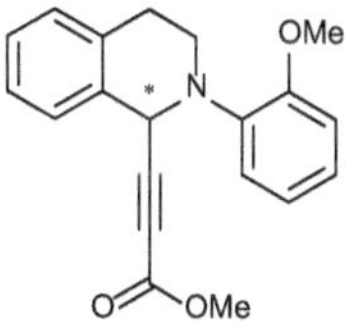
7o 72%, 62% *ee*, 40 min	**7p** 71%, 52% *ee*, 40 min	**7q** 65%, 34% *ee*, 40 min	**7r** 60%, 79% *ee*, 50 min

[a]Reactions were carried out with **1** (1.0 mmol), **6** (1.1 mmol), DDQ (1.0 mmol), Pybox-1 (10 mol%), silica gel (0.5 g) and two copper balls ($d = 12.0$ mm) at 30 Hz for 10 min followed by a 5 min pause until the finish of reaction. Isolated yields based on **1**.
[b]Enantiomeric excess determined by HPLC analysis using a chiral stationary phase.

yields. Noticeably, higher *ee*'s were obtained by using propiolates as pronucleophiles, while aliphatic alkynes give moderate yield with low stereoselectivity.

As in previous work, 6,7-dimethoxytetrahydroisoquinolines still show low activity in this asymmetric CDC reaction, which requires further activation. By adding a catalytic amount of $Cu(OTf)_2$ to the system, an improvement in yields was observed, but the enantioselectivity remained low (Table 5.6).

5.3.2 Effects on Enantioselectivity

It is widely accepted that high temperatures disfavor asymmetric reactions. Under ball milling conditions, higher frequency usually leads to a higher reaction temperature, which may cause a decrease in enantioselectivity. As an uncommon technique, it is hard to control the change of temperature during the ball milling procedure. To avoid overheating, the strategy of intermittent milling[16a,17a,b] was taken to cool down the reaction. The effect of frequency on reaction enantioselectivity was then investigated in detail.

Interestingly, high frequency gave the best result after 30 min, while lower frequencies required prolonged reaction time and gave lower enantioselectivity (Table 5.7, entries 1–4). Noticeably, when the reaction was performed without a pause at 30 Hz, lower enantioselectivity was obtained after 30 min, while the yield was improved to 75% (Table 5.7, entry 5).

Further research was conducted to explain the phenomenon. Owing to the good conductivity of heat of the stainless steel vial, the inside temperature was probably similar to the outside temperature. After the reaction was completed, the outside temperature was measured (about 25–27 °C at 30 Hz; 23–26 °C at 25 Hz; 22–24 °C at 20 Hz, <20 °C at 15 Hz). Since an intermittent

Table 5.6 Asymmetric CDC reaction of 6,7-dimethoxytetrahydroisoquinolines with alkynes by ball milling.[a,b]

1d: $R^1 = R^2 =$ OMe, Ar = Ph　　**6a**: R = Ph

1g: $R^1 = R^2 =$ OMe, Ar = 4-$MeOC_6H_4$　　**6f**: R = CO_2Me

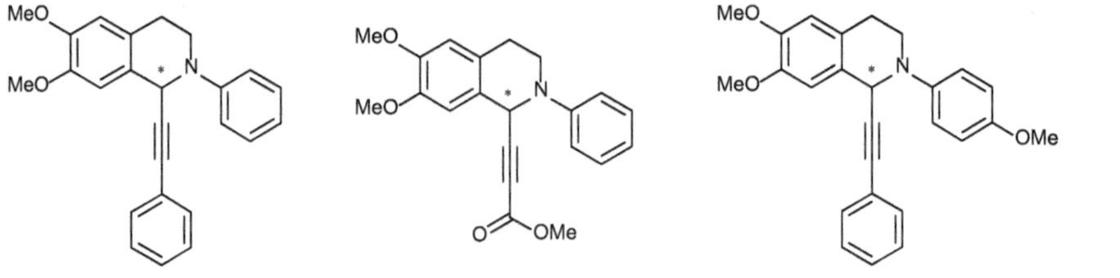

7l	**7s**	**7t**
72%, 22% *ee*, 50 min	70%, 22% *ee*, 40 min	56%, 20% *ee*, 150 min

[a]Reactions were carried out with **1** (1.0 mmol), **6** (1.1 mmol), DDQ (1.0 mmol), $Cu(OTf)_2$ (5 mol%), Pybox-1 (10 mol%), silica gel (0.5 g) and two copper balls ($d = 12.0$ mm) at 30 Hz for 10 min followed by a 5 min pause until the finish of reaction. Isolated yields based on **1**.
[b]Enantiomeric excess determined by HPLC analysis using a chiral stationary phase.

Table 5.7 Effects of frequency on $C(sp^3)-H$ and $C(sp)-H$ bond coupling reactions.[a] Reproduced from ref. 24. Copyright 2013, with permission from Elsevier.

Entry	Frequency (Hz)	Time (min)	Yield (%)[b]	*ee* (%)[c]
1	30	30	69	74
2	25	60	62	69
3	20	60	42	70
4	15	80	—	—
5[d]	30	30	75	43

[a]Unless otherwise noted, reactions were carried out with **1a** (1.0 mmol), **6a** (1.1 equiv.), DDQ (1.0 equiv.), PyBox-1 (10 mol%), 0.5 g of silica gel and two copper balls ($d = 12.0$ mm) at a certain frequency for 10 min followed by a 5 min pause until the finish of reaction.
[b]Isolated yields based on **1a**.
[c]Determined by HPLC analysis using a chiral stationary phase.
[d]The reaction was carried out without pause.

milling cycle of 10 min followed by a 5 min pause was adopted during the reaction process, the heat generated by the mill was insignificant for each 10 min cycle. If the reaction was carried out without a pause in milling, the conversion was accelerated and the product's *ee* was decreased. In this case, the inside temperature would be relatively high (the measured outside temperature was about 38–40 °C).

5.4 Summary and Outlook

In conclusion, ball milling promoted CDC and asymmetric CDC reactions provide an efficient method for constructing C–C bonds directly from different C–H bonds, a method that does not require the presence of solvents or a long reaction time. In some of these reactions, even additional unrecoverable metal catalysts could be avoided by exploring a unique milling media. Furthermore, the asymmetric CDC reaction under these unusual conditions showed a possibility for construction of optically active compounds at high speed, and which could be easily performed without critical temperature control.

References

1. (a) Z. P. Li and C. J. Li, *Org. Lett.*, 2004, **6**, 4997; (b) Z. P. Li and C. J. Li, *J. Am. Chem. Soc.*, 2004, **126**, 11810; (c) Z. P. Li and C. J. Li, *J. Am. Chem. Soc.*, 2005, **127**, 3672; (d) Z. P. Li and C. J. Li, *J. Am. Chem. Soc.*, 2005, **127**, 6968; (e) Z. P. Li and C. J. Li, *Eur. J. Org. Chem.*, 2005, 3173; (f) Y. Zhang, H. Fu, Y. Jiang and Y. F. Zhao, *Org. Lett.*, 2007, **9**, 3813; (g) O. Basle and C. J. Li, *Org. Lett.*, 2008, **10**, 3661; (h) L. L. Chu, X. G. Zhang and F. L. Qing, *Org. Lett.*, 2009, **11**, 2197; (i) Y. M. Shen, M. Li, S. Z. Wang, T. G. Zhan, Z. Tan and C. C. Guo, *Chem. Commun.*, 2009, 953.
2. O. Baslé and C. J. Li, *Green Chem.*, 2007, **9**, 1047.
3. M. Y. Niu, Z. M. Yin, H. Fu, Y. Y. Jiang and Y. F. Zhao, *J. Org. Chem.*, 2008, **73**, 3961.
4. X. Xu and X. Li, *Org. Lett.*, 2009, **11**, 1027.
5. A. J. Catino, J. M. Nichols, B. J. Nettles and M. P. Doyle, *J. Am. Chem. Soc.*, 2006, **128**, 5648.
6. (a) C. M. R. Volla and P. Vogel, *Org. Lett.*, 2009, **11**, 1701; (b) M. Ghobrial, K. Harhammer, M. D. Mihovilovic and M. Schnuürch, *Chem. Commun.*, 2010, **46**, 8836.
7. A. Sud, D. Sureshkumarzn and M. Klussmann, *Chem. Commun.*, 2009, 3169.
8. A. H. Yu, Z. Gu, D. Chen, W. M. He, P. Tan and J. N. Xiang, *Catal. Commun.*, 2009, **11**, 162.
9. X. Z. Shu, Y. F. Yang, X. F. Xia, K. G. Ji, X. Y. Liu and Y. M. Liang, *Org. Biomol. Chem.*, 2010, **8**, 4077.
10. P. Liu, C. Y. Zhou, S. Xiang and C. M. Che, *Chem. Commun.*, 2010, **46**, 273.
11. (a) B. Rodriguez, A. Bruckmann, T. Rantanen and C. Bolm, *Adv. Synth. Catal.*, 2007, **349**, 2213; (b) J. Mack, D. Fulmer, S. Stofel and N. Santos, *Green Chem.*, 2007, **9**, 1041; (c) Y. W. Dong, G. W. Wang and L. Wang, *Tetrahedron*, 2008, **64**, 10148; (d) X. Y. Zhu, Z. H. Li, C. Jin, L. Xu, Q. Q. Wu and W. K. Su, *Green Chem.*, 2009, **11**, 163; (e) D. A. Fulmer, W. C. Shearouse, S. T. Medonza and J. Mack, *Green Chem.*, 2009,

11, 1821; (f) F. Schneider, A. Stolle, B. Ondruschka and H. Hopf, *Org. Process. Res. Dev.*, 2009, **13**, 44; (g) R. Thorwirth, A. Stolle and B. Ondruschka, *Green Chem.*, 2010, **12**, 985.
12. (a) D. Tewari, B. Saffari, C. Cowan, A. C. Wallick, M. Z. Koontz and B. J. Monk, *Gynecol. Oncol.*, 2006, **102**, 421; (b) J. Zhu, J. Lu, Y. Zhou, Y. Li, J. Cheng and C. Zheng, *Bioorg. Med. Chem. Lett.*, 2006, **16**, 5285; (c) R. Gitto, R. Ficarra, R. Stancanelli, M. Guardo, L. De Luca, M. L. Barreca, B. Pagano, A. Rotondo, G. Bruno, E. Russo, G. De Sarro and A. Chimirri, *Bioorg. Med. Chem.*, 2007, **15**, 5417; (d) C. J. Suckling, J. A. Murphy, A. I. Khalaf, S. Zhou, D. E. Lizos, A. N. Van Nhien, H. Yasumatsu, A. McVie, L. C. Young, C. McCraw, P. G. Waterman, B. J. Morris, J. A. Pratt and A. L. Harvey, *Bioorg. Med. Chem. Lett.*, 2007, **17**, 2649; (e) X. H. Deng, T. L. Jimmy, J. Liu, H. McAllister, C. Schubert and N. S. Mani, *Org. Process Res. Dev.*, 2007, **11**, 1043; (f) K. Tsoyi, H. J. Kim, J. S. Shin, D. H. Kim, H. J. Cho, S. S. Lee, S. K. Ahn, H. S. Yun-Choi, J. H. Lee, H. G. Seo and K. C. Chang, *Cell Signal*, 2008, **20**, 1839.
13. W. K. Su, J. B. Yu, Z. H. Li and Z. J. Jiang, *J. Org. Chem.*, 2011, **76**, 9144.
14. A. Domling and K. Khoury, *ChemMedChem*, 2010, **5**, 1420.
15. T. Fujii and M. Ohba, *The Alkaloids*, ed. G. A. Cordell, Academic Press, New York, 1998, vol. 51, p. 271.
16. (a) B. Rodríguez, A. Bruckmann and C. Bolm, *Chem. Eur. J.*, 2007, **13**, 4710; (b) H. Watanabe, E. Matsui, Y. Ishiyama and M. Senna, *Tetrahedron Lett.*, 2007, **48**, 8132; (c) P. R. Patil and K. P. R. Kartha, *J. Carbohydr. Chem.*, 2008, **27**, 279; (d) R. Trotzki, M. M. Hoffmann and B. Ondruschka, *Green Chem.*, 2008, **10**, 767; (e) V. Declerck, P. Nun, J. Martinez and F. Lamaty, *Angew. Chem., Int. Ed.*, 2009, **48**, 9318; (f) T. Szuppa, A. Stolle, B. Ondruschka and W. Hopfe, *Green Chem.*, 2010, **12**, 1288; (g) T. Szuppa, A. Stolle, B. Ondruschka and W. Hopfe, *ChemSusChem*, 2010, **3**, 1181.
17. (a) B. Rodriguez, T. Rantanen and C. Bolm, *Angew. Chem., Int. Ed.*, 2006, **45**, 6924; (b) T. Rantanen, I. Schiffers and C. Bolm, *Org. Process Res. Dev.*, 2007, **11**, 592; (c) J. G. Hernandez and E. Juaristi, *J. Org. Chem.*, 2011, **76**, 1464; (d) P. Chauhan and S. S. Chimni, *Beilstein J. Org. Chem.*, 2012, **8**, 2132; (e) J. G. Hernández, V. García-López and E. Juaristi, *Tetrahedron*, 2012, **68**, 92.
18. (a) M. Chrzanowska and M. D. Rozwadowska, *Chem. Rev.*, 2004, **104**, 3341; (b) S. M. Lu, Y. Q. Wang, X. W. Han and Y. G. Zhou, *Angew. Chem., Int. Ed.*, 2006, **45**, 2260.
19. (a) Y. Ukaji, Y. Shimizu, Y. Kenmoku, A. Ahmed and K. Inomata, *Chem. Lett.*, 1997, 59; (b) Y. Ukaji, Y. Shimizu, Y. Kenmoku, A. Ahmed and K. Inomata, *Bull. Chem. Soc. Jpn*, 2000, **73**, 447; (c) S. Wang and C. T. Seto, *Org. Lett.*, 2006, **8**, 3979; (d) K. Funabashi, H. Ratni, M. Kanai and M. Shibasaki, *J. Am. Chem. Soc.*, 2001, **123**, 10784; (e) T. Itoh, M. Miyazaki, H. Fukuoka, K. Nagata and A. Ohsawa, *Org. Lett.*, 2006, **8**, 1295; (f) S. I. Murahashi, Y. Imada, T. Kawakami, K. Harada, Y. Yonemushi and N. Tomita, *J. Am. Chem. Soc.*, 2002, **124**, 2888;

(g) T. Itoh, M. Yokoya, K. Mizauchi, K. Nagata and A. Ohsawa, *Org. Lett.*, 2003, **5**, 4301; (h) M. S. Taylor, N. Tokunaga and E. N. Jacobsen, *Angew. Chem., Int. Ed.*, 2005, **44**, 6700; (i) K. Frisch, A. Langda, S. Saaby and K. A. Jørgensen, *Angew. Chem., Int. Ed.*, 2005, **44**, 6058; (j) A. M. Taylor and S. Schreiber, *Org. Lett.*, 2006, **8**, 143.
20. (a) K. Ito, S. Akashi, B. Saito and T. Katsuki, *Synlett*, 2003, 1809; (b) C. Shi and I. Ojima, *Tetrahedron*, 2007, **63**, 8563.
21. Z. P. Li, P. D. MacLeod and C. J. Li, *Tetrahedron: Asymmetry*, 2006, **17**, 590.
22. G. Zhang, Y. X. Ma, S. L. Wang, Y. H. Zhang and R. Wang, *J. Am. Chem. Soc.*, 2012, **134**, 12334.
23. J. M. Zhang, B. Tiwari, C. Xing, X. K. Chen and Y. R. Chi, *Angew. Chem., Int. Ed.*, 2012, **51**, 3649.
24. J. B. Yu, Z. H. Li, K. Y. Jia, Z. J. Jiang, M. L. Liu and W. K. Su, *Tetrahedron Lett.*, 2013, **54**, 2006.

CHAPTER 6

Amino Acids and Peptides in Ball Milling[†]

THOMAS-XAVIER MÉTRO, EVELINA COLACINO, JEAN MARTINEZ AND FRÉDÉRIC LAMATY*

Institut des Biomolécules Max Mousseron, UMR 5247 CNRS–UM1–UM2–ENSCM, Université Montpellier 2, Place E. Bataillon, 34095 Montpellier Cedex 5, France
*Email: frederic.lamaty@univ-montp2.fr

6.1 Introduction

For many years, pharmaceutical companies have focused their attention on the development of drugs based on the biological activity of small molecules. More recently, peptides have been recognized as efficient active pharmaceutical ingredients and new delivery systems have moved them forward in the modern therapeutic arsenal. Peptides also serve as pharmacological tools.[1]

Peptides have many advantages over small drugs, the major ones being their high potency and selectivity. They can also be investigated over a broad range of targets, providing generally a better binding with fewer side-effects and avoiding accumulation in tissues. Moreover, at a time when small drugs and their metabolites have been recognized as pollutants in the environment, peptides are considered as less eco-toxic since their degradation pathways in nature lead to the generation of more innocuous molecules such as amino acids.[2]

[†]The contribution of Thomas-Xavier Métro and Evelina Colacino to this work is equivalent.

RSC Green Chemistry No. 31
Ball Milling Towards Green Synthesis: Applications, Projects, Challenges
Edited by Brindaban Ranu and Achim Stolle

Published by the Royal Society of Chemistry, www.rsc.org

An inconvenience can be their poor metabolic stability and oral availability. In this regard, new practical delivery technologies have been investigated such as nasal spray or micro-needles. Furthermore, chemists have been involved in the design and synthesis of mimicking original molecules and preparing new structures that are more resistant to degradation by endogenous enzymes.

About 100 peptidic drugs have now reached the pharmaceutical market and many peptides are now in the pipeline of pharmaceutical companies. They are usually made of a 5–10 amino acid sequence but, in some cases, larger peptides up to 50 amino acids have been synthesized and commercialized.[3] One can cite as an example Fuzeon (enfuvirtide), a 36-amino-acid antiretroviral.

As a consequence, the market for therapeutic bulk peptides is expected to grow rapidly in the next few years. It has been evaluated as more than $40 billion. Some of these molecules are becoming blockbusters. Indeed, peptides such as Copaxone® (for multiple sclerosis therapy) and hormone-related products such as leuprolide, octreotide, and goserelin (Figure 6.1) have reached annual sales of more than $1 billion each.[4]

The synthesis of peptides is now very well established with three major approaches, in solution, in the solid phase or using recombinant techniques for the larger peptides. The chemical preparation of these molecules consists of assembling amino acids by stepwise successive reactions consisting of a coupling reaction with a protected amino acid followed by a deprotection step. The synthesis of peptides has undergone strong developments with the discovery of solid phase supported synthesis.[5] This technique, based on the use of an insoluble polymeric support to anchor a first amino acid, allows a stepwise synthesis, including washings to eliminate soluble excess of coupling and deprotection reagents and side products (Scheme 6.1). One of the major advantages is the possibility to fully automate such a process.[6,7]

Nevertheless, while being extremely practical and efficient, these methods make use of large amounts of solvent. To produce 1 kg of peptide it is thought that 5000 kg of solvent are needed. From this point of view, there is a need to explore new methods for the scale up of peptide production that would avoid or decrease the use of solvents, all the more so given that the solvents recovered from the reaction and the washings are loaded with toxic compounds used during the coupling or the deprotection step.[8]

The building blocks that are used to make peptides are protected amino acids. Consequently, the preparation of amino acids and their protected derivatives are of the upmost importance in this area. Furthermore, amino acids and their derivatives may exert biological activities on their own.[9,10] They are also important starting materials arising from the chiral pool for the preparation of heterocycles.[11]

We describe in this chapter the application of ball-milling in peptide synthesis, including the preparation of amino acids and their protected derivatives.

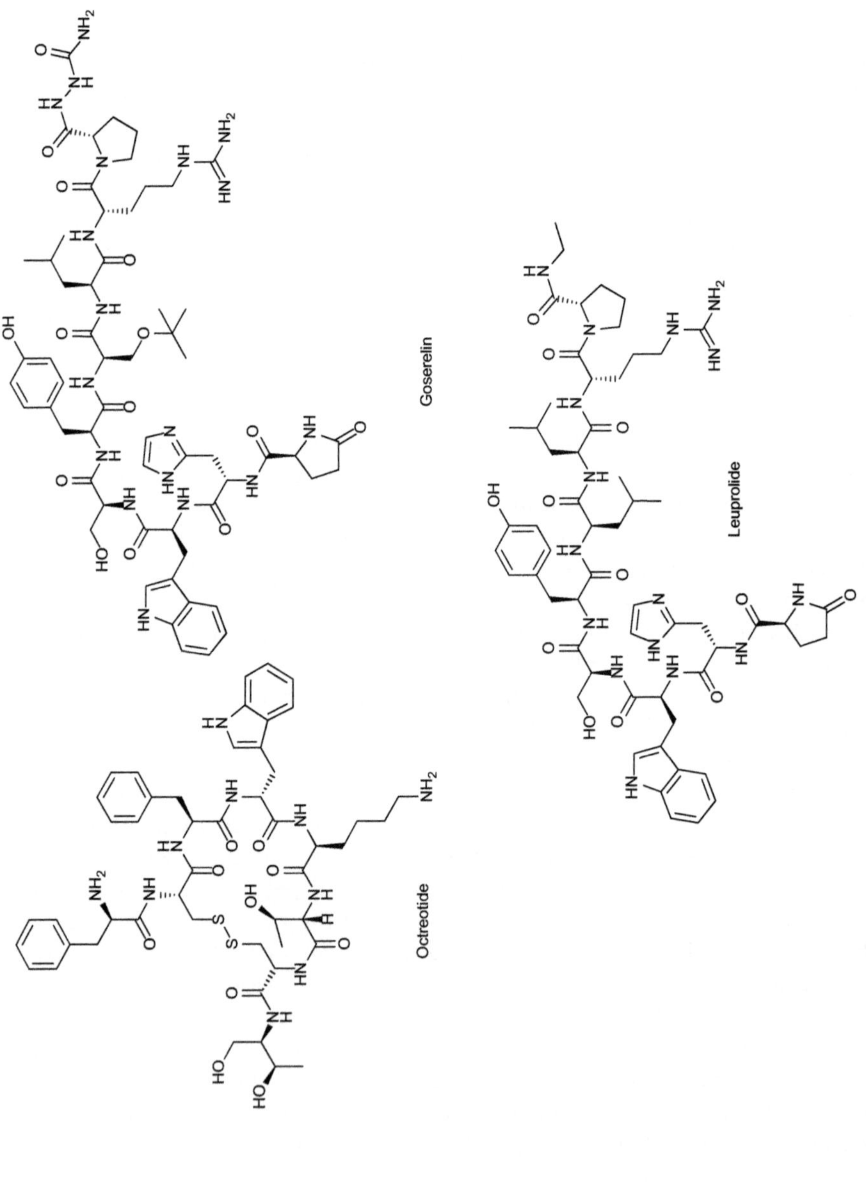

Figure 6.1 Examples of therapeutic peptides.

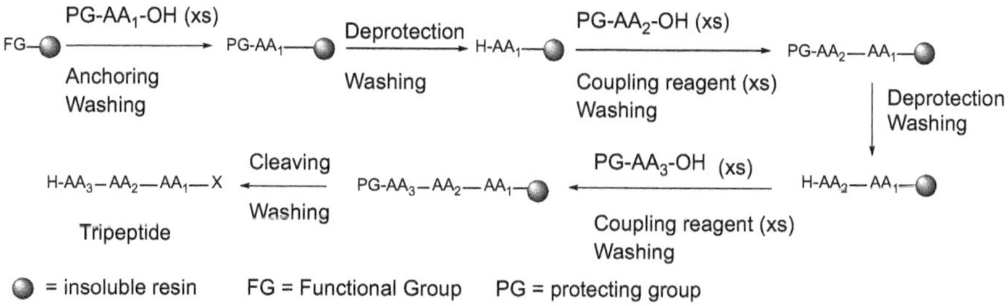

Scheme 6.1 Schematic synthesis of a tripeptide on solid-phase.

6.2 Mechanochemical Synthesis and Derivatization of Amino Acids

Amino acids are interesting targets for solid-state reactions, due to their intrinsic properties such as their zwitterionic nature and high melting points. However, their reactivity and uses in mechanochemical processes have not yet been fully explored and exploited.

6.2.1 Synthesis of Amino Acid Derivatives

Historically, the first example of their use in organic mechanochemistry dates back to 2000, when L-cysteine[12–14] (**1**) (together with its hydrochloride monohydrate derivative) and L-proline[13] (**3**) were tested for their solid-state reactivity in the presence of stoichiometric quantities of paraformaldehyde (Scheme 6.2).

Solid paraformaldehyde $(HCOH)_n$ (**6**) polymer is a handling-friendly and convenient alternative to access gaseous formaldehyde monomer (**7**) (formed *in situ* during mechanochemical milling by complete breakage of the weak polymer chain bonds). Thus, the solid-state condensation in the ball-mill with amino (or ammonium) group led to the corresponding methylene iminium salts,[15] (*R*)-**1** · HCl and (*S*)-**5** quantitatively. However, they are extremely reactive and they can be easily trapped by nucleophiles, such as the thiol group on L-cysteine[12] (**1**) – leading to L-thiazolidine (or its hydrochloride)[12] (*R*)-**2** after removing *in vacuo* the water of the reaction – or water,[13] leading to large-scale quantities (200 g)[16] of stable *N*/O-hemiacetal (*S*)-**4**.[16,17]

(*S*)-Proline (**3**) also served for the waste free, large scale and quantitative synthesis of azomethine ylide **11**[14,16,18] using stoichiometric milling with ninhydrin, *via* a three-step solid-state cascade reaction (substitution/elimination/decarboxylation) without the need of purification,[17] outperforming the synthesis in solution (82% yield) (Scheme 6.3).

6.2.2 Oxidation Reactions

As an alternative to a plethora of methods for the synthesis of disulfides in solution, the aerobic solid-state oxidation of thiol to symmetrical organodisulfides under ball milling was achieved using L-cysteine (**1**) with iodine

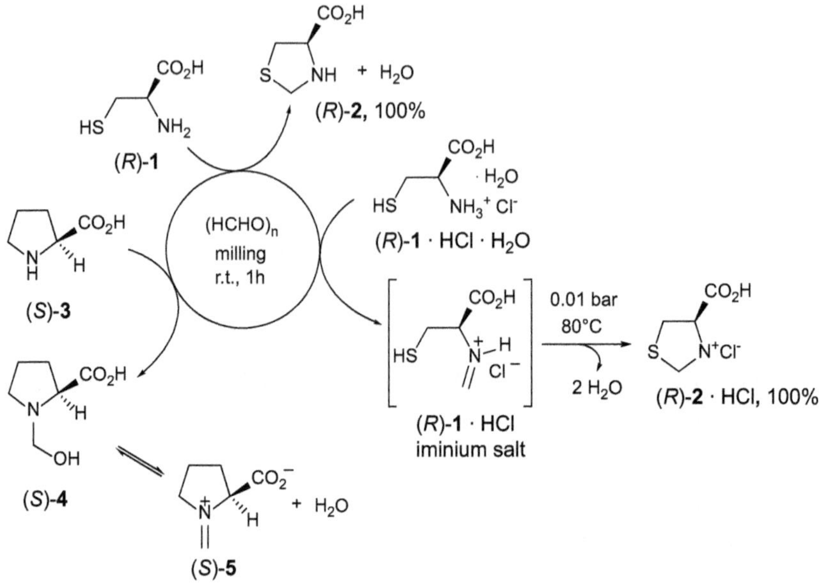

Scheme 6.2 Quantitative methylene iminium cysteine and proline salts synthesis (and trapping).

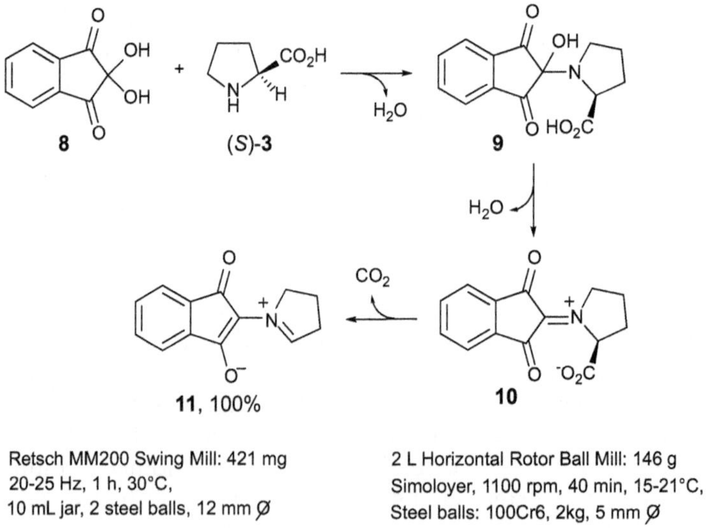

Scheme 6.3 Cascade synthesis of azomethine ylide **11**.

(quantitatively)[14] or over neutral aluminium oxide[19] (grinding auxiliary) in the absence of any metal-catalyst, base or solvent (Scheme 6.4). In the last case, the product was isolated by washing the reaction mixture with ethanol, followed by evaporation of the solvent, while the grinding auxiliary could be recovered, dried and recycled.

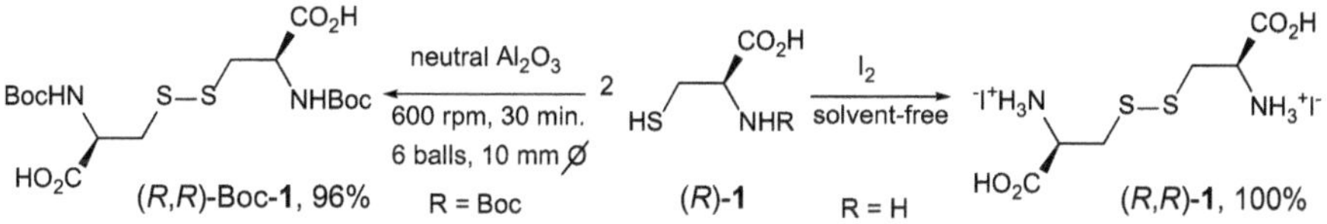

Scheme 6.4 Solid-state oxidation to organo-disulfides.

6.2.3 Asymmetric Synthesis of Amino Acids

Ball milling was also successfully applied to the preparation of amino acid derivatives in their chiral form, which remains an ongoing challenge for organic chemists.[20] For example, the metal-free aminohalogenation[21,22] of electron-deficient olefins was described from cinnamate ($R^2 = OMe$) and cinnamide ($R^2 = NEt_2$) **12**, promoted by (diacetoxyiodo)benzene [$PhI(OAc)_2$], with high regio- and diastereo-selectivity (Scheme 6.5).

Commercially available and inexpensive chloramine-T trihydrate ($TsNClNa \cdot 3H_2O$) or tosylamide/*N*-bromosuccinimide ($TsNH_2$/NBS)[23] system were used as nitrogen and halogen sources, respectively. Although the reaction mechanism is not yet known, it was proposed that the oxidation of chloro- **13** or bromo-derivative **14** by $PhI(OAc)_2$ led to a similar intermediate *N*-acetoxy-*N*-halogen-*p*-toluenesulfonamide **B**, *via* intermediate **A** after releasing iodobenzene. The electrophilic attack of **B** on α,β-unsaturated enones **12** afforded the highly reactive aziridinium intermediate **C** that underwent fast ring opening *via* S_N2 attack of the more electrophilic β-position by the *in situ* formed nearby halogen anion. High regio- and diastereoselectivity (up to *anti/syn* > 99 : 1) characterized the formation of intermediate **D**-(±), which reacted with the suitable halogeno-derivative (**13** or **14**) to afford the final product, regenerating the intermediate **B**. The mechanism involving the formation of a bridged halogenium ion intermediate (instead of aziridinium) was excluded because a reversed regio-stereoselectivity would have been observed (dominant formation of the *syn*-diastereoisomer).

Chiral amino esters were also prepared by an asymmetric alkylation reaction starting from the Schiff base of glycine in the presence of chiral ammonium phase-transfer catalyst (PTC) **20** under basic conditions[24] (Scheme 6.6).

The transamination reaction between benzophenone imine (**17**) and stoichiometric quantities of glycine *tert*-butyl ester hydrochloride (**18**) led to Schiff base **19** in nearly quantitative yield (and up to two-gram scale), after washing the milling powder with water to eliminate ammonium chloride salt. The reaction time was shorter compared to the synthesis in solution (overnight) and no chlorinated solvent was needed. For the enantioselective reaction, the first step was a solid-state deprotonation occurring at the interface of the Schiff base and KOH – performed best among the various bases explored with M_2CO_3 or MOH (M = Na, K, Cs) – and leading to a

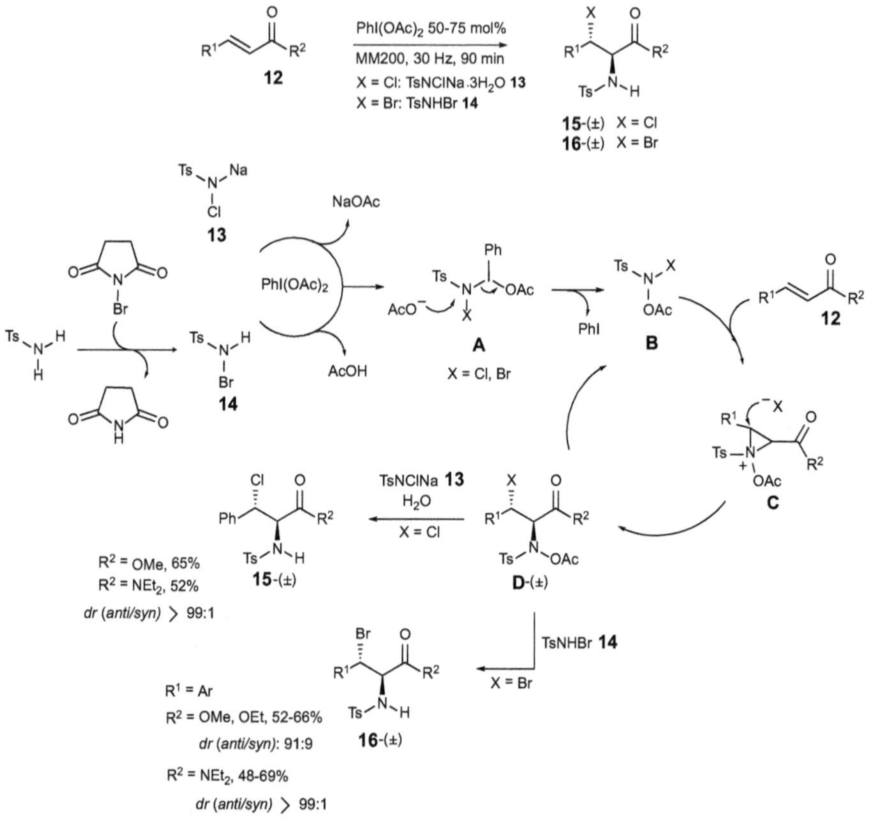

Scheme 6.5 Diastereoselective aminohalogenation and plausible reaction mechanism.

potassium enolate, which remained stable as shown by spectroscopic analysis. Glycine *t*-butyl ester Schiff base was selected for the alkylation reaction due to its higher stability to hydrolysis. In the second step, interaction with the chiral ammonium salt derived from cinchonidine **20** induced the reaction of the electrophile R^1X on a preferential face. Although excellent yields and exclusive monoalkylation were obtained, the enantiomeric excess (*ee* up to 75%) values were lower than those obtained in solution under PTC conditions. The enantiomeric excess could not be improved by increasing the amount of catalyst or varying the quantity of base. The ball-milling frequency showed little influence on the measured enantiomeric excess during the preparation of allylglycine *t*-butyl ester: at 10 or 20 Hz identical *ee*s (64%) were obtained, with a further decrease of optical purity at 30 Hz (58% *ee*). Despite the difficulties associated with temperature control and a high concentration of reactants, the results demonstrated that asymmetric synthesis in a ball-mill is not a chimera.

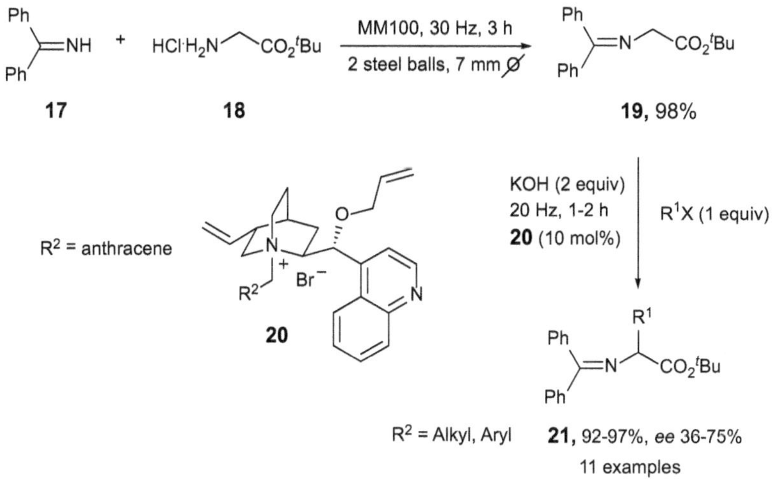

Scheme 6.6 Solvent-free synthesis of glycine Schiff base and enantioselective alkylation under PTC-conditions.

6.2.4 Synthesis of Unsaturated Amino Acids

Unsaturated unnatural amino acid derivatives **23** were also obtained using the ball-milling technique by performing a Horner–Wadsworth–Emmons reaction[25] or a Heck–Jeffery protocol,[26,27] and used as platform to access 2-carbonylindoles **26** (Scheme 6.7).[28]

High yields of Boc-protected unsaturated amino-esters were obtained by grinding amino-phosphoryl acetate **22** and various aromatic or aliphatic aldehydes in a planetary ball-mill based on a Horner–Wadsworth–Emmons reaction, with a conversion close or equal to 100% (Scheme 6.7).[25] The (*Z*)/(*E*) selectivity was always very high (100 : 0) with exclusive formation of the (*Z*)-isomer using aromatic aldehydes and hindered *ortho*-substituted substrates, except in one case (with naphthaldehyde the selectivity was lower, (*Z*)/(*E*) 82 : 18). Linear aliphatic aldehydes were also tested: in most cases high yields (up to 90%) and 100% selectivity in favor of the (*Z*)-isomer were obtained, while branched or sterically hindered aldehydes were less suitable, with incomplete conversion of starting material, while ketones were unreactive. A comparative experiment starting from aminophosphoryl acetate **22** and 3,5-dimethoxybenzaldeyde (with Cs_2CO_3 as base) was carried out in a flask with a stir bar. After 7 h, incomplete conversion (43%) was obtained, as well as by heating the mixture at 50 °C (45% conversion). Under the same conditions, mechanochemical activation led to full conversion and 94% yield for unsaturated Boc-protected amino ester **23** (R^1 = 3,5-dimethoxyphenyl), showing the importance of ball-milling activation for this transformation.

Exclusive (*Z*)-selectivity was also observed when dehydrophenylalanine derivatives **23** were prepared using solvent- and phosphine-free Heck–Jeffery protocol[26,27] in a planetary mill under phase transfer conditions involving

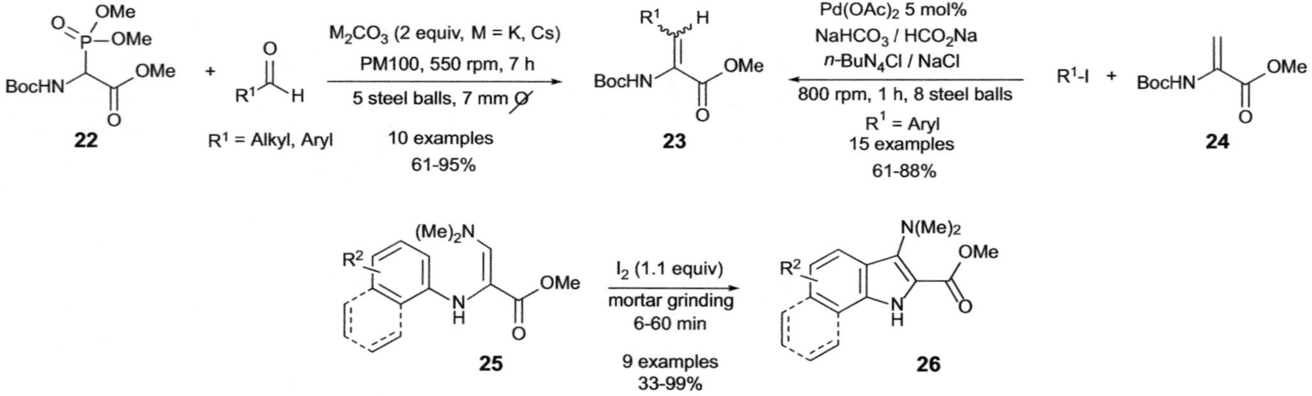

Scheme 6.7 Solvent-free synthesis of unsaturated amino esters and 2-carboxy-substituted indoles.

tetrabutylammonium chloride (Scheme 6.7). Various aryl iodo-anilines were coupled successfully (up to 5 mmol scale with 78% yield), in the presence of NaCl (5 mg mg^{-1} aryl halide) as a grinding aid,[29,30] while bromide derivatives failed to give the coupling products. The cross-coupling reaction outcome was influenced by steric and electronic effects of substituents on the aromatic ring: electron-withdrawing groups or the proximity of a heteroatom to the halide led to poor results. The results obtained under ball-milling conditions were compared to those obtained under usual conditions: (i) on heating with or without stirring, (ii) under microwave irradiation, (iii) under high static pressure conditions (200 kg cm^{-2}), using an ordinary hydraulic press for making IR-tablets. The reaction mixture was heated in all cases, at 80 °C, a temperature that matches that generated during milling at full speed (800 rpm) for 1 h. The yields were always lower compared to ball-mill experiments (up to 33%), demonstrating that the conditions created during ball-milling by combination of pressure (rotation of the steel balls), heat, grinding and stirring were not easy to obtain under usual conditions, affirming the power of milling for the positive outcome of the reaction. This trend was also confirmed for the iodine-promoted intramolecular cyclization of substituted 2-anilinoenaminones **25** to prepare 2-carbonylated-3-dimethylamino-indoles **26** by solvent-free manual grinding in a mortar (Scheme 6.7).[28] The reactions were generally faster (6–60 min instead of 12 h) and higher yielding (often quantitative) compared to synthesis in solution using acetonitrile or in the presence of Lewis acids.

6.2.5 Synthesis of Protected Amino Acids

Ball-milling technology was also applied to the solvent-free synthesis of carbamate *N*-protected α- and β-amino acids[31] as Boc-, Z-, and Fmoc-derivatives (Scheme 6.8).

The reactions were performed using two different planetary ball-mill apparatus and process parameters were also investigated: (i) stainless steel or tungsten carbide (WC) grinding jar material, (ii) number of grinding balls, (iii) rotation speed, (iv) mode of operation under cycled or continuous milling, and (v) grinding additives.

In the one-pot/two-step protocol, the first step relied on the inhibition of the reactivity of α- or β-carboxylic acid function through the *in situ* formation of a transient potassium internal salt, and then stoichiometric amounts of the suitable protecting group (Boc_2O, Z-OSu or Fmoc-OSu) were introduced in the second step (Scheme 6.8). This eco-friendly methodology was general for all the protecting groups tested in the study, and gave good to excellent yields of amino acid derivatives without any purification and in a scale from 50 mg up to 1 g of final product. The *N*-protected amino acid derivatives were not soluble in water: in the case of *N*-Fmoc- and *N*-Z-derivatives, after acidification, the pure product precipitated off the solution and was recovered, while liquid–liquid extraction was necessary for *N*-Boc-protected amino acids. The only by-products were water, CO_2 and *t*-BuOH (eliminated by

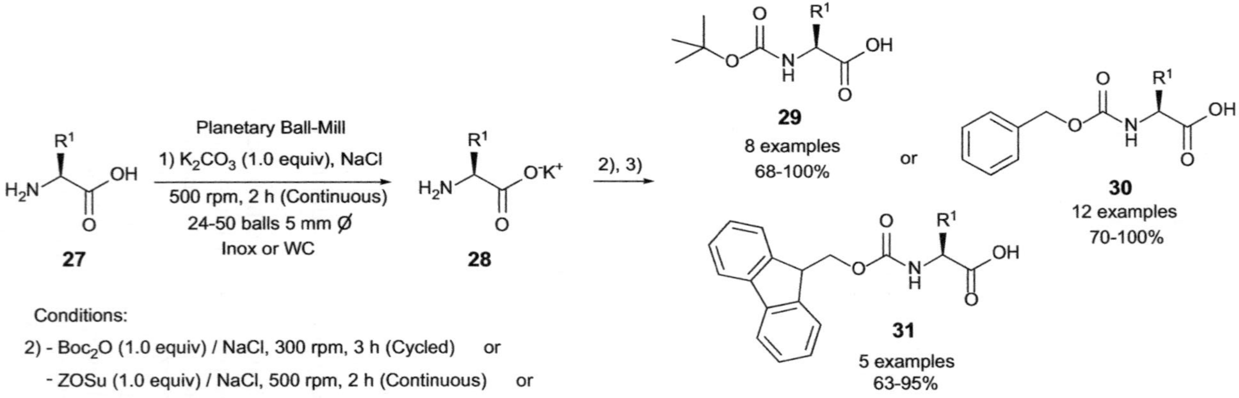

Conditions:

2) - Boc_2O (1.0 equiv) / NaCl, 300 rpm, 3 h (Cycled) or
- ZOSu (1.0 equiv) / NaCl, 500 rpm, 2 h (Continuous) or
- FmocOSu (1.0 equiv) / NaCl, 500-750 rpm, 2-3 h (Continuous or Cycled)

3) Aqueous acidic work-up, then filtration/precipitation

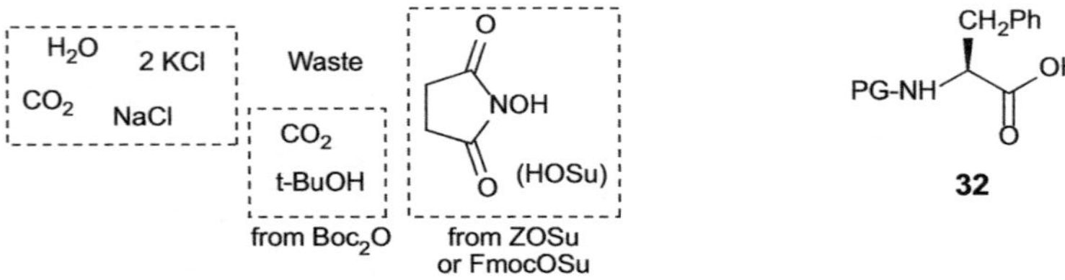

E-factor for the synthesis of **32**

PG	In Solution	Solvent-free
Boc	62	265
Z	20	8
Fmoc	288	6

Scheme 6.8 Two step/one-pot *N*-protection of amino acids in a planetary ball-mill.

evaporation), together with water soluble compounds such as inorganic salts (MCl, with M = Na, K), traces of unreacted amino acid hydrochloride or hydroxysuccinimide (SuOH), which were washed away during aqueous acidic precipitation/filtration work-up (Scheme 6.8).

In the case of Boc-protection, cycled milling mode (at 300 rpm during three cycles of 1 h each, with 10 min pause between cycle) was the right approach to avoid both the formation of *N*-diacylated by-products and to promote a fast kinetics for *N*-protection, while diminishing the kinetics of Boc_2O degradation, which were otherwise observed under continuous milling, at higher rotation speeds (500 or 650 rpm cycled or not) and independently of the number of balls used (6, 12 or 24 stainless steel balls). In the case of Z- and Fmoc-protection, if the highly reactive Z-Cl, Z_2O or Fmoc-Cl were not suitable because of their fast decomposition, the hydroxysuccinimidyl derivatives Z-OSu and Fmoc-OSu, respectively, were used. In the case of *N*-Fmoc derivatives, yields and selectivity were higher at increased rotation speed (up to 750 rpm) and number of balls (up to 50), while the modulation of the energetic conditions applied to the grinding jar, using different milling materials, had no influence. Notice that in the case of Fmoc-protection the formation of side-products usually observed in solution (*i.e.* Fmoc-β-Ala-OH or dipeptides Fmoc-β-Ala-AA-OH *via* the Lossen rearrangement)[32,33] was avoided under mechanochemical activation. From the point of view of environmental impact, especially in the case of *Z*- and Fmoc-protection, the *E*-factors[34] were greatly improved compared to classical syntheses in solution (Scheme 6.8). In the case of Boc-protection, the nature of waste for solution or solvent-free synthesis was the same, but liquid–liquid extraction (instead of precipitation) work-up was needed to recover the product, with a negative impact on the *E*-factor value.

In a similar approach, a vibrational ball-mill was used for the carbamoylation reaction of α-, β- or quaternary amino esters (Scheme 6.9),[35] sharing the same advantages as previously illustrated for the *N*-protection of amino acid derivatives in the planetary ball-mill[31] (Scheme 6.8) in terms of: (i) ease of recovery of pure final products, (ii) entity and nature of waste, and (iii) full conversion and yield. In addition, Z-OSu and Fmoc-OSu could be replaced by the more convenient chloride derivatives. Independently of the incoming protecting group, the reaction kinetics depended on the nature of the *C*-terminal ester, which proved to be slower when *t*-butyl instead of methyl esters were used. The number of balls in the milling jar also seemed to have an impact on the outcome of the reaction: one ball instead of two led to incomplete conversion of substrates and much lower yields.

Notably, mechanochemical activation was particularly suitable for the *N*-Boc carbamoylation of hindered substrate such as H-Aib-OMe, affording impressive improved yield (68%) – three times more than in solution.[36] However, very low conversions were obtained for the *N*-carbamoylation reaction in the vibrational ball-mill from amino acids (instead of planetary milling, Scheme 6.8), or in the planetary ball-mill from amino esters (instead of vibrational apparatus, Scheme 6.9), probably because of the different

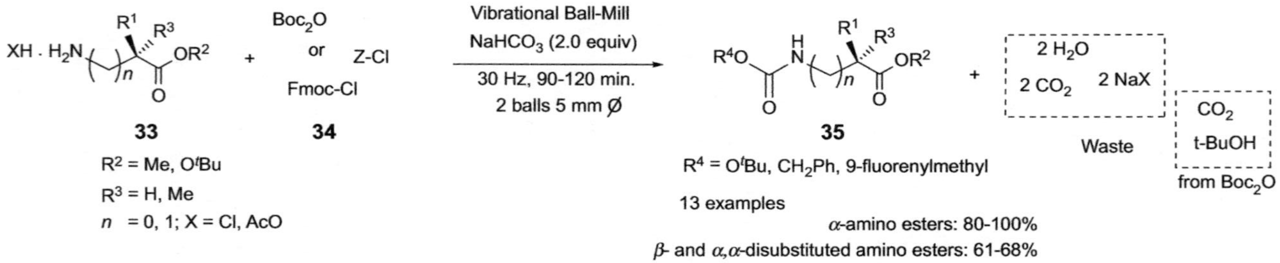

Scheme 6.9 *N*-Carbamoylation of amino-ester derivatives in a vibrational ball-mill.

stress phenomena (horizontally vibrating or circularly shaking milling) and quantity of energy delivered to the sample, working at 30 Hz or at 450 rpm (which corresponded to 7.5 Hz), showing that the two apparatus were not necessarily interconvertible when performing the same reaction.

The same trend was observed during the solvent-less esterification of the *C*-terminal position of amino acids,[35] which is effective in the planetary and not in the vibrational ball-mill apparatus (Scheme 6.10).

Cycled milling was suitable for the reaction of various dialkyl dicarbonates such as Boc_2O, Z_2O, Moc_2O [$(ROCO)_2O$, with R = O^{t}Bu, Bn, Me, respectively], carbonates (RO_2CO, with R = succinimide, Me) or alkyl chloroformates (ROCOCl, with R = Bn, Me, Et, allyl) with *N*-protected α-amino acids, in the presence of DMAP as base. As already illustrated for the *N*-protection of amino acids[31] (Scheme 6.8), continuous milling provoked the premature decomposition of activating agents **37** used in stoichiometric quantity. The decarboxylative esterification proceeded *via* formation of a mixed carboxylic-carbonic anhydride **A** (Scheme 6.10), which was converted into acylpyridinium derivative **B** by nucleophilic attack of DMAP. The evolution of carbon dioxide provided the driving force of the reaction.

Compared to solution synthesis, the preparation of *tert*-butyl esters from Boc_2O presented the advantages of shorter reaction times and reduced quantity of waste, also avoiding the use of expensive Boc_2O in excess, Lewis acids, solvents or *t*-BuOH to speed up the reaction. Benzyl or methyl ester derivatives using the highly reactive dicarbonates Z_2O, Moc_2O or dimethyl carbonate were not obtainable under solvent-less mechanochemical activation because of their fast decomposition. Alternatively, the corresponding chloroformates reacted straightforwardly, and with the succinimidyl esters prepared from *N,N*′-disuccinimidyl carbonate (DSC). In all cases, the pure products were recovered after simple and clean filtration/precipitation work-up, as already illustrated (Schemes 6.8 and 6.9), with lower environmental impact compared to solution syntheses.

For the spot-to-spot solvent-free mechanochemical derivatizations of *N*- or *C*-protected amino acids (Schemes 6.9 and 6.10, respectively) the temperature of the mixtures was 23–25 °C, after vibrational or planetary milling, ruling out a temperature increase into the milling jar[37] as responsible for the good yields.

The solvent-free derivatization of the *N*- or *C*-terminal position of amino acids was also reported with *N,N*′-carbonyldiimidazole (CDI) (Scheme 6.11).[38] In a first step, carboxylic acids were efficiently transformed into acylimidazoles by treatment with CDI as activating agent. Subsequent addition of amine in a ball-mill reactor furnished, through a highly stereoselective process, the amino acid derivatives **40** and **41** when using enantiopure HCl · NH_2-Phe-O^{t}Bu or Boc-NH-Phe-OH, respectively (Scheme 6.11).

Notably, consumption of both the carboxylic acid in the first step and the acyl-imidazole in the second step were monitored by infrared analyses. Precipitation of pure final compounds in water, followed by filtration, afforded the amides **40** and **41** without the need of any organic solvent.

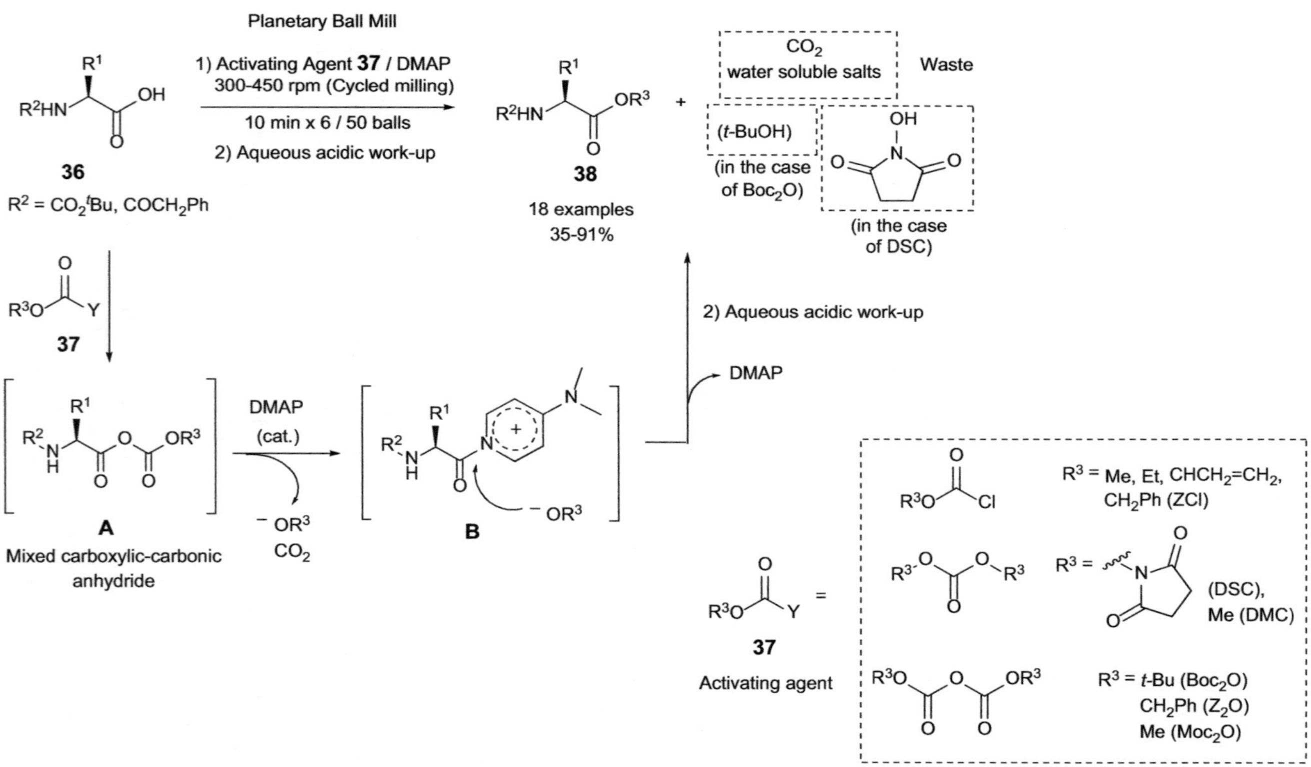

Scheme 6.10 *C*-Terminal esterification for protection/activation of *N*-protected amino acids in a planetary ball-mill.

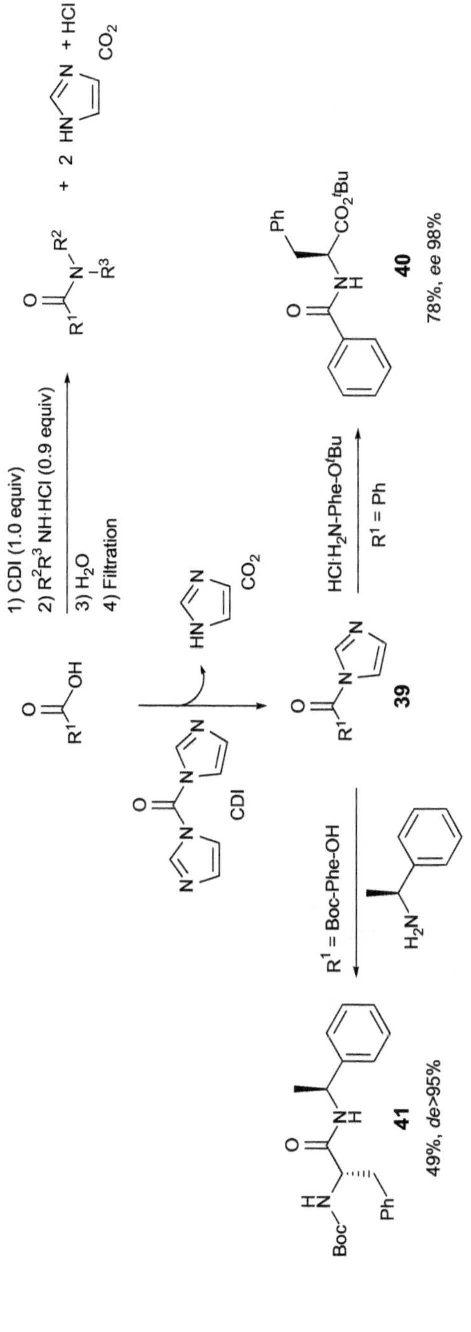

Scheme 6.11 CDI-mediated mechanosynthesis of amides.

A similar approach was developed by Margetić and coworkers to produce amino amides by using a mechanochemical process (Scheme 6.12).[39] Boc-Alanine was activated in the reaction media by addition of *N*-ethyl-*N*′-(3-dimethylaminopropyl)carbodiimide hydrochloride (EDC · HCl) in the presence of 2 equivalent of dimethylaminopyridine (DMAP) as the base, 20 equivalents of NaCl as a solid grinding assistant and small quantities of nitromethane as a liquid grinding auxiliary. Subsequent reaction with *p*-anisidine or 4-choroaniline provided the corresponding amino amides in 87% and 88% yield respectively.

6.3 Mechanosynthesis of Peptides

6.3.1 Synthesis of Di- and Tripeptides

In 2009, motivated by the desire to mitigate the environmental impact of classical peptide synthesis approaches, Lamaty and coworkers envisioned performing solvent-free peptide synthesis based on the use of the ball-milling technology.[40] Initial investigations were carried out on the coupling of the Boc-protected phenylalanine *N*-carboxyanhydride (Boc-Phe-NCA) with alanine methyl ester hydrochloride salt (HCl · H-Ala-OMe) in the presence of $NaHCO_3$ in a vibrating ball-mill (Scheme 6.13).

After 1 h of vigorous agitation (30 Hz) in a vibrating ball-mill, reaction media was recovered from the milling jar by using both EtOAc and water. Classical treatments of the organic phase such as water washings, drying and concentration furnished the pure dipeptide Boc-Phe-Ala-OMe in 79% yield without epimerization as observed by HPLC analysis. The low environmental impact of this reaction has to be noticed. The usual highly toxic solvents such as DMF, CH_2Cl_2, NMP or THF were not required and volatile and corrosive bases such as Et_3N and DIPEA were replaced by innocuous $NaHCO_3$. In addition, this approach furnished non-toxic CO_2 and NaCl as the only byproducts.

As one could argue that this reaction could have taken place during reaction media recovery with EtOAc and water, the authors performed solid-state IR and CP/MAS ^{13}C NMR analysis of the reaction media before EtOAc recovery. These analysis revealed the disappearance of the characteristic signals of Boc-Phe-NCA ($\nu = 1817$ and $1872\ cm^{-1}$ in IR and $\delta = 147.6$ in CP/MAS ^{13}C NMR) and the emergence of the typical signals of the desired product Boc-Phe-Ala-OMe ($\nu = 1624$ and $1655\ cm^{-1}$ in IR and $\delta = 156.0$ and 177.8 in CP/MAS ^{13}C NMR). As these analytical samples were not involved in any solubilization process, these results proved that the reaction clearly occurred in the solid state.

Another particularity of these reaction conditions has been revealed by studying the kinetics of the reaction. As previously described by others on solid state reactions,[41] measuring the conversion of Boc-Phe-NCA into Boc-Phe-Ala-OMe at different times of reaction revealed apparent zero-order kinetics (Figure 6.2). Notably, the reaction order seems to be independent of the frequency applied to the reaction media.

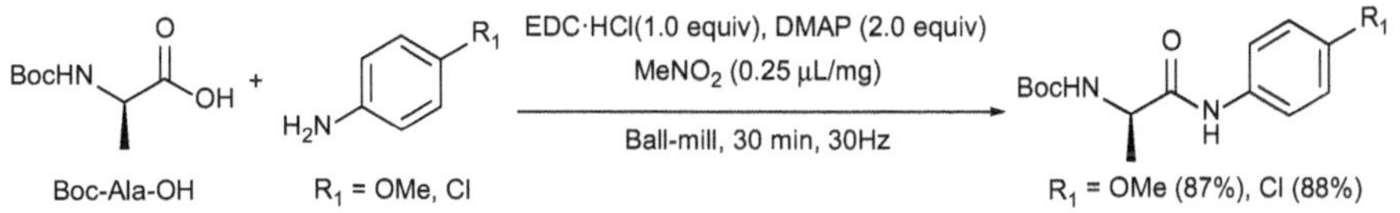

Scheme 6.12 EDC-mediated mechanosynthesis of amides.

This approach was applied to the coupling of various α-UNCAs with α-aminoesters, furnishing numerous α,α-dipeptides with high yields (Table 6.1). When starting from Boc-Phe-NCA with alanine or leucine esters as the nucleophiles, high conversions and yields were obtained (Table 6.1, entries 1–3). Under the same conditions, treatment of Boc-Phe-NCA with phenylalanine methyl ester hydrochloride salt gave a low conversion of 58%, which was supposed by the authors to be related with the physicochemical state of the reaction mixture (Table 6.1, entry 4). Indeed, due to its insolubility in water, Boc-Phe-Phe-OMe could be recovered as pure material by simple trituration in aqueous media followed by filtration and drying. It was isolated in 55% yield, which can be considered as very satisfying compared to the relatively low conversion. Notably, the production of Boc-Phe-Phe-OMe under these conditions fits particularly well the 12 principles of Green Chemistry[42] as it was produced without the use of any organic solvent from the reaction to the product recovery. In addition, aminoesters can also be used as their acetate salt as AcOH · H-Gly-O*t*Bu was transformed into Boc-Phe-Gly-O*t*Bu with high conversion and yield (Table 6.1, entry 5). Bulky electrophiles such as Boc-Val-NCA can also lead efficiently to the corresponding dipeptides with yields ranging from 85% to 100% (Table 6.1, entries 6–10). The urethane protecting group can also be switched to the widely used Fmoc group as treating Fmoc-Val-NCA with various amino-esters furnished the corresponding dipeptides with high conversions (Table 6.1, entries 11–15).

The same approach is applicable to the synthesis of tripeptides as reacting the dipeptide HCl · H-Ala-Gly-OMe with Boc-Val-NCA produced Boc-Val-Ala-Gly-OMe in 89% yield (Scheme 6.14).

6.3.2 Scale-up of Peptide Synthesis

The scalability of this approach was also studied. While the reaction conditions were optimized for a vibrating ball-mill, performing the synthesis of a dipeptide on a large scale was realized on a planetary ball-mill by using a 250 mL jar equipped with an in-line temperature and pressure monitoring system. The latter was filled with 3.85 g of Boc-Phe-NCA, 1.84 g of HCl · H-Ala-OMe and 1.7 g of $NaHCO_3$ and agitated for 2 h in the planetary ball-mill at 350 rpm, furnishing 4.3 g of the dipeptide Boc-Phe-Ala-OMe in an excellent yield of 94% (Scheme 6.15).[43]

Measurement of the temperature indicated a slow but regular increase from 26 to 33 °C during the first 60 min of the reaction (Figure 6.3). In addition, the pressure increased during 50 min from atmospheric pressure

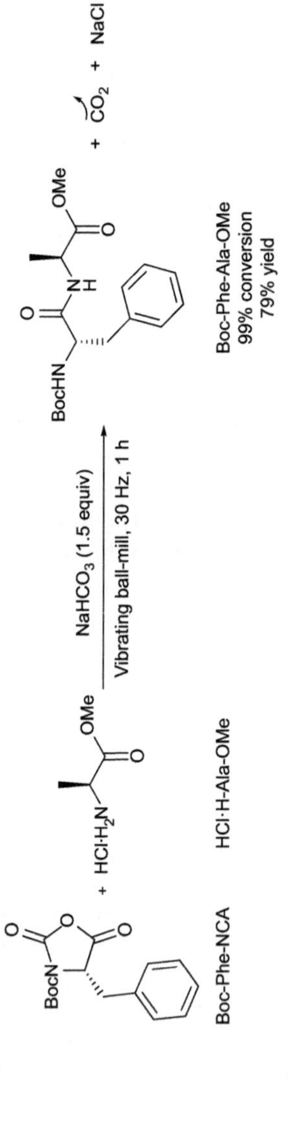

Scheme 6.13 Mechanosynthesis of Boc-Phe-Ala-OMe.

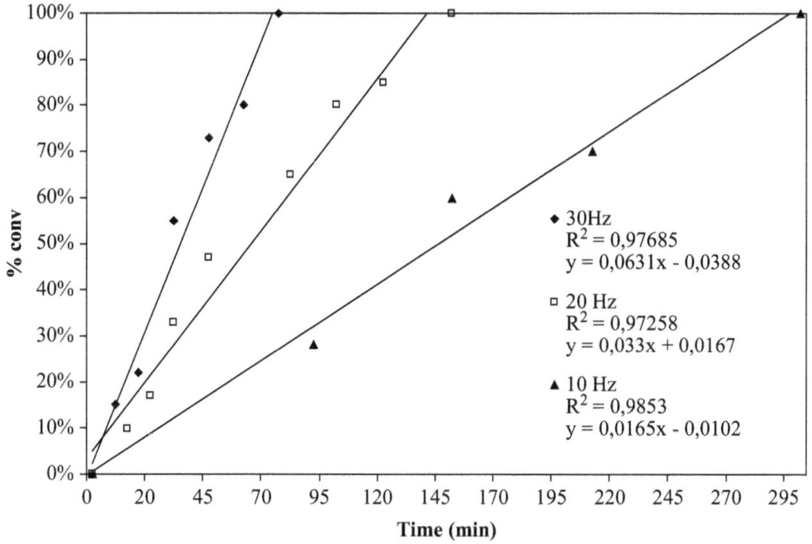

Figure 6.2 Conversion of Boc-Phe-NCA into Boc-Phe-Ala-OMe as a function of time.

to a plateau at 3 bar that lasted 15 min, until the reactor was opened ($t = 60$ min). This pressure increase is in complete accordance with the formation of CO_2 that occurs during the reaction. Opening the jar to follow completion of the reaction resulted in a pressure drop to atmospheric pressure. An additional hour of agitation at 350 rpm resulted in an insignificant rise in pressure and temperature indicating the end of the reaction, which could be confirmed by ^{1}H NMR analysis of the crude.

This approach was then applied to the synthesis of the dipeptide aspartame, a nutritive sweetener that is produced at the multi-ton scale every year. After preparation of Boc-Asp(O*t*Bu)-NCA, the latter was treated with phenylalanine methyl ester hydrochloride during 1 h milling at 30 Hz to produce Boc-Asp(O*t*Bu)-Phe-OMe with an excellent yield of 97% (Scheme 6.16). Boc and *t*Bu protecting groups were then removed by gaseous HCl treatment in the absence of any solvent to furnish aspartame hydrochloride in quantitative yield. Notably, these two first reactions are performed with a close to quantitative global yield while producing either volatile or water-soluble side-products (CO_2, isobutylene and NaCl). Finally, aspartame hydrochloride was solubilized in water and neutralization to the isoelectric point (pH 5.0) with an aqueous solution of Na_2CO_3 furnished aspartame as a white powder in 84% yield.

6.3.3 Synthesis of α,β- and β,β-Dipeptides

By using a very similar method, Juaristi and coworkers published a year later the solvent-free, ball-mill mediated synthesis of α,β- and β,β-dipeptides.[44] Indeed, β-UNCAs were used in place of the α-UNCAs utilized by Lamaty and

Table 6.1 Mechanosynthesis of α,α-dipeptides from α-UNCAs and α-aminoesters.

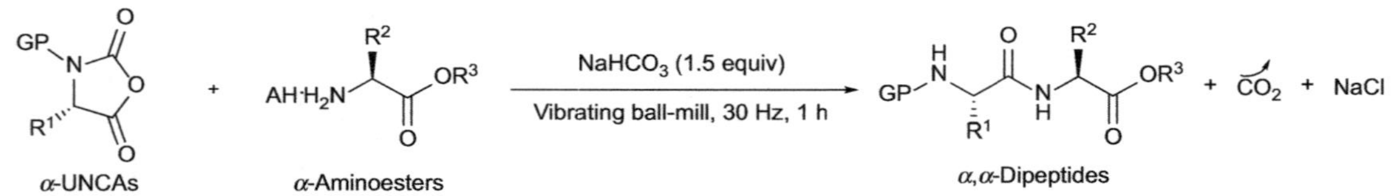

Entry	α-UNCA	α-Aminoesters	α,α-Dipeptides	Conversion (%)	Yield (%)
1	Boc-Phe-NCA	HCl · H-Ala-O*t*Bu	Boc-Phe-Ala-O*t*Bu	100	73
2		HCl · H-Leu-OMe	Boc-Phe-Leu-OMe	85	—
3		HCl · H-Leu-O*t*Bu	Boc-Phe-Leu-O*t*Bu	100	70
4		HCl · H-Phe-OMe	Boc-Phe-Phe-OMe	58	55
5		AcOH · H-Gly-O*t*Bu	Boc-Phe-Gly-O*t*Bu	100	90
6	Boc-Val-NCA	HCl · H-Leu-OMe	Boc-Val-Leu-OMe	100	87
7		HCl · H-Leu-O*t*Bu	Boc-Val-Leu-O*t*Bu	97	85
8		HCl · H-Ala-OMe	Boc-Val-Ala-OMe	100	100
9		HCl · H-Ala-O*t*Bu	Boc-Val-Ala-O*t*Bu	100	100
10		HCl · H-Phe-OMe	Boc-Val-Phe-OMe	100	88
11	Fmoc-Val-NCA	HCl · H-Leu-OMe	Fmoc-Val-Leu-OMe	90	—
12		HCl · H-Leu-O*t*Bu	Fmoc-Val-Leu-O*t*Bu	92	—
13		HCl · H-Ala-OMe	Fmoc-Val-Ala-OMe	100	76
14		HCl · H-Ala-O*t*Bu	Fmoc-Val-Ala-O*t*Bu	78	—
15		HCl · H-Phe-OMe	Fmoc-Val-Phe-OMe	93	—

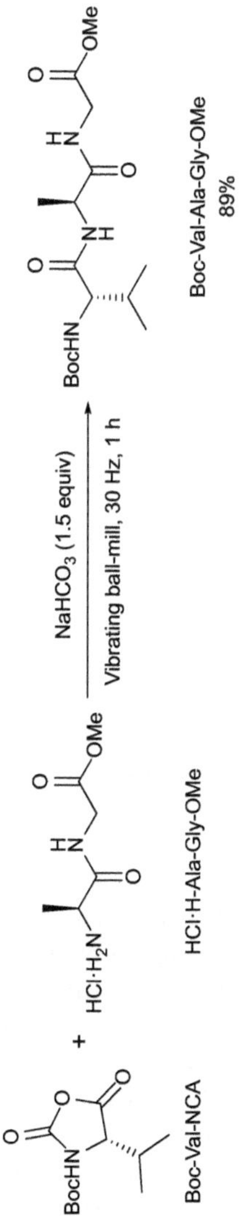

Scheme 6.14 Mechanosynthesis of the tripeptide Boc-Val-Ala-Gly-OMe.

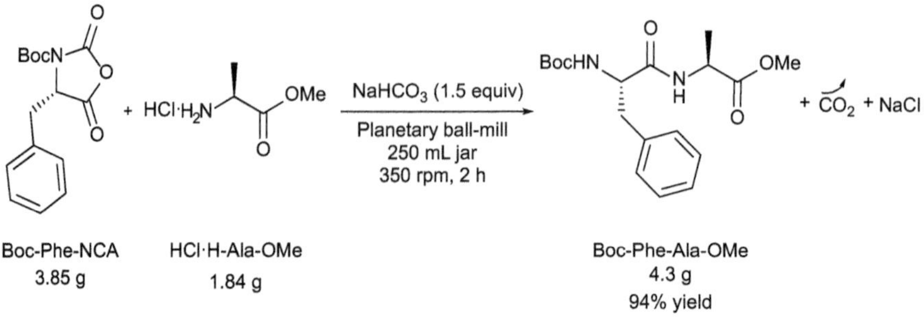

Scheme 6.15 Synthesis of Boc-Phe-Ala-OMe on a 4.3 g scale.

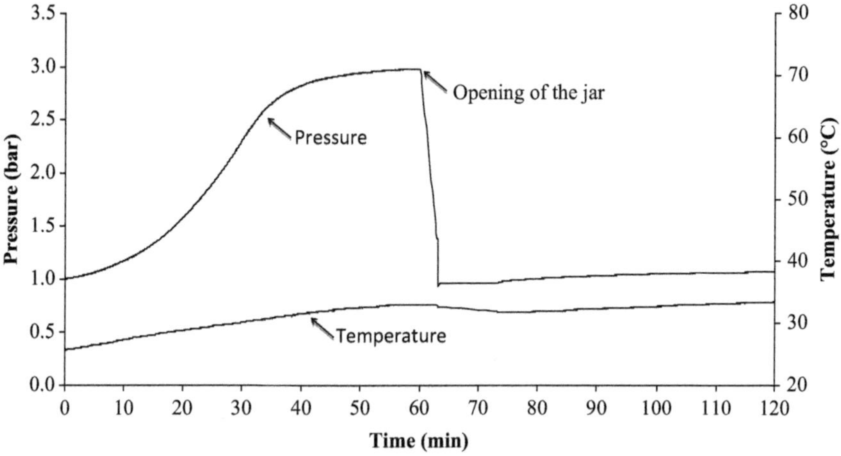

Figure 6.3 Temperature and pressure measurement during the synthesis of Boc-Phe-Ala-OMe at the 4.3 g scale.

coworkers. While known to present slightly lower reactivity in solution than α-UNCAs, β-UNCAs could be easily transformed into α,β- and β,β-dipeptides, showing the high versatility and capacity of the ball-milling approach.

Indeed, treating Boc-β-Ala-NCA with α-aminoester hydrochlorides such as glycine, alanine, leucine, or phenylalanine methyl ester hydrochlorides in the presence of $NaHCO_3$ in a vibrating ball-mill during 2 h furnished the corresponding α,β-dipeptides with yields ranging from 83% to 88% yield (Scheme 6.17). Utilizing bulkier nucleophiles such as valine or isoleucine methyl ester hydrochlorides did not hamper the reaction as Boc-β-Ala-Val-OMe and Boc-β-Ala-Ile-OMe were isolated in 82% and 80% respectively. Starting from Boc-β-Ala-NCA also allowed the synthesis of a precursor of the natural L-carnosine (H-β-Ala-His-OH), a mammalian dipeptide presenting biological activity.[44] Indeed, when 2HCl · H-His-OMe was reacted in a ball-mill with Boc-β-Ala-NCA, Boc-β-Ala-His-OMe was produced in 91% yield. Finally, substituted β-UNCA could also be used as substrates under the above conditions as Boc-protected (*S*)-β^3-carboxyhomoglycine *N*-carboxy

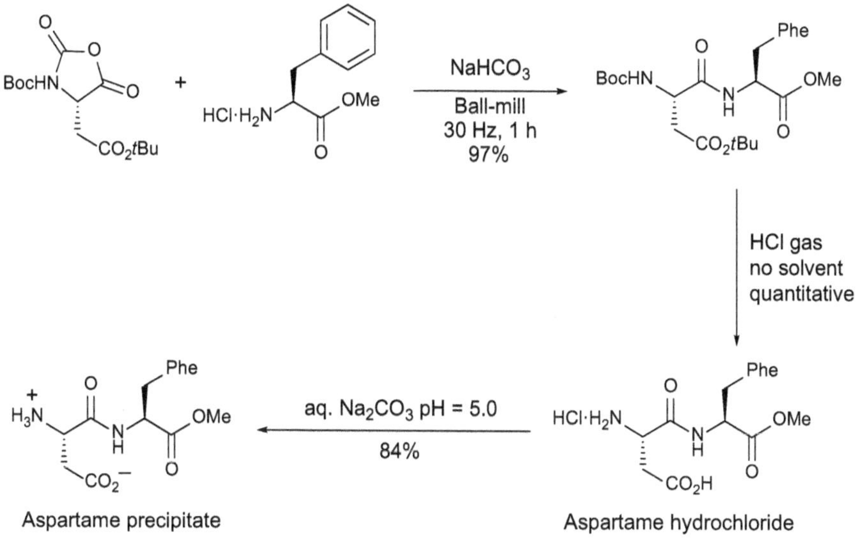

Scheme 6.16 Synthesis of aspartame by mechanochemistry.

anhydride (Boc-(*S*)-β^3-Chg(OMe)-NCA) led to Boc-(*S*)-β^3-Chg(OMe)-Ala-OMe in 79% yield when reacted with HCl · H-Ala-OMe (Scheme 6.17).

β,β-Dipeptides were also produced efficiently by treating β-UNCAs with β-aminoesters. Surprisingly, the yields were higher when using β-aminoesters than when using α-aminoesters. Indeed, treatment of HCl · H-β-Ala-OMe and HCl · H-(*S*)-β^3-hPhg-OMe [(*S*)-β^3-homophenylglycine methyl ester hydrochloride] with Boc-β-Ala-NCA furnished Boc-β-Ala-β-Ala-OMe and Boc-β-Ala-(*S*)-β^3-hPhg-OMe with 96% and 91% yield, respectively (Scheme 6.18). When treated with the same β-aminoesters, Boc-(*S*)-β^3-hPhg-NCA led to the production of Boc-(*S*)-β^3-hPhg-β-Ala-OMe and Boc-(*S*)-β^3-hPhg-(*S*)-β^3-hPhg-OMe with excellent yields of 93% and 94%. Finally, treatment of Boc-(*S*)-β^3-Chg(OMe)-NCA with HCl · H-β-Ala-OMe furnished Boc-(*S*)-β^3-Chg(OMe)-β-Ala-OMe with a high yield of 91%.

6.3.4 Synthesis of Peptides with a Longer Amino Acid Sequence

Subsequent development of the methodology was provided by Lamaty and coworkers with the aim of applying the later approach to the synthesis of longer peptide sequences.[45] As this objective requires the production of dipeptides on a large scale, efforts were directed to increase the productivity of the method. Indeed, whatever the size of the jar, the previously described operating procedure was performed using a relatively low milling load. Milling load can be defined as "the sum of the mass of the reactants per free volume in the jar".[45] In other words, increasing the milling load will result in improving the productivity of the process.

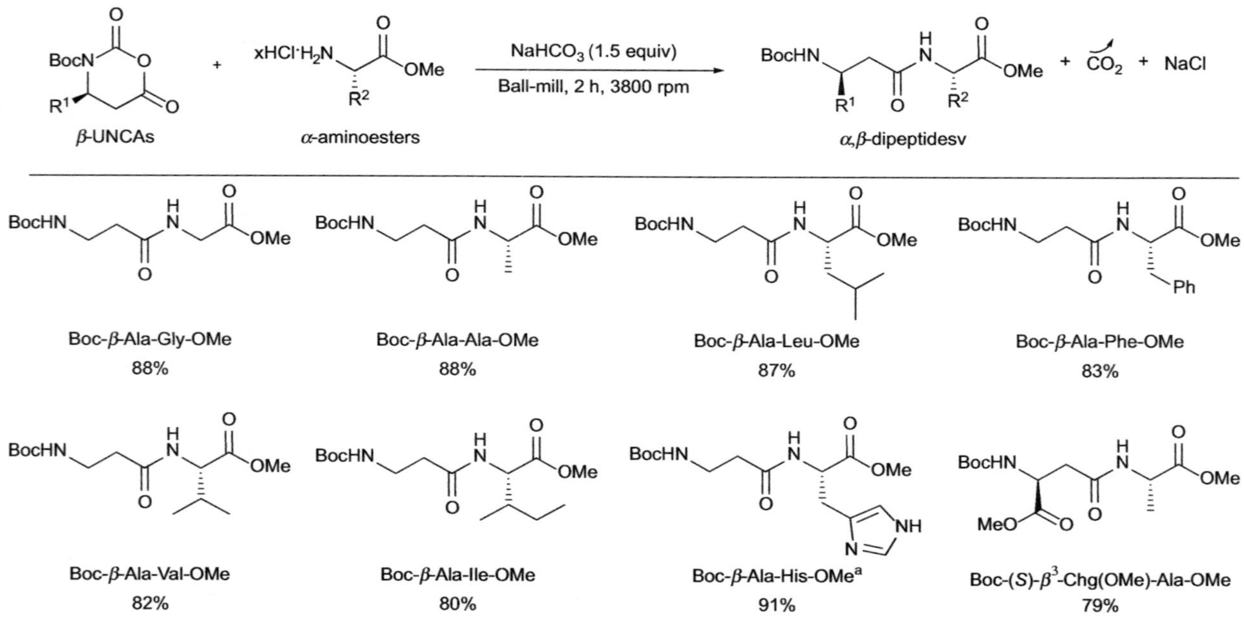

[a] 2 equiv of $NaHCO_3$ were used.

Scheme 6.17 Mechanosynthesis of α,β-dipeptides from β-UNCAs and α-aminoesters.

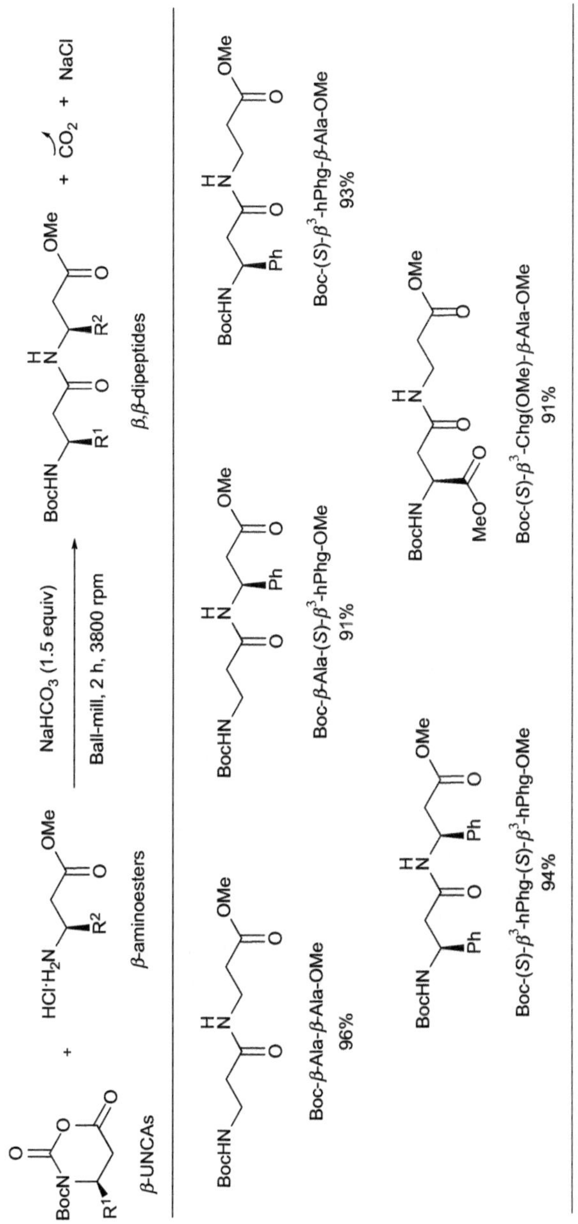

Scheme 6.18 Mechanosynthesis of β,β-dipeptides from β-UNCAs and β-aminoesters.

Indeed, in a typical experiment involving a 10 mL jar filled with one 10 mm diameter metallic ball, Boc-Phe-NCA, HCl · H-Leu-OMe and $NaHCO_3$, up to a milling load of 5.9 mg mL^{-1}, gave a satisfying conversion of 59% reached after 10 min of agitation (Scheme 6.19, Figure 6.4). Unfortunately, when the milling load was increased up to 22.5 mg mL^{-1}, conversion into Boc-Phe-Leu-OMe dropped to 17% (Figure 6.4), which was attributed by the authors to the "highly viscous and sticky reaction media". This observation was hypothesized to be related to mass transfer limitations resulting in a low speed of formation of Boc-Phe-Leu-OMe. To improve mass transfer of the reactants in the jar, the authors envisioned the use of small amounts of a liquid as a grinding assistant. Thus, adding 1.4 μL of EtOAc per mg of reactants to the reaction mixture ($\eta = 1.4$ μL mg^{-1}; η stands for the ratio of added liquid volume to the mass of solid reactants)[46] led to 91% conversion into Boc-Phe-Leu-OMe after the same grinding time. One could argue that adding EtOAc to the solid reactants and placing this reaction mixture under classical agitation could be enough to overcome the mass transfer limitations. The latter possibility was rejected by the fact that when the same reaction mixture was placed under classical magnetic agitation, the speed of the reaction was far slower than under ball-milling agitation. Indeed, when placed under classical agitation, the conversion into Boc-Phe-Leu-OMe reached 12% after 40 min (Figure 6.4). The latter experiment proved unambiguously that it is the combination of ball-milling with the presence of a grinding assistant that is responsible for the fast conversion into the dipeptide. Besides, switching from EtOAc to a predominant solvent used in peptide synthesis such as DMF did not match the performance of the ball-mill approach as 37% of conversion was reached after 40 min of agitation.

Another positive effect of using a liquid grinding assistant is the influence on Boc-Phe-NCA hydrolysis. Indeed, while conversion of Boc-Phe-NCA into Boc-Phe-OH reached 23% after 20 min of grinding with a milling load of 22.5 mg mL^{-1}, it declined to 2% after the same amount of time if EtOAc was present in the reaction media (Figure 6.5).

These conditions were then successfully applied to the synthesis of a wide range of di- to pentapeptides (Table 6.2). First of all, when previously optimized conditions were applied to the production of Boc-Phe-Leu-OMe, the latter could be isolated with 95% yield (Table 6.2, entry 1). No

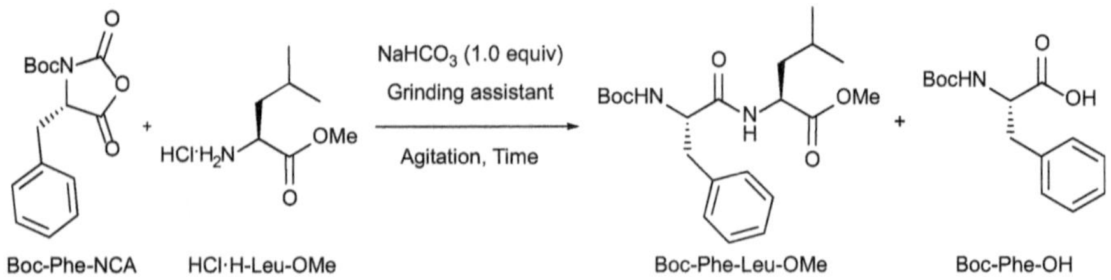

Scheme 6.19 Liquid-assisted mechanosynthesis of Boc-Phe-Leu-OMe.

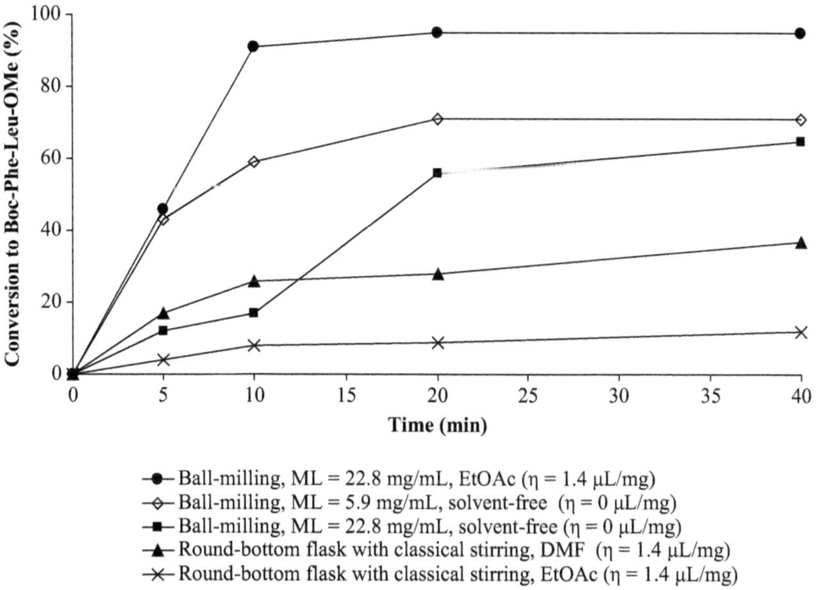

Figure 6.4 Influence of the composition of the reaction mixture and the type of agitation on the conversion into Boc-Phe-Leu-OMe.[45]

epimerization occurred during the coupling as no traces of Boc-D-Phe-Leu-OMe could be observed by HPLC analysis. Other NCAs such as Boc-Leu-NCA, Boc-Val-NCA or Boc-Ile-NCA reacted efficiently with differently substituted α-amino esters (Table 6.2, entries 2–5). Indeed, replacing either the hydrochloride amino ester salt with a *p*-tosylate salt or the methyl ester with a benzyl ester did not affect the efficiency of the reaction as *p*-TsOH · H-Leu-OBn and HCl · H-Pro-OBn were transformed into Boc-Leu-Leu-OBn and Boc-Leu-Pro-OBn with 93% and 90% yield, respectively (Table 6.2, entries 2 and 3). Similarly, sterically challenging Boc-Val-NCA and Boc-Ile-NCA reacted with HCl · H-Phe-OMe and HCl · H-Ile-OMe to produce Boc-Val-Phe-OMe and Boc-Ile-Ile-OMe with high yields (Table 6.2, entries 4 and 5). Functionalities other than *N*-carboxyanhydride can be used to activate α-amino acid substrates. Indeed, *N*-hydroxysuccinimide esters of Boc-protected glycine, tyrosine or phenyalanine (Boc-AA-OSu) led to the efficient production of the corresponding peptides (Table 6.2, entries 6–11). This α-amino acid activation presents the advantage of producing the low toxicity and water-soluble *N*-hydroxysuccinimide (HOSu) as a side-product during the course of the reaction. Afterwards, HOSu can easily be eliminated during extractions. To evaluate the efficacy of liquid-assisted grinding on the elimination of the mass transfer limitations, the milling load of the reaction producing Boc-Phe-Phe-OMe from Boc-Phe-OSu and *p*-TsOH · H-Leu-OBn was raised substantially to 188.8 mg mL^{-1}. While being slightly slower than with a milling load of 28.4 mg mL^{-1}, performing the reaction with a milling load of 188.8 mg mL^{-1} allowed close to 800 mg

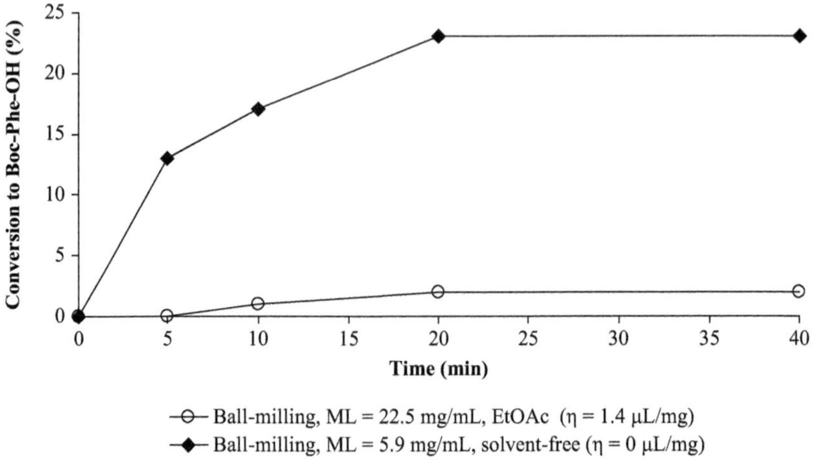

Figure 6.5 Influence of the composition of the reaction mixture on the hydrolysis of Boc-Phe-NCA.[45]

of dipeptide to be obtained in a small 10 mL jar with 80% yield (Table 6.2, entry 11). Producing this amount of peptide enabled the synthesis of much longer peptides by using this approach. On this perspective, the Boc protecting groups of the mechanosynthesized dipeptides were cleaved using gaseous HCl treatment. This procedure has the advantages of being solvent-free and of furnishing the deprotected peptides as hydrochloride salts in quantitative yields.

The corresponding dipeptides were then engaged in another coupling step with activated α-amino acid derivatives. A wide variety of tripeptides were isolated with good to excellent yields starting from either Boc-AA-NCA or Boc-AA-OSu (Table 6.2, entries 12–19). Thus, hydrophobic Boc-Leu-Leu-Leu-OBn and Boc-Phe-Phe-Phe-OMe were produced in 79% and 86% yield respectively (Table 6.2, entries 12 and 13). This approach perfectly tolerates the presence of an heteroatom in the side chain of the reactants as Boc-Lys(Boc)-Gly-Pro-OBn and Boc-Trp-Val-Phe-OMe were isolated in 79% and 86% (Table 6.2, entries 14 and 15). Interestingly, sterically challenging Boc-Ile-NCA furnished both Boc-Ile-Leu-Pro-OBn and Boc-Ile-Ile-Ile-OMe with 86% yield when treated with HCl · H-Leu-Pro-OBn and HCl · H-Ile-Ile-OMe respectively (Table 6.2, entries 16 and 17). Similarly, less encumbered Boc-Gly-OSu also produced Boc-Gly-Gly-Phe-OMe and Boc-Gly-Phe-Leu-OBn with good yields (Table 6.2, entries 18 and 19). The latter peptide could be produced at the gram scale as 937 mg of this tripeptide (94% yield) could be isolated by performing the corresponding reaction with a high milling load of 152.7 mg mL^{-1} (Table 6.2, entry 19). After having been deprotected by gaseous HCl, two tripeptides could be engaged in another mechano-mediated peptide bond coupling. The corresponding tetrapeptide Boc-Leu-Leu-Leu-Leu-OBn and Boc-Gly-Gly-Phe-Leu-OBn were both isolated with an excellent yield of 96% (Table 6.2, entries 20 and 21). Once again, increasing

Table 6.2 Mechanosynthesis of di- to pentapeptides.

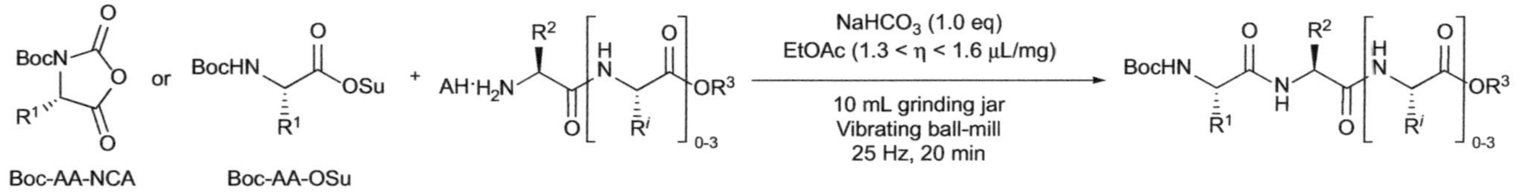

Entry	Boc-AA-NCA or Boc-AA-OSu	AH · H-(AA)$_x$-OR^3	Mechanosynthesized peptides	Yield (%)[a]
1	Boc-Phe-NCA	HCl · H-Leu-OMe	Boc-Phe-Leu-OMe	95[b]
2	Boc-Leu-NCA	*p*-TsOH · H-Leu-OBn	Boc-Leu-Leu-OBn	93
3	Boc-Leu-NCA	HCl · H-Pro-OBn	Boc-Leu-Pro-OBn	90
4	Boc-Val-NCA	HCl · H-Phe-OMe	Boc-Val-Phe-OMe	84
5	Boc-Ile-NCA	HCl · H-Ile-OMe	Boc-Ile-Ile-OMe	95
6	Boc-Gly-OSu	HCl · H-Phe-OMe	Boc-Gly-Phe-OMe	96
7	Boc-Gly-OSu	HCl · H-Pro-OBn	Boc-Gly-Pro-OBn	97
8	Boc-Tyr(Bn)-OSu	HCl · H-Leu-OMe	Boc-Tyr(Bn)-Leu-OMe	98[c]
9	Boc-Phe-OSu	HCl · H-Phe-OMe	Boc-Phe-Phe-OMe	92
10	Boc-Phe-OSu	HCl · H-Leu-OMe	Boc-Phe-Leu-OMe	83[b]
11	Boc-Phe-OSu	*p*-TsOH · H-Leu-OBn	Boc-Phe-Leu-OBn	90[d], 80[e]
12	Boc-Leu-NCA	HCl · H-Leu-Leu-OBn	Boc-Leu-Leu-Leu-OBn	79
13	Boc-Phe-OSu	HCl · H-Phe-Phe-OMe	Boc-Phe-Phe-Phe-OMe	86
14	Boc-Lys(Boc)-OSu	HCl · H-Gly-Pro-OBn	Boc-Lys(Boc)-Gly-Pro-OBn	79

Table 6.2 (*Continued*)

Entry	Boc-AA-NCA or Boc-AA-OSu	AH · H-(AA)$_x$-OR^3	Mechanosynthesized peptides	Yield (%)[a]
15	Boc-Trp-NCA	HCl · H-Val-Phe-OMe	Boc-Trp-Val-Phe-OMe	86
16	Boc-Ile-NCA	HCl · H-Leu-Pro-OBn	Boc-Ile-Leu-Pro-OBn	86
17	Boc-Ile-NCA	HCl · H-Ile-Ile-OMe	Boc-Ile-Ile-Ile-OMe	86
18	Boc-Gly-OSu	HCl · H-Gly-Phe-OMe	Boc-Gly-Gly-Phe-OMe	74
19	Boc-Gly-OSu	HCl · H-Phe-Leu-OBn	Boc-Gly-Phe-Leu-OBn	93[f] (94)[g]
20	Boc-Leu-NCA	HCl · H-Leu-Leu-Leu-OBn	Boc-Leu-Leu-Leu-Leu-OBn	96
21	Boc-Gly-OSu	HCl · H-Gly-Phe-Leu-OBn	Boc-Gly-Gly-Phe-Leu-OBn	96[h] (90)[i]
22	Boc-Tyr(Bn)-OSu	HCl · H-Gly-Gly-Phe-Leu-OBn	Boc-Tyr(Bn)-Gly-Gly-Phe-Leu-OBn	88[j]

[a]Isolated yield. $1.1 < \eta < 1.6$ μL mg^{-1} unless otherwise noted.
[b]>98% diastereomeric excess determined by chiral HPLC.
[c]Completed in 1 h.
[d]ML = 28.4 mg mL^{-1}. $\eta = 1.1$ μL mg^{-1}.
[e]ML (milling load) = 188.8 mg mL^{-1}. $\eta =$ of 1.1 μL mg^{-1}. Completed in 2 h.
[f]ML = 22.9 mg mL^{-1}.
[g]ML = 152.7 mg mL^{-1}. Completed in 40 min with *t*BuOAc as the grinding auxiliary.
[h]ML = 22.2 mg mL^{-1}, $\eta = 1.1$ μL mg^{-1}.
[i]ML = 60.6 mg mL^{-1}, $\eta = 1.1$ μL mg^{-1}. Completed in 40 min with *t*BuOAc as the grinding aid.
[j]Completed in 1 h.

milling load from 22.2 to 60.6 mg mL^{-1} did not dramatically hamper the efficacy of the reaction as the tetrapeptide Boc-Gly-Gly-Phe-Leu-OBn was produced in 96% and 90% yield, respectively. Finally, after solvent-free deprotection of the Boc group, the tetrapeptide HCl · H-Gly-Gly-Phe-Leu-OBn was reacted with Boc-Tyr(Bn)-OSu and furnished, after 1 h of grinding, the pentapeptide Boc-Tyr(Bn)-Gly-Gly-Phe-Leu-OBn in 88% yield (Table 6.2, entry 22). The synthesis of the last peptide involved four mechano-mediated peptide bond couplings realized in a row intersected by three solvent-free Boc deprotections. To the best of our knowledge, this is the longest linear organic synthesis to have been performed through ball-milling technology. This approach has allowed strong minimization of the use of organic solvents while avoiding the predominant toxic solvents such as DMF or DCM encountered in classical peptide synthesis. In addition, these reaction conditions require no more than stoichiometric quantities of non-hazardous reactants and transient Boc deprotections are realized without any solvent.

This approach was finally applied to the synthesis of the natural peptide Leu-enkephalin. The latter is an endogenous neurotransmitter presenting agonist activity at μ and δ opioid receptors, which was first reported and synthesized in 1975.[47] Thus, benzylic groups of Boc-Tyr(Bn)-Gly-Gly-Phe-Leu-OBn were cleaved by pallado-catalyzed hydrogenation to furnish Boc-Tyr-Gly-Gly-Phe-Leu-OH in 77% yield (Scheme 6.20). Then, the Boc protecting group was removed by gaseous HCl to produce Leu-enkephalin hydrochloride salt in quantitative yield. By using this strategy, Leu-enkephalin was efficiently produced in nine steps with a 46% overall yield.

Another interesting approach was developed in 2012 by Margetić[39] and coworkers to produce dipeptides by using a mechanochemical process. Instead of starting from pre-activated *N*-carboxyanhydrides or hydroxysuccinimide esters of α-amino acids substrates, Margetić and coworkers utilized Boc-protected α-amino acids as starting material. These α-amino acids were activated in the reaction media by addition of *N*-ethyl-*N*′-(3-dimethylaminopropyl)carbodiimide hydrochloride (EDC · HCl) in the presence of 2 equivalents of dimethylaminopyridine (DMAP) as the base, 20 equivalents of NaCl as a solid grinding assistant and small quantities of nitromethane as a liquid grinding auxiliary. Whereas originally optimized for coupling benzoic acid with *p*-anisidine (Section 6.2.5), these conditions were applied to the synthesis of a limited range of dipeptides with yields ranging from 70% to 81% (Scheme 6.21). It is worth mentioning that the dipeptides were recovered without using any organic solvent. Based on the low water-solubility of the products, a simple aqueous work-up followed by filtration allowed the isolation of the pure dipeptides. Thus, Boc-Gly-Gly-OBn was isolated with 70% yield starting from Boc-Gly-OH and *p*Ts · H-Gly-OBn. When treated with *p*Ts · H-Gly-OBn, α-substituted Boc-L-Ala-OBn and Boc-D-Ala-OBn furnished Boc-L-Ala-Gly-OBn and Boc-D-Ala-Gly-OBn with slightly improved yields of 78% and 79%, respectively. Similar yields were obtained when Boc-L-Ala-L-Ala-OBn and its diastereoisomer Boc-D-Ala-L-Ala-OBn were synthesized (80% and 81% respectively).

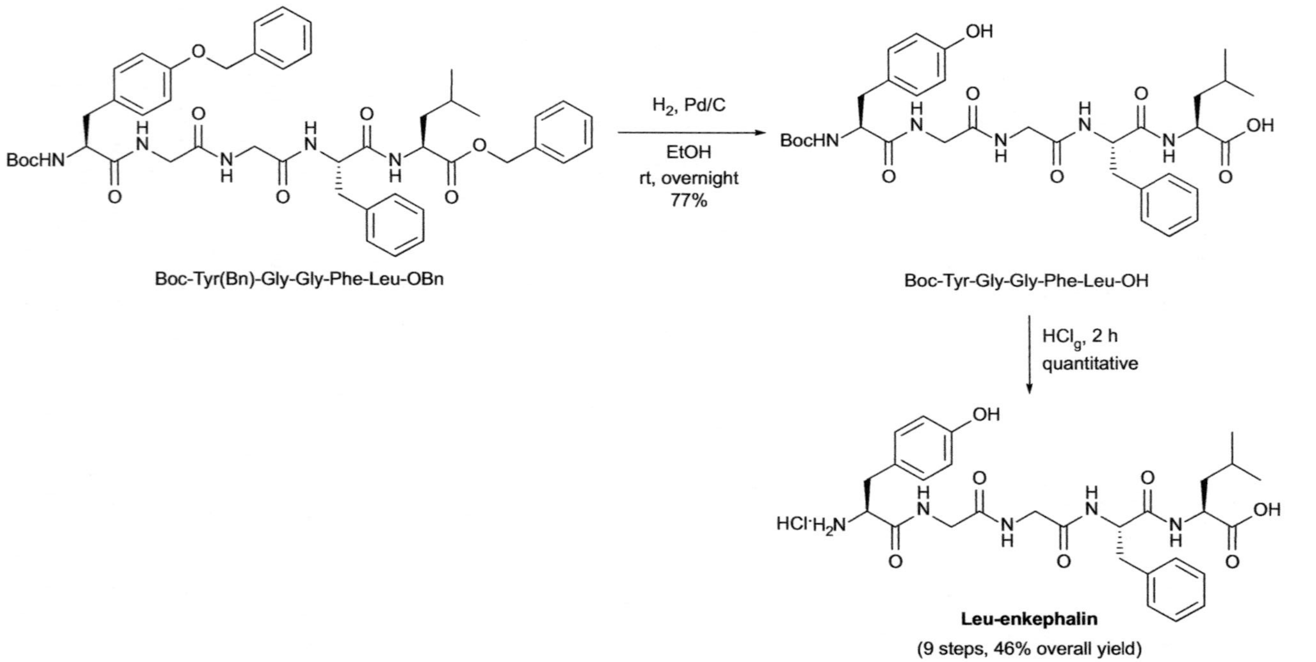

Scheme 6.20 Synthesis of Leu-enkephalin.

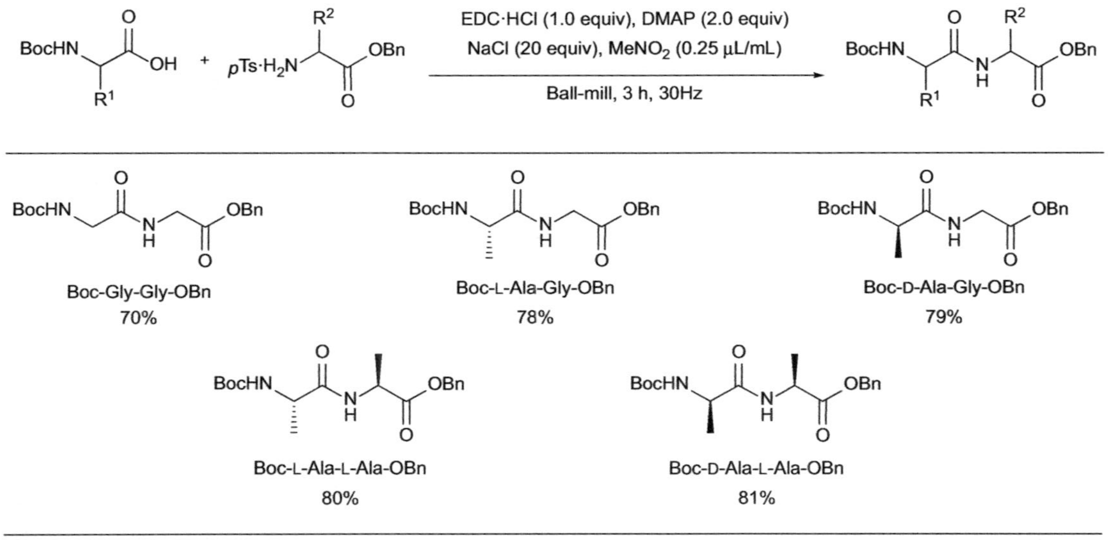

Scheme 6.21 Mechanosynthesis of dipeptides from non-preactivated α-amino acids.

This approach is advantageous in that it enables the production of dipeptide avoiding the use of pre-activated α-amino acids as substrates. Nevertheless, this particularity may be useful only in the very limited number of cases where the pre-activated α-amino ester is hardly synthesizable or not commercially available. Besides, the approach developed by Margetić and coworkers still requires the use of 2 equivalents of the highly toxic organic base DMAP. In addition, one can regret the use of an explosive liquid as grinding assistant such as nitromethane, which unfortunately hampers the environmental impact of the process.

In contrast, a very interesting point in this approach is that the peptide recovery is realized in the absence of any organic solvent. Organic solvent-free recovery and purification have been a scarcely studied theme in fine organic chemistry, but promising methods have been published recently.[38,40,48–51] Mechanochemical peptide syntheses would gain even more interest in the organic chemist's community if solutions addressing this challenge would emerge.

6.4 Conclusion

Performing a reaction in a ball-mill is an efficient way of reaching more sustainable conditions in organic synthesis, especially if combined with innocuous solvent-assisted grinding and organic solvent-free recovery or purification. These approaches were highlighted in this chapter, demonstrating that amino acid derivatization or protection and peptide synthesis could greatly benefit from such processes.

References

1. P. Vlieghe, V. Lisowski, J. Martinez and M. Khrestchatisky, *Drug Discovery Today*, 2010, **15**, 40–56.
2. J. O. Straub, in *Green and Sustainable Pharmacy*, Springer, Berlin, 2010, pp. 127–133.
3. D. J. Craik, D. P. Fairlie, S. Liras and D. Price, *Chem. Biol. Drug Des.*, 2013, **81**, 136–147.
4. A. M. Thayer, *Chem. Eng. News*, 2011, **89**, 13–20.
5. R. B. Merrifield, *J. Am. Chem. Soc.*, 1963, **85**, 2149–2154.
6. M. Amblard, J.-A. Fehrentz, J. Martinez and G. Subra, *Mol. Biotechnol.*, 2006, **33**, 239–254.
7. V. Mäde, S. Els-Heindl and A. G. Beck-Sickinger, *Beilstein J. Org. Chem.*, 2014, **10**, 1197–1212.
8. A. El-Faham and F. Albericio, *Chem. Rev.*, 2011, **111**, 6557–6602.
9. S. Bouifraden, C. Drouot, M. El Hadrami, F. Guenoun, L. Lecointe, N. Mai, M. Paris, C. Pothion, M. Sadoune, B. Sauvagnat, M. Amblard, J. L. Aubagnac, M. Calmes, P. Chevallet, J. Daunis, C. Enjalbal, J. A. Fehrentz, F. Lamaty, J. P. Lavergne, R. Lazaro, V. Rolland,

M. L. Roumestant, P. Viallefont, Y. Vidal and J. Martinez, *Amino Acids*, 1999, **16**, 345–379.
10. R. Saladino, G. Botta and M. Crucianelli, *Mini-Rev. Med. Chem.*, 2012, **12**, 277–300.
11. M. I. Calaza and C. Cativiela, *Amino Acids, Pept. Proteins Org. Chem.*, 2011, **3**, 83–114.
12. G. Kaupp, J. Schmeyers and J. Boy, *Tetrahedron*, 2000, **56**, 6899–6911.
13. G. Kaupp, *CrystEngComm*, 2009, **11**, 388–403.
14. G. Kaupp, *Top. Curr. Chem.*, 2005, **254**, 95–183.
15. G. Kaupp, J. Schmeyers and J. Boy, *J. Prakt. Chem.*, 2000, **342**, 269–280.
16. G. Kaupp, *CrystEngComm*, 2006, **8**, 794–804.
17. G. Kaupp, M. Reza Naimi-Jamal and J. Schmeyers, *Chem. Eur. J.*, 2002, **8**, 594–600.
18. G. Kaupp, *CrystEngComm*, 2003, **5**, 117–133.
19. T. Chatterjee and B. C. Ranu, *RSC Adv.*, 2013, **3**, 10680–10686.
20. V. A. Soloshonok and K. Izawa, *Asymmetric Synthesis and Application of alpha-Amino-Acids*, American Chemical Society, Washington DC, ACS Symposium Series, vol. 1009, 2009.
21. G.-W. Wang and X.-L. Wu, *Adv. Synth. Catal.*, 2007, **349**, 1977–1982.
22. G.-W. Wang, X.-L. Wu and J.-J. Xia, *Org. Biomol. Chem.*, 2008, **6**, 548–553.
23. A. J. Catino, J. M. Nichols, R. E. Forslund and M. P. Doyle, *Org. Lett.*, 2005, 7, 2787–2790.
24. P. Nun, V. Pérez, M. Calmès, J. Martinez and F. Lamaty, *Chem. Eur. J.*, 2012, **18**, 3773–3779.
25. A. Baron, J. Martinez and F. Lamaty, *Tetrahedron Lett.*, 2010, **51**, 6246–6249.
26. E. Tullberg, D. Peters and T. Frejd, *J. Organomet. Chem.*, 2004, **689**, 3778–3781.
27. E. Tullberg, F. Schacher, D. Peters and T. Frejd, *Synthesis*, 2006, 1183–1189.
28. A. V. Jerezano, E. M. Labarrios, F. E. Jiménez, M. d. C. Cruz, D. C. Pazos, R. U. Gutiérrez, F. Delgado and J. Tamariz, *ARKIVOC*, 2014, **3**, 18–53.
29. S. F. Nielsen, D. Peters and O. Axelsson, *Synth. Commun.*, 2000, **30**, 3501–3509.
30. M. O. Rasmussen, O. Axelsson and D. Tanner, *Synth. Commun.*, 1997, **27**, 4027–4030.
31. L. Konnert, A. Gauliard, F. Lamaty, J. Martinez and E. Colacino, *ACS Sustain. Chem. Eng.*, 2013, **1**, 1186–1191.
32. E. Hlebowicz, A. J. Andersen, L. Andersson and B. A. Moss, *J. Pept. Res.*, 2005, **65**, 90–97.
33. A. Isidro-Llobet, X. Just-Baringo, A. Ewenson, M. Alvarez and F. Albericio, *Biopolymers*, 2007, **88**, 733–737.
34. R. A. Sheldon, *Pure Appl. Chem.*, 2000, **72**, 1233–1246.
35. L. Konnert, F. Lamaty, J. Martinez and E. Colacino, *J. Org. Chem.*, 2014, **79**, 4008–4017.

36. I. Lee, D. Pyeon, M. Shin, J. Hwang, C. Park, S. Kim and H. Chae, *Pat. Appl.* WO2013/019091.
37. E. Colacino, P. Nun, F. M. Colacino, J. Martinez and F. Lamaty, *Tetrahedron*, 2008, **64**, 5569–5576.
38. T.-X. Métro, J. Bonnamour, T. Reidon, J. Sarpoulet, J. Martinez and F. Lamaty, *Chem. Commun.*, 2012, 11781–11783.
39. V. Štrukil, B. Bartolec, T. Portada, I. Đilović, I. Halasz and D. Margetić, *Chem. Commun.*, 2012, **48**, 12100–12102.
40. V. Declerck, P. Nun, J. Martinez and F. Lamaty, *Angew. Chem., Int. Ed.*, 2009, **48**, 9318–9321.
41. G. Kaupp, *CrystEngComm*, 2003, **5**, 117–133.
42. P. T. Anastas and J. C. Warner, *Green Chemistry: Theory and Practice*, Oxford University Press, New York, 1998.
43. P. Nun, Micro-ondes et activation mécanique: application en synthèse organique et peptidique, PhD Thesis, Université Montpellier 2, France, 2009.
44. J. G. Hernández and E. Juaristi, *J. Org. Chem.*, 2010, **75**, 7107–7111.
45. J. Bonnamour, T.-X. Métro, J. Martinez and F. Lamaty, *Green Chem.*, 2013, **15**, 1116–1120.
46. T. Friščić, S. L. Childs, S. A. A. Rizvi and W. Jones, *CrystEngComm*, 2009, **11**, 418–426.
47. J. Hughes, T. W. Smith, H. W. Kosterlitz, L. A. Fothergill, B. A. Morgan and H. R. Morris, *Nature*, 1975, **258**, 577–579.
48. Z. Zhang, Y.-W. Dong, G.-W. Wang and K. Komatsu, *Synlett*, 2004, 61–64.
49. Z. Zhang, Y.-W. Dong, G.-W. Wang and K. Komatsu, *Chem. Lett.*, 2004, **33**, 168–169.
50. M. Ferguson, N. Giri, X. Huang, D. Apperley and S. L. James, *Green Chem.*, 2014, **16**, 1374–1382.
51. I. Huskić, I. Halasz, T. Friščić and H. Vančik, *Green Chem.*, 2012, **14**, 1597–1600.

CHAPTER 7

Ball-milling Mechanochemical Synthesis of Coordination Bonds: Discrete Units, Polymers and Porous Materials

TOMISLAV FRIŠČIĆ

Department of Chemistry and FRQNT Centre for Green Chemistry and Catalysis, McGill University, 801 Sherbrooke St. W., Montreal, H3A 0B8, Canada
Email: tomislav.friscic@mcgill.ca

7.1 Introduction

While mechanochemical reactions[1] and processing are well established in the areas of inorganic chemistry and materials science (*e.g.* syntheses of ferrites and perovskites,[2] nanoparticles[3] and alloys[4]), as well as in the syntheses of organic compounds,[5,6] the application of ball milling for the synthesis of metal–organic materials is relatively new. Pioneering reports on the synthesis of coordination polymers date from early 2000s,[1] and mechanochemical synthesis of discrete metal–organic complexes, although noted in 1970s,[7] have remained largely unexplored until the twenty-first century. Despite a late beginning, the area of metal–organic mechanochemistry has developed with explosive velocity and now encompasses a vast diversity of materials, from molecular cages[8] and squares, microporous metal–organic frameworks (MOFs),[9] luminescent coordination polymers[10] and

RSC Green Chemistry No. 31
Ball Milling Towards Green Synthesis: Applications, Projects, Challenges
Edited by Brindaban Ranu and Achim Stolle

Published by the Royal Society of Chemistry, www.rsc.org

organic light-emitting diode (OLED) materials[11] to organometallics,[12] pharmaceutical derivatives[13] and metallodrugs.[14] Notably, the first techniques[15] for *in situ* monitoring of the mechanisms of mechanochemical reactions using X-ray powder diffraction and Raman spectroscopy[16] have focused largely on the assembly of coordination polymers and MOFs by milling.[17] This chapter will illustrate, through selected examples, the current state of affairs in using and understanding mechanochemical reactions for the syntheses and transformations of metal–organic compounds, with particular focus on extended metal–organic structures, such as coordination polymers and open MOFs. For a deeper overview of this field, the reader is advised to also consult some of the existing expert reviews on the topic of mechanochemistry and mechanochemical synthesis of coordination compounds.[1,18,19]

7.2 Benefits of Mechanochemical Synthesis in making Coordination Bonds

The rapid growth of mechanochemical metal–organic synthesis can be rationalized by several benefits that emerge from the applications of mechanochemistry to the synthesis of coordination bonds. While the early explorations were largely inspired by the ability to achieve transformations of metal–ligand bonds in the absence of solvents[7] it quickly became clear that mechanochemical reactions can also enable significant enhancements in reaction rates. For example, Otera and co-workers[8] noted the significant acceleration of the formation of platinum- and palladium-based molecular squares and cages by grinding the corresponding *cis*-protected metal(II) nitrate with bridging ligands, such as 4,4′-dipyridyl (bipy) and tris(4-pyridyl)triazine. Grinding of these building blocks yielded the expected supramolecular structures within minutes, as opposed to weeks required for conventional solution-based processes (Figure 7.1a).

Another emergent advantage of the mechanochemical environment over the conventional solution synthesis is the enhanced control over the stoichiometric composition and phase purity of the product.[20] One of the first systematic explorations of the ability to precisely control the stoichiometric composition of a mechanochemically obtained metal–organic product addressed the self-assembly of zinc chloride with a small ditopic ligand cyanoguanidine (cnge).[20] The cnge molecule provides two different potential metal-binding sites in the form of nitrile and imine nitrogen atoms (Figure 7.1b), enabling the formation of at least two types of complexes: those in which cnge acts as a monodentate ligand by utilizing only one type of metal bonding site, and those in which cnge acts as a bridging ligand by engaging both types of binding sites. These two binding modes are expected to yield materials with different metal : ligand stoichiometric ratios. Indeed, complexation of $ZnCl_2$ and cnge is known to provide the discrete (0-dimensional, 0D) tetrahedral complex $Zn(cnge)_2Cl_2$ in which the ligands are

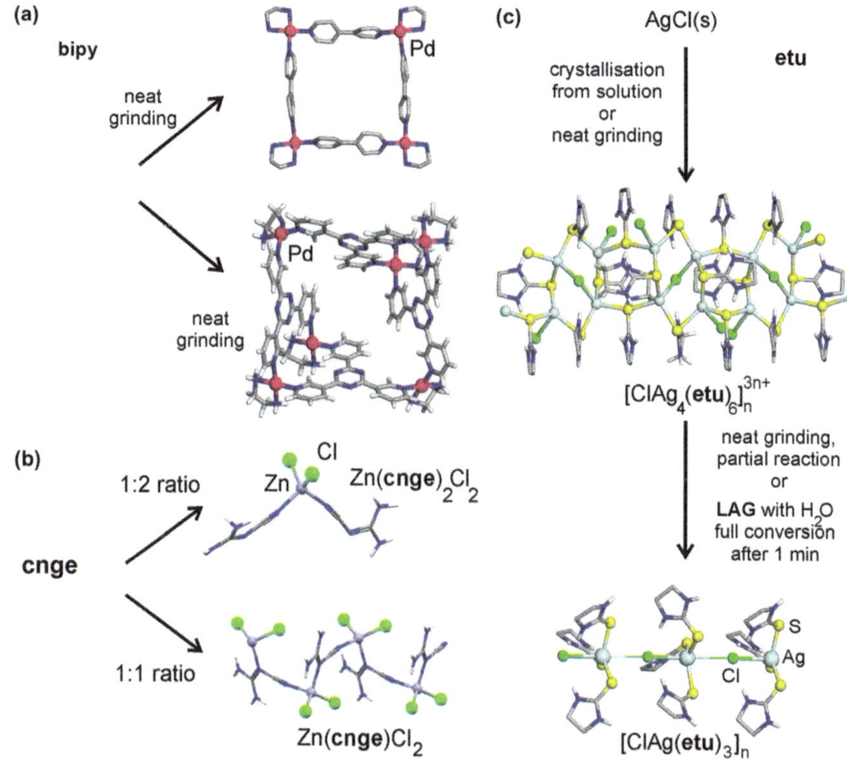

Figure 7.1 (a) Mechanochemical synthesis of metal–organic molecular squares and cages.[8] (b) Milling $ZnCl_2$ and cnge in a 1 : 1 ratio produced the coordination polymer $Zn(cnge)Cl_2$, while milling the same reactants in a 1 : 2 stoichiometric ratio selectively yields the discrete complex $Zn(cnge)_2Cl_2$.[20] (c) Acceleration of the mechanochemical synthesis of the coordination polymer $AgCl(etu)_3$ by switching from neat grinding to LAG with a small amount of water.[21]

attached to the metal ion *via* nitrile groups only, as well as the 1D coordination polymer $Zn(cnge)Cl_2$ in which pairs of zinc ions are bridged by cnge. In solution, $ZnCl_2$ and cnge typically yield a mixture of these two products. In contrast, mechanochemistry provides almost perfect selectivity: milling of $ZnCl_2$ and cnge in a 1 : 1 stoichiometric ratio yields exclusively the polymer $Zn(cnge)Cl_2$, while milling a mixture of $ZnCl_2$ and cnge in a respective 1 : 2 ratio yields the monomeric complex as the sole product (Figure 7.1b). Such excellent control over the stoichiometric composition of the product in mechanochemical reactions was recognized and exploited by the Bowmaker group for the discovery and synthesis of previously unknown complexes of silver halides with ethylenethiourea (etu).[21] For example, whereas solution synthesis is known to produce only the previously known complex $AgCl(etu)_{1.5}$, mechanochemical synthesis readily provided either $AgCl(etu)_{1.5}$ or the previously unknown $AgCl(etu)_3$ depending on the employed ratio of AgCl and etu. Indeed, investigations of mechanochemical co-crystallisation,[22] as well as of inorganic[23] and organic syntheses,[24] indicate that excellent control over reaction stoichiometry is a general property of mechanochemical reactions (Figure 7.1c).

Mechanochemical reactions can also provide coordination complexes with stereoselectivity that is either different, or superior, to that observed in solution. A notable example is the neat grinding reaction of platinum(II) chloride with triphenylphosphine (PPh_3), which yields exclusively the *cis*-complex $Pt(PPh_3)_2Cl_2$.[25] This contrasts with thermal reactions, which also provide the *trans*-isomer. Besides demonstrating superior reaction selectivity by grinding, this outcome also infers a difference between mechanochemical and thermally-induced reactions. Another example of improved control over reaction stereochemistry by switching from solution to mechanochemical synthesis is the oxidative addition of bromine onto the organometallic complex tricarbonyl(cyclopentadienyl)rhenium(I), $CpRe(CO)_3$ (Cp = cyclopentadienide anion).[26] Variation of mechanochemical reaction parameters enabled the reaction to be either optimized for selective and quantitative synthesis of the *trans*-isomer of the complex $CpRe(CO)_2Br_2$ or to yield the *cis*-isomer as the major (70%) product. The analogous reaction in solution normally gives the two isomers in *ca.* 50 : 50 ratio.

However, probably the most important practical advantages of mechanochemistry in coordination chemistry, which can be identified as the major driving force for the rapid development of this area, address the synthesis of microporous MOFs. The rapid development of MOFs,[27] particularly for applications in hydrogen storage,[28] carbon dioxide sequestration,[29] light harvesting,[30] gas separation[31] and molecular sensing,[32] as well as their recent commercialization and large-scale manufacture (*e.g.* the Basolite® series of products)[33] has highlighted the need to develop new techniques for their discovery and synthesis in a scalable, atom- and energy-efficient manner. Considerations[33,34] of safety, cost and environmental impact dictate that thermally sensitive, toxic and often costly organic solvents and metal precursors should be avoided. It is particularly desirable to replace the hazardous or corrosive metal nitrates and chlorides with safer sulfates or oxides.[33,34] Indeed, using metal oxides as starting materials is expected to greatly improve synthetic procedures, as the only reaction byproduct would in many cases be water and, as oxides are primary products in the course of mineral and metal processing, such starting materials are inexpensive. Unfortunately, low solubilities of metal sulfates and oxides, especially in organic solvents, present a severe limitation to their use in solvent-based MOF syntheses. Such limitations do not exist in mechanochemical procedures in which the solvent is either not present or is added in sub-stoichiometric, catalytic amount.[35] Consequently, mechanochemistry permits the synthesis of MOFs from slightly soluble substances, such as metal oxides or carbonates, at room temperature. The recently introduced mechanochemical techniques of liquid-assisted grinding[35] and ion- and liquid-assisted grinding[36] are particularly attractive for such clean syntheses of coordination polymers and microporous MOFs, as they allow the initially unreactive or low-yielding synthetic procedures to be optimized for the rapid (minutes to hours) and quantitative synthesis.

7.3 Methods for the Mechanosynthesis of Coordination Bonds

Currently, there are four distinct experimental methodologies for mechanochemical synthesis and transformations of coordination bonds: neat grinding, liquid-assisted grinding (LAG)[35] or kneading[37] (also known under other names, such as solvent-drop grinding[38] or solvent-assisted grinding[39]), ion- and liquid-assisted grinding (ILAG),[36] and neat grinding followed by annealing.[40] While these procedures can be conducted using conventional equipment, *e.g.* mortar and pestle or automated mills, recent reports have investigated alternative experimental arrangements, notably vortex grinding[41] or the use of a lysis mill.[42]

7.3.1 Neat Grinding

Neat grinding involves simply grinding a physical mixture of reactants. Grinding can often be performed manually, using a mortar and a pestle, although such approaches may lead to results that are difficult to reproduce due to a lack of control over numerous experimental variables, such as the strength of the experimenter, intensity of grinding, atmosphere, humidity and temperature. The use of an electronically controlled laboratory mechanical ball mill enables the reactions to be conducted under well-controlled and reproducible conditions. The effect of the surrounding atmosphere on neat grinding reactions in open air is often recognized by the formation of unexpected hydrated products, due to absorption of moisture from the air. For example, if grinding of silver acetate and 1,4-diaminocyclohexane (dace)[46] is conducted in air, the product is a hydrated coordination polymer $Ag(CH_3COO)(dace)\cdot nH_2O$ even if both reactants are anhydrous.

Neat grinding has also been applied for the synthesis of 1D and 2D coordination polymers, as well as three-dimensional (3D) porous MOFs. An extensive array-based study[43] of neat grinding reactions between diverse organic ligands and metal precursors conducted by the James group indicates that neat grinding reactivity is favored with solid reactants that either have low melting points and/or are solvated. Although mechanochemical reactions by neat grinding are considered to occur in a dry environment, it is noteworthy that most successful neat grinding reactions often involve hydrated metal salt precursors. It is likely that, when using hydrated acetates or chlorides as reagents, water or acid that is liberated during mechanochemical reaction can act as a liquid phase that facilitates the mechanochemical reaction (as in LAG), and/or templates the formation of a porous structure.[44,45]

7.3.2 Grinding–Annealing

Heating of a monomeric metal complex initially obtained by neat grinding can lead to a subsequent coordination-driven polymerization to form a new product. For example, grinding of 3-cyanoacetylacetone (HCNacac) with acetates of divalent transition metals leads to the formation of discrete monomeric complexes with axially coordinated water ligands.[40] Annealing of these primary products results in the removal of coordinated water and assembly of extended frameworks (Figure 7.2a).

A different type of reactivity obtained by a neat grinding–annealing sequence is illustrated by heating pyridinium and imidazolium tetrachlorometallates(II) made by grinding of corresponding organic hydrochlorides and metal chlorides. Thermal annealing of the mechanochemically prepared tetrachlorometallates(II) leads to the reversible loss of hydrogen chloride and the formation of 1D or 2D coordination polymers consisting of metal chloride units connected to pyridine (or imidazole) ligands (Figure 7.2b).[47]

7.3.3 Liquid-assisted Grinding and Kneading

Liquid-assisted grinding (LAG)[35,36] and the related kneading[19,37] technique were first introduced to mechanosynthesis in the context of forming solid-state hydrogen-bonded molecular complexes (also known as co-crystals).[48] LAG utilizes a small quantity of a liquid phase to assist or enable the mechanochemical reaction. In co-crystal synthesis, LAG is known to reliably result in quantitative yields, significantly increased reaction rates, and a greater scope for reactants and products.[38] A systematic study revealed that the increase in the rate of mechanochemical reaction is almost independent of the thermodynamic solubility of reactants in the liquid additive used for LAG.[49] This apparent lack of relationship between reactant solubility and reaction acceleration is a principal reason why LAG might be preferred over alternative names, such as solvent-drop grinding or solvent-assisted grinding. Mechanisms through which a liquid can accelerate mechanochemical reactions and produce a highly crystalline product are not yet clear, but are most likely related to enhanced diffusion on the surface of the reactant solids, and the plasticizing effect of the liquid phase. Products obtained by LAG are often highly crystalline,[48] suggesting a means to avoid or minimize amorphization that usually results from neat grinding. Whereas LAG has been extensively applied for co-crystal synthesis, similar observations have been made in the context of coordination chemistry. An example is the mechanochemical complexation of etu with silver halides (Figure 7.1c).[21] Solution reaction of AgCl and etu yields only $AgCl(etu)_{1.5}$, composed of 1D cationic tapes $[ClAg_4(etu)_6]^{3+}$ surrounded by Cl^- anions. Neat grinding of this cationic polymer with a further 1.5 equivalents of etu results in slow partial formation of a complex $AgCl(etu)_3$. However, LAG with a small amount of water quantitatively yields $AgCl(etu)_3$ within 1 min (Figure 7.1c).

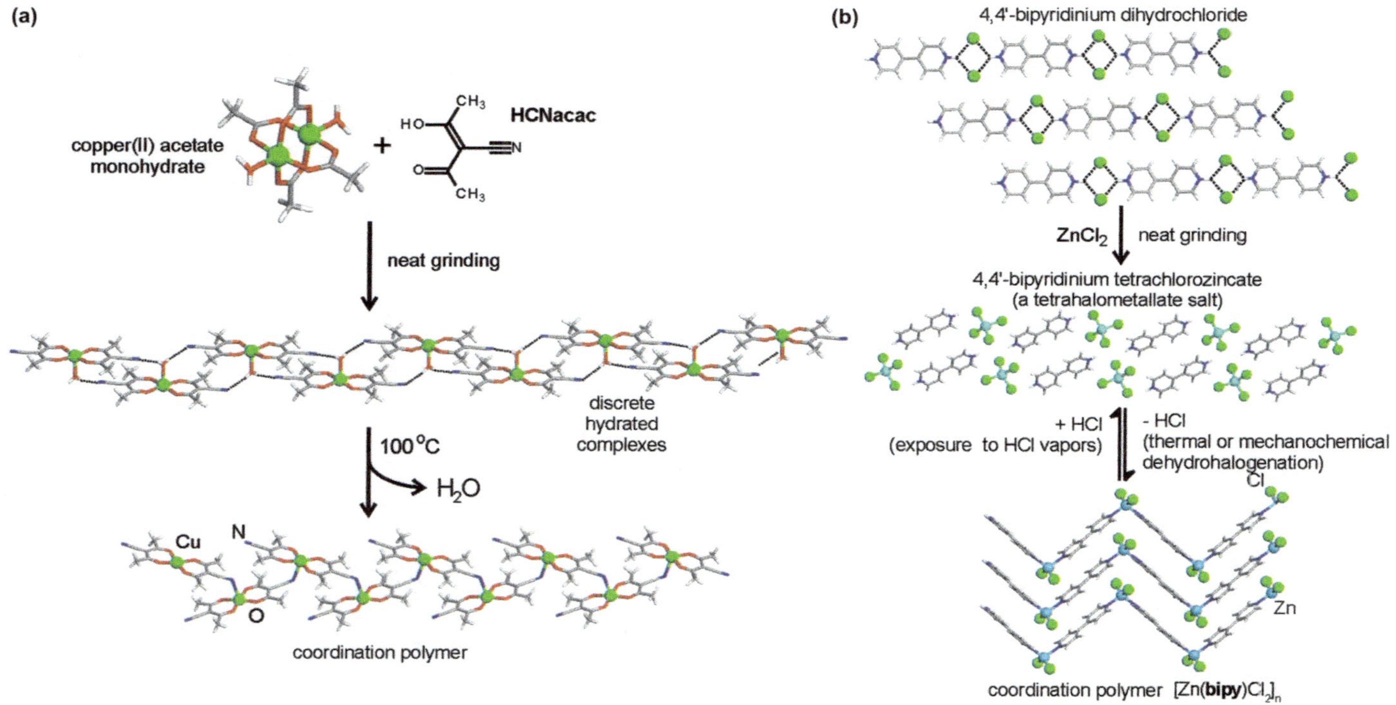

Figure 7.2 Synthesis of coordination polymers by a grinding–annealing technique: (a) dry grinding of copper(II) acetate monohydrate with HCNacac produces a hydrated monomeric complex $Cu(H_2O)(CNacac)_2$ that, upon heating to 100 °C in the solid state, loses water to form a coordination polymer of composition $Cu(CNacac)_2$; (b) grinding of 4,4′-bipyridinium dihydrochloride with zinc chloride yields the salt 4,4′-dipyridinium tetrachlorozincate. Heating of the solid salt leads to a loss of HCl and the formation of the coordination polymer $Zn(bipy)Cl_2$.

Similarly, while AgI and etu do not react by neat grinding, addition of a small amount of water quantitatively yields the complex $AgI(etu)_2$.

Whereas LAG is often described as grinding with "drops" of liquid, a recent attempt to quantify the effect of the added liquid on mechanosynthesis has led to the introduction of the parameter η, the ratio of the liquid additive (in μL) to the weight of solid reactants (in mg).[9,49] Neat grinding corresponds to $\eta = 0$ μL mg^{-1}, while LAG reactivity takes place in the η range between 0 and 2 μL mg^{-1}. Higher η values correspond to slurry reactions (*ca.* 2–12 μL mg^{-1}) and to reactions in solution ($\eta > 12$ μL mg^{-1}). LAG is particularly interesting for the synthesis of porous materials and inclusion compounds, as the liquid can become incorporated into the final product as a guest and even template the product structure.[50] One example is the formation of a metal–organic host–guest complex by kneading of copper(II) chloride and dace with dimethyl sulfoxide (DMSO) or water. Kneading leads to the *in situ* formation of the 1D coordination polymer $Cu(dace)Cl_2$, which forms a host–guest inclusion compound with the kneading liquids, for example $Cu(dace)Cl_2 \cdot n$DMSO (Figure 7.3a).[51] The inclusion compounds consist of layers of juxtaposed chains of $Cu(dace)Cl_2$, separated by layers of included guest (Figure 7.3b). Thermal desolvation of the inclusion compound provides the non-solvated polymer $Cu(dace)Cl_2$, which can reversibly include various organic molecules upon kneading and suspension overnight. That the same inclusion compounds can be prepared directly by kneading of $CuCl_2$ and dace in only a few cases suggests that pre-assembled $Cu(dace)Cl_2$ layers are a prerequisite for inclusion, making this system a simple analogue of layered clay minerals with inclusion capabilities.[19,51]

The ability of LAG to enhance mechanochemical reactions and lead to molecular inclusion has been exploited to screen for coordination polymers and porous MOFs using reactants that do not readily react under neat grinding. For example, LAG of ZnO and fumaric acid (H_2fum) with bridging ligands bipy or *trans*-1,2-bis(4-pyridyl)ethylene leads to a one-pot neutralization and self-assembly process to yield the anticipated pillared MOFs (Figure 7.3c).[36,52] The reaction yield is dependent on the amount of the added liquid that acts as a space-filling agent: quantitative yield is only achieved in the presence of sufficient DMF to fill the pores of the MOF. Mechanosynthesis of the porous MOFs can also be conducted with other space-filling liquids, including environmentally more acceptable methanol, ethanol or 2-propanol.[52] Importantly, such LAG reactions provided rapid access to gram amounts of pillared MOFs, each within 30 min and at room temperature.[36,52] Thermogravimetric analysis of the mechanochemically prepared pillared MOFs revealed the loss of included guest upon heating and subsequent X-ray powder diffraction analysis revealed the retention of the initial pillared structure upon guest loss, indicating that the mechanochemically synthesized material exhibits structural robustness comparable to that of solvothermally-synthesized samples.

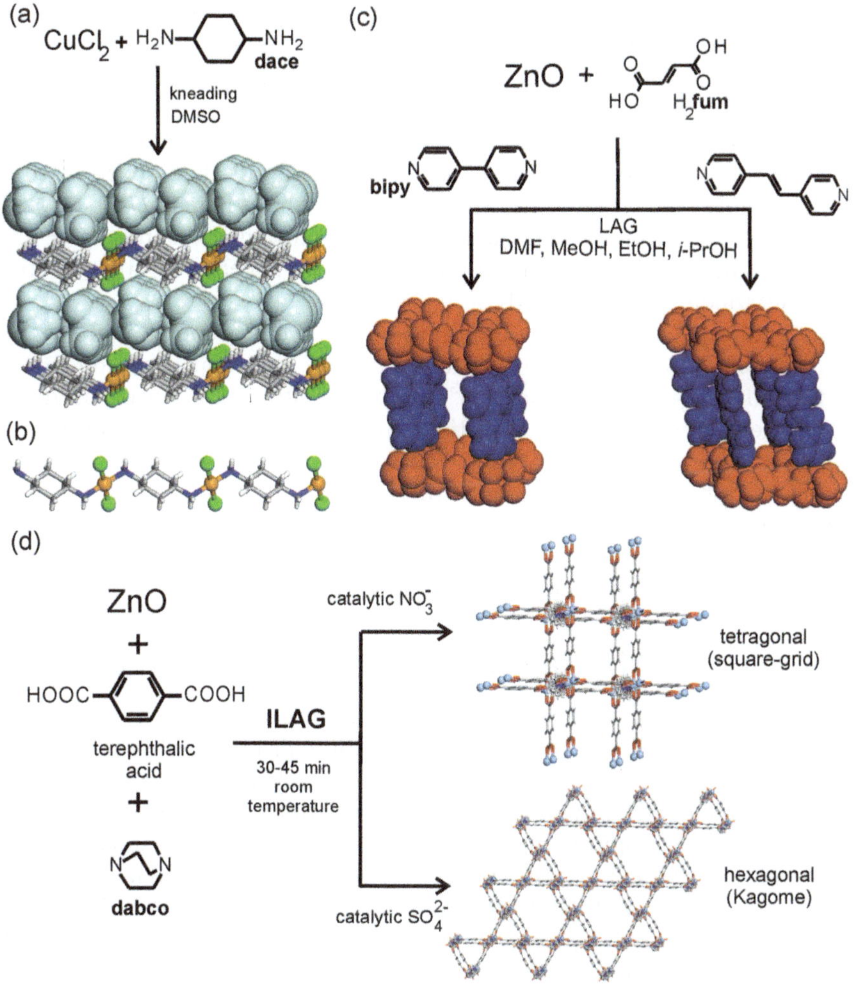

Figure 7.3 (a) Mechanosynthesis of a metal–organic clay mimic material by kneading with DMSO[51] and (b) a single chain of the coordination polymer host Cu(dace)Cl_2 as found in the structure of the resulting DMSO inclusion compound. Solvent molecules included between the layers of Cu(dace)Cl_2 are shown using a space-filling model. Mechanosynthesis of pillared porous MOFs from a metal oxide using: (c) LAG and (d) ILAG with catalytic addition of nitrate or sulfate salts.[36,52,53]

7.3.4 Ion- and Liquid-assisted Grinding (ILAG)

The ILAG methodology[36,53] represents an improvement of the LAG technique that is particularly suited for the synthesis of metal–organic materials from metal oxides. ILAG utilizes catalytic amounts of simple ionic salts to induce, enhance and direct mechanochemical reactivity. For example, attempts to synthesize the large-pore pillared MOFs based on zinc terephthalate and diazabicyclo[2.2.2]octane (dabco), by milling zinc oxide, terephthalic acid, dabco and a space-filling agent, were unsuccessful without the addition of small amounts of ionic salts (few molar percent with respect to the metal). The addition of small amounts of nitrate salts (*e.g.* KNO_3 or NH_4NO_3) to the reaction mixture led to the quantitative formation of the porous framework based on square-grid zinc terephthalate sheets within 45 min (Figure 7.3d).[53] Ionic additives containing the sulfate ion (*e.g.* Na_2SO_4 or $(NH_4)_2SO_4$) selectively and quantitatively produced the supramolecular isomer of this MOF, based on zinc terephthalate sheets with a hexagonal (Kagome) topology (Figure 7.3d).

These unexpected structure-directing effects were accompanied by the inclusion of salts in the MOF porous structure, as revealed by solid-state NMR and infrared spectroscopy.[53] Subsequently, ILAG was used to enable the mechanochemical oxidation of cobalt and ruthenium complexes, and for the conversion of ZnO into ZIFs,[54] a family of porous MOFs with topologies analogous to those of zeolites (Figure 7.3b).[55]In the context of pharmaceutically-relevant materials, ILAG also enabled a rapid and environmentally-friendly synthesis of the metallodrug bismuth subsalicylate,[56] active component of the popular gastrointestinal drug Pepto-Bismol®.

7.4 Characterization of Mechanochemical Products

Low solubility, possibilities of solvate formation, polymorphism or dissociation/association upon dissolution make it difficult to reliably characterize products of metal–organic mechanosynthesis in any other way except in the solid form obtained immediately from the reaction. In that way mechanosynthesis of coordination compounds (and, indeed, of co-crystals held by non-covalent interactions) is different from organic mechanosynthesis where the product and reaction mixtures can often be analyzed by conventional solution NMR spectroscopy after extraction from the solid reaction mixture. Consequently, products of metal–organic mechanosynthesis are characterized primarily by X-ray powder diffraction (XRPD), thermogravimetric analysis (TGA), reflectance FTIR spectroscopy and solid-state NMR spectroscopy. Although the latter two methods can immediately provide extensive information on the chemical composition of the product, full structural characterization very often depends on the ability to obtain an identical product by single-crystal growth from solution. In such cases, structural characterization is readily achieved *via* single-crystal X-ray diffraction. In cases where reaction in solution provides a different product than mechanosynthesis, it is, in principle, possible to obtain single crystals from solution by seeding it with fine powder of the grinding product.[57] Such a seeding strategy provides single crystals of the grinding product through heterogeneous nucleation.

If crystallization from solution consistently fails to provide single crystals of the desired phase, a possible alternative is crystal structure solution from XRPD data.[58] Recent advances in laboratory technology and software have made structure determination from XRPD data, especially if aided by information gathered through spectroscopic, thermal or elemental analysis, a very viable and almost routine method for structural characterization of molecular materials.[59] An example of complete structure determination from XRPD data is the porous interpenetrated pillared MOF with composition $Zn_2(fum)_2(bipy)$, prepared by mechanochemical synthesis from zinc acetate, H_2fum and bipy.[60] Determination of the crystal structure for the related framework $Zn_2(fum)_2(bpe)$ from XRPD data was facilitated by structural similarity to its copper(II) analogue, by the Rietveld refinement of the structural coordinates of the copper(II) analogue to the diffraction pattern measured for the LAG product.

7.5 Synthetic Strategies

The experimental techniques for the mechanosynthesis of coordination complexes, polymers and MOFs listed in Section 7.3 are readily applicable to a diversity of chemical transformations that form new metal–ligand bonds. Analysis of existing literature suggests that such chemical transformations can be divided into six distinct categories:

(1) reactions of ligand addition in which the addition of a multi-topic ligand onto the metal center leads to the formation of a new discrete or extended (1-, 2- or 3D) structure;
(2) ligand exchange reactions, which involve the mechanochemical replacement of ligands on a metal center;
(3) acid–base, proton-transfer reactions in which the formation of a new metal–organic structure is associated with the protonation and expulsion of a ligand attached to a metal precursor, such as acetate, chloride, carbonate or oxide;
(4) dehydration reactions in which the loss of coordinated water on a metal centre results in a change in the structure of a metal–organic material, *e.g.* polymerization of a small molecule complex;
(5) oxidative addition is one of the fundamental reactions of organometallic chemistry and corresponds to a process in which the coordination number of a central metal atom is increased, along with its oxidation state;
(6) increasingly popular one-pot and multi-component reactions, which can involve some or all of the above listed processes, or couple them with mechanochemical covalent bond formation or redox reactivity.

7.5.1 Ligand Addition Reactions

Mechanochemical addition of neutral ligands onto a metal-containing building block has been extensively used for the construction of discrete complexes, including organometallics, as well as coordination polymers. Early examples of mechanochemical ligand exchange leading to a discrete coordination complex were noted by Thabet and co-workers in 1972, who highlighted the rapid (within minutes) complexation of anhydrous cobalt(II) chloride by manual grinding with phosphines, phosphine and arsine oxides, as well as organic ligands like dimethylglyoxime.[7] Possibly the first report[1] of a mechanochemical synthesis of a coordination polymer is the transformation of the pre-synthesized 1D polymer of zinc bromide and pyrazine, $ZnBr_2$(pyrazine), into a 2D square grid framework $ZnBr_2$(pyrazine)$_2$ by grinding with extra pyrazine.[61] A bridging ligand that is often used for mechanosyntheses of coordination polymers is bipy. For example, neat grinding of zinc chloride and bipy readily yields the 2D coordination polymer Zn(bipy)Cl_2, which, interestingly, adopts only one out of three known polymorphic forms of the coordination polymer.[47,62] In contrast, the analogous cobalt-based polymer Co(bipy)Cl_2 was not obtainable by neat grinding of anhydrous $CoCl_2$ and bipy. Instead, this polymer could be synthesized either by a ligand exchange reaction of $CoCl_2 \cdot 6H_2O$ and bipy, suggesting that the water produced by desolvation of the reagents plays a role in mechanochemical reactivity (Figure 7.4a), or by LAG reaction of $CoCl_2$ and bipy. Another example of a ligand addition leading to a coordination polymer by grinding was presented by Pichon and James,[43] who utilized neat grinding of copper(II) acetylacetonate or hexafluoroacetylacetonate with bipy to construct 1D polymers held together by long axial Cu–N bonds (Figure 7.4b).

Reactions[20] of metal chlorides and cnge have already been noted as illustrations of excellent control of reaction stoichiometry arising from mechanochemical synthesis. These reactions, however, can also serve to illustrate how the mechanosynthesis of coordination polymers can be

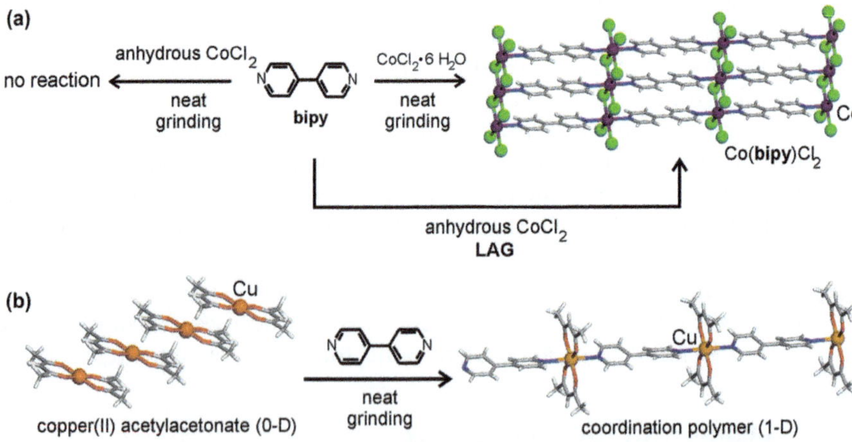

Figure 7.4 Examples of the mechanochemical synthesis of coordination polymers by ligand addition reactions: (a) illustration of the difference in mechanochemical reactivity of bipy towards anhydrous $CoCl_2$ and $CoCl_2 \cdot 6H_2O$[47] and (b) the synthesis of a 1D coordination polymer from copper(II) acetylacetonate and bipy.[43]

manipulated by varying the conditions of neat milling, specifically the impact force and temperature. While milling of $ZnCl_2$ with one or two equivalents of cnge selectively provides the 1D polymer Zn(cnge)Cl_2 or the discrete (0D) complex Zn(cnge)$_2$$Cl_2$, respectively, milling of $CdCl_2$ and cnge in a 1 : 1 stoichiometric ratio provides the 3D structure Cd(cnge)Cl_2. However, neat milling of $CdCl_2$ and cnge in the 1 : 2 stoichiometric ratio provides only a mixture of Cd(cnge)Cl_2 with excess ligand. The expected 1D polymer Cd(cnge)$_2$$Cl_2$ could be obtained mechanochemically, but only through harsher grinding conditions, *i.e.* by the use of heavier grinding media or by milling at a higher temperature. The difficulty in forming 1D Cd(cnge)$_2$$Cl_2$ was related to the higher dimensionality and, hence, kinetic stability of Cd(cnge)Cl_2.[20]

7.5.2 Ligand Exchange

An example of a mechanochemical ligand exchange reaction is ball milling of *cis*-$Pt(PPh_3)_2Cl_2$ with potassium carbonate, giving rise to the neutral carbonato complex *cis*-$Pt(PPh_3)_2CO_3$ (Figure 7.5a).[25] The Coville group has conducted extensive studies of solvent-free ligand exchange reactions on organometallic complexes[12] and, in some cases, established that the reactions are mediated by a low melting eutectic phase (Figure 7.5b).[63,64]

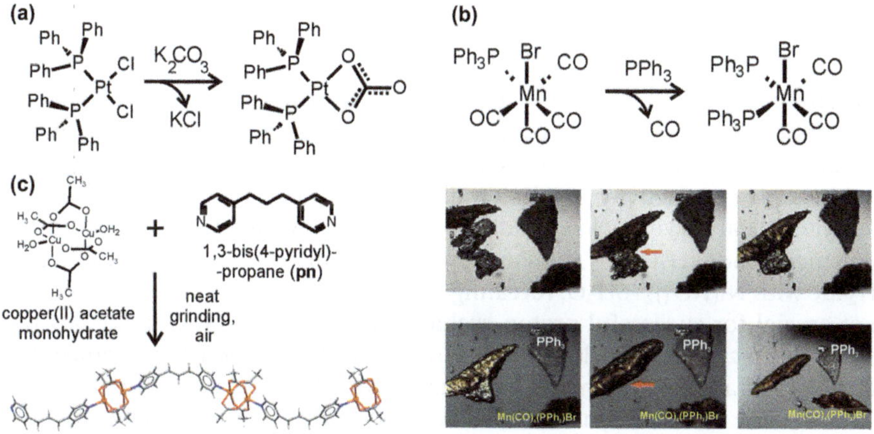

Figure 7.5 Mechanochemical reactions of ligand exchange: (a) synthesis of *cis*-$Pt(PPh_3)_2CO_3$ by grinding of mechanochemically obtained *cis*-$Pt(PPh_3)_2Cl_2$ with K_2CO_3.[25] (b) Thermal microscopy reveals that the ligand exchange reaction between touching crystals of $Mn(CO)_4(PPh_3)Br$ and PPh_3 (on the left of each image) proceeds through an intermediate eutectic melt phase although the temperature remains below the melting point of any of the reactants.[63] For comparison, the right side of each image shows a separate crystal of PPh_3 and the organometallic reactant. Adapted with permission from ref. 63. (c) Mechanochemical replacement of water ligands on the copper(II) acetate paddlewheel unit with the bridging ligand pn yields a 1D coordination polymer.[65]

The first reported mechanosynthesis of a coordination polymer by design appears to have been based on a ligand exchange reaction involving the binuclear copper(II) acetate paddlewheel cluster: manual grinding of copper(II) acetate monohydrate with 1,3-bis(4-pyridyl)propane (pn) produced the water inclusion compound of a 1D polymer by mechanochemical replacement of copper-coordinated water molecules with bridging pn ligands (Figure 7.5c).[65] The product was isostructural to the product of an analogous reaction in refluxing methanol, which yielded a methanol inclusion compound. Isostructurality indicated that the water produced by mechanochemical ligand exchange probably became incorporated into the mechanochemical product as a guest. This was also supported by the ^{13}C MAS-NMR spectrum of the mechanochemical product, which was identical to the one of the methanol solvate, with the exception of NMR signals of the guest methanol molecules.

During manual grinding of silver(I) acetate (AgOAc) and dabco, the neutral dabco ligand displaces the anionic acetate ligands coordinated to Ag^+ with simultaneous absorption of water from air to form the product $Ag(dabco)_2(OAc)\cdot 5H_2O$. The structure consists of cationic $[Ag(dabco)_2(H_2O)]_n^{n+}$ chains with additional water molecules and acetate ions acting as inclusion guests.[66] The formation of a hydrated product upon grinding in air illustrates the influence that the surrounding atmosphere can have on mechanosynthesis in the open mortar-and-pestle setup.

7.5.3 Acid–Base Reactions

Reactions mediated by proton transfer are a powerful tool in mechanosynthesis of coordination compounds, particularly reactions of metal acetates with organic acids, which yield acetic acid as a byproduct. An example of such reactivity involves milling of hydrated nickel(II) acetate and acetylenedicarboxylic acid (H_2adc) to form the hydrated 3D coordination polymer $Ni(adc)(H_2O)_2$.[43] The same product was also obtained by grinding of H_2adc with $Ni(NO_3)_2\cdot 6H_2O$, revealing that very basic anions like acetates are not crucial for successful proton exchange in the solid state. Analogous reaction with zinc acetate led to the formation of a previously unknown 3D polymer $Zn(adc)(H_2O)_2$ that is isostructural to the Ni(II) analogue, demonstrating how mechanosynthesis can be used to discover new metal–organic materials. A similar approach was used for the synthesis of coordination polymers of alkaline earth metals, specifically by grinding–annealing of magnesium or calcium acetate with H_2adc[67] to yield Ca(adc) and $Mg(adc)\cdot 2H_2O$ products, isostructural to Sr(adc) and $Mn(adc)\cdot 2H_2O$.

The acidic ligand in reactions with metal acetates does not need to be a carboxylic acid, as demonstrated by grinding of transition metal acetates with HCNacac (Figure 7.2a).[40] Acetates of Fe(II), Co(II) and Ni(II) directly yield non-porous 3D frameworks. However, hydrated acetates of Mn(II), Cu(II) or Zn yield previously unknown discrete complexes $Mn(CNacac)_2\cdot 2H_2O$, $Cu(CNacac)_2\cdot H_2O$ and $Zn(CNacac)_2\cdot H_2O$, respectively. As mentioned previously, upon heating to 100 °C, these complexes lose water and form coordination polymers.

A mechanochemical alternative to the synthesis of coordination polymers from 4,4′-bipyridinium salts of $FeCl_4^{2-}$, $CoCl_4^{2-}$ and $ZnCl_4^{2-}$ by thermal dehydrohalogenation (Figure 7.2b) is milling with an external base, such as KOH. While adding an external base is likely to be energetically advantageous to thermal dehydrohalogenation, this advantage is somewhat offset by the need to remove the byproduct salt (*e.g.* KCl).[68,69]

7.5.4 Synthesis of Coordination Polymers and MOFs from Metal Oxides and Carbonates

So far, all discussed examples of acid–base reactions used to construct coordination polymers involve soluble metal salts as precursors. However, as mechanochemical reactivity is independent of the solubility of reactants, the use of slightly soluble but cleaner and readily available carbonate or oxide reactants is also possible.[36,52,66] With oxide reactants, the highly basic nature of the oxide ion provides a strong driving force for the mechanochemical reaction. For metal carbonates, formation of CO_2 gas serves as an additional reaction driving force.

Perhaps the first example of mechanochemical reactivity of metal oxides against organic ligands was reported by Fernández-Bertrán, who obtained coordination polymers of Ag(I), Zn, Cd and Hg(II) by manually grinding the respective metal oxides with imidazole.[70] Analogous reactions with oxides of other transition metals, Pb(II) or Mg could not be achieved, suggesting that the nd^{10} configuration of the metal ion was important for the reactivity. Construction of coordination polymers from metal carbonates was utilized in the synthesis of the 2D framework $Co(bipy)Cl_2$ by LAG of cobalt(II) carbonate with bipyridinium chloride (Figure 7.6a).[68] The product was identical to the one constructed by kneading of $CoCl_2$ and bipy with water. Similarly, LAG of the basic zinc carbonate with bipyridinium chloride gave a mixture of two polymorphs of the $[Zn(bipy)Cl_2]_n$ polymer. This contrasts with the neat grinding reaction of $ZnCl_2$ and bipy, which yields only one polymorph, thus suggesting a surface templating effect in the reaction.

The reaction of ZnO with H_2fum demonstrates rapid screening for coordination polymers using LAG.[36,52] Different liquid phases were found to steer the reaction towards the formation of different products (Figure 7.6b). Quantitative yields, obtainable within 30–45 min milling, allowed the first structural characterization of anhydrous zinc fumarate and a previously unknown zinc fumarate dihydrate directly from powder X-ray diffraction data. Grinding of equimolar amounts of ZnO and H_2fum with three and four equivalents of water selectively produced the tetrahydrate and the pentahydrate forms of zinc fumarate, respectively, known to form concomitantly from solution. Formation of differently hydrated coordination polymers was subsequently correlated to the activity of water in the grinding liquid,

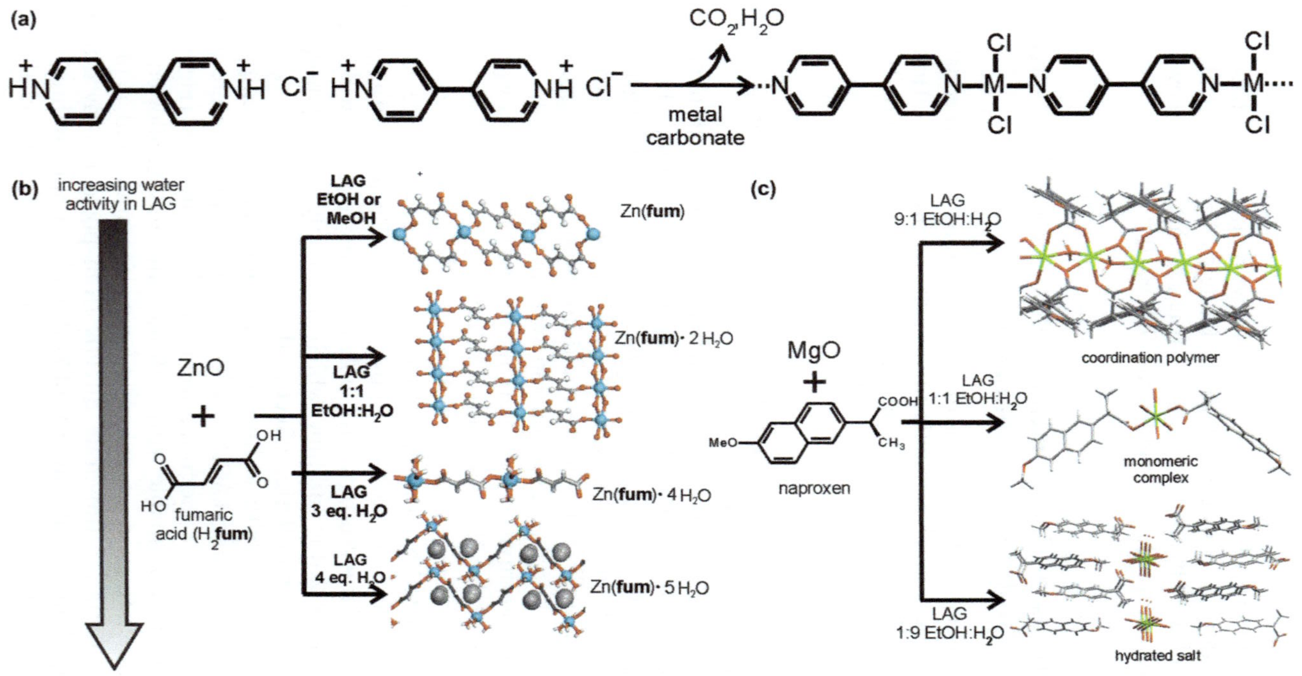

Figure 7.6 (a) Mechanosynthesis of coordination polymers by acid–base reactions involving a metal carbonate. Controlling the outcome of LAG mechanosynthesis by varying water activity in the grinding liquid: (b) synthesis of anhydrous and hydrated zinc fumarate coordination polymers from ZnO and (c) screening for new forms of magnesium naproxen directly from MgO.

suggesting a rational strategy to mechanochemically screen for hydrated coordination polymers.[71] Formation of solid hydrates is of particular importance in developing pharmaceutical formulations and new solid forms of pharmaceuticals. Consequently, control of water activity in LAG was exploited in the pharmaceutical context, as a strategy to explore the phase space of hydrated magnesium complexes of the drug naproxen: deliberate variation of water activity in the grinding liquid enabled the LAG discovery of new magnesium derivatives of naproxen formed directly from MgO and the drug (Figure 7.6c).[72]

7.5.5 Mechanochemical Dehydration

Liquid-assisted grinding offers an interesting alternative to grinding–annealing for achieving the formal dehydration of small molecule complexes and their association into extended structures. As demonstrated by Wang and co-workers,[73] mechanochemical milling of zinc oxide or copper(II) acetate monohydrate in the presence of water leads to the formation of corresponding hydrated metal isonicotinates, as small discrete complexes of composition $M(ina)_2(H_2O)_4$ (where M = Zn or Cu). A second round of milling, but now in the presence of methanol, a solvent well known to greatly reduce the activity of water, leads to dehydration of the monomeric tetrahydrate complexes and their reversible polymerization into either the non-porous triply interpenetrated MOF $Zn(ina)_2$ with a diamondoid topology or the open 2D framework $Cu(ina)_2$ (Figure 7.7a). Thus, it appears that LAG with a liquid exhibiting a particularly high affinity for water can be used as a means of non-thermal dehydration and polymerization of small molecule complexes into MOFs. Notably, such mechanochemical LAG dehydration is readily reversed by milling the two MOFs with water, yielding again the $M(ina)_2(H_2O)_4$ complexes.

7.5.6 One-pot Multi-component Strategies

Whereas applications of mechanochemical reactivity are, to a very large extent, focused on a particular chemical reaction, there is growing interest in the potential of mechanochemistry for conducting multi-component, multi-step reactions. Such potential, if realized, would further enhance the synthetic benefits of mechanochemical reactions by allowing rapid and solvent-free assembly of increasingly complex products directly from the simplest reactants. In the context of metal–organic materials, an early investigation of multi-component, one-pot reactivity involved the synthesis and screening for inclusion behavior of wheel-and-axle lattice hosts based on cobalt(II) and nickel(II) dibenzoylmethanates.[74] Milling of either nickel(II) or cobalt(II) dibenzoylmethanate dihydrate with nicotinamide and a suitable guest liquid quantitatively yielded metal–organic inclusion compounds in which a small molecule guest was trapped through weak van der Waals forces within an extended host structure based on coordination bonds and

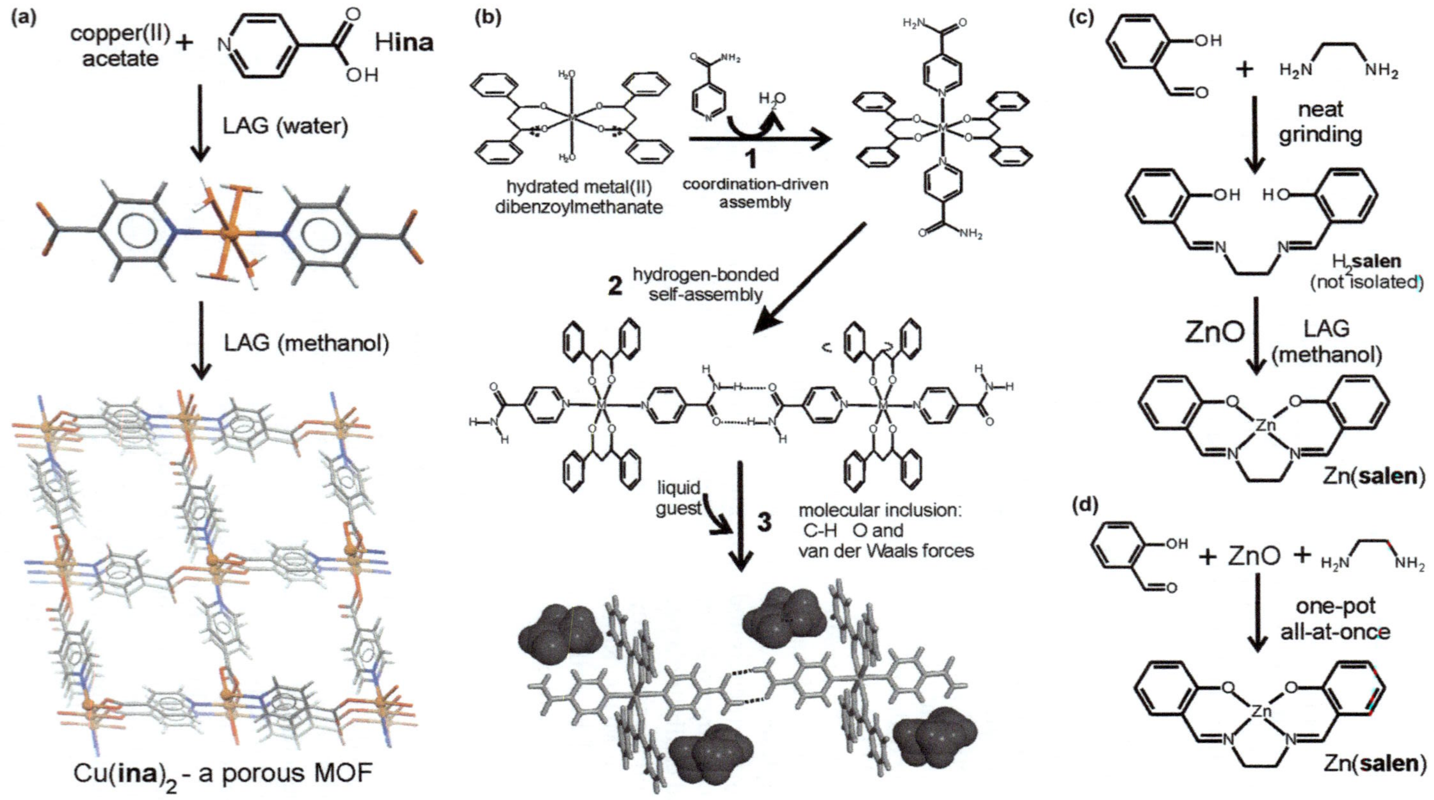

Figure 7.7 (a) Mechanochemical LAG reaction of copper(II) acetate and isonicotinic acid yields discrete hydrated complexes that can be dehydrated to form the porous MOF Cu(ina)$_2$ by LAG with methanol.[73] (b) Mechanochemical synthesis of a complex metal–organic material by combining three levels of self-assembly in a one-pot multi-component mechanochemical reaction: formation of coordination bonds, hydrogen bond-driven self-assembly and formation of a lattice inclusion compound *via* weak non-covalent interactions.[74] (c) One-pot synthesis of the Zn(salen) complex by sequential formation of covalent bonds followed by coordination bonds and (d) analogous process conducted quantitatively in a one-pot "all-at-once" manner.[76]

hydrogen bonding, thus demonstrating a one-pot mechanochemical process that combines three levels of molecular self-assembly (Figure 7.7b). A more complex system, involving five distinct reactants, was demonstrated by Chow *et al.* who milled magnesium oxide, salicylic acid, 15-crown-5, acetonitrile and water all at once to obtain an acetonitrile inclusion compound of a complex salt involving a simple mononuclear metal–organic cation and a trinuclear metal–organic anion.[75] This one-pot reaction demonstrated how mechanochemical acid–base neutralization can be coupled to the formation of coordination bonds, hydrogen bonds and molecular inclusion.

That the mechanochemical syntheses of coordination bonds and covalent bonds can be combined in a one-pot reaction was first shown by Ferguson and co-workers, who explored the synthesis of the discrete mononuclear complex of zinc with the popular bis(salicylidene)ethylenediamine (H_2salen) ligand (Figure 7.7c,d).[76] One-pot milling of zinc oxide, salicylaldehyde and ethylenediamine quantitatively yielded the expected complex in a process that combined acid–base neutralization with the formation of covalent and coordination bonds. In addition to this three-component one-pot mechanochemical reaction, the same complex could readily be obtained by a consecutive (also known as telescopic) two-step reaction strategy in which the ligand was first mechanochemically synthesized from salicylaldehyde and ethylenediamine, followed by the addition of zinc oxide and further milling. The consecutive reaction strategy was also applicable for the preparation of analogous complexes of nickel(II) and copper(II), by replacing zinc oxide in the second milling step with either nickel(II) acetate tetrahydrate or copper(II) acetate monohydrate, respectively.[76] A systematic exploration of multicomponent sequential and one-pot reactions based on mutually orthogonal chemical reactions was recently described by Hernández and co-workers, who targeted organometallic complexes of rhenium(I) based on the $Re(CO)_3$ fragment relevant for catalysis and design of model radiopharmaceuticals.[77] A judicious choice of oxidative addition and ligand exchange reactions allowed the mechanochemical one-pot synthesis of a complex organometallic fluoride complex from altogether five components, using the metal carbonyl as the simplest and most readily accessible metal precursor.[77]

7.6 Functional Metal–Organic Materials

This section highlights selected examples that illustrate the use of mechanosynthesis for making functional and, in some cases, commercially relevant metal–organic materials, specifically microporous MOFs, pharmaceuticals and luminescent materials.

7.6.1 Microporous Metal–Organic Frameworks

The possibility of forming a porous materials by mechanochemistry was first demonstrated[1] in the acid–base reaction of copper(II) acetate monohydrate and isonicotinic acid (Hina), which provided the open copper(II) isonicotinate MOF $Cu(ina)_2$.[45] The porous framework was obtained quantitatively by neat milling, with the byproducts water and acetic acid most likely included in its pores (Figure 7.8a). The same synthetic strategy was used to construct the industrially relevant framework HKUST-1 (pore diameter *ca.* 9 Å) by neat grinding of copper(II) acetate with trimesic acid.[76] As noted previously, application of the LAG and ILAG techniques enabled the facile and rapid, topologically specific transformation of zinc oxide into microporous pillared MOFs based on zinc fumarate or terephthalate sheets.[36,52,53]

Mechanochemistry was also applied for the synthesis of ZIFs directly from a metal oxide.[54] Whereas grinding ZnO with solid imidazole is of very limited scope for making ZIFs, advanced techniques such as LAG or ILAG enable rapid and quantitative synthesis of a variety of close-packed and open ZIFs, including the commercially highly relevant ZIF-8. The grinding liquid and salt catalysts in ILAG can both enhance the reactivity of ZnO and imidazole ligands, as well as exert control over the final product topology. This is illustrated in Figure 7.8b in the case of ZIFs based on 2-ethylimidazole, where different choices of the ionic additive and the grinding liquid enabled the selective formation of zeolite rho (RHO), analcime (ANA) and β-quartz (qtz) frameworks.[54] Mechanosynthesis of ZIFs proceeds through a stepwise mechanism in which the most porous structures are formed first and subsequently transform into close-packed structures *via* intermediate

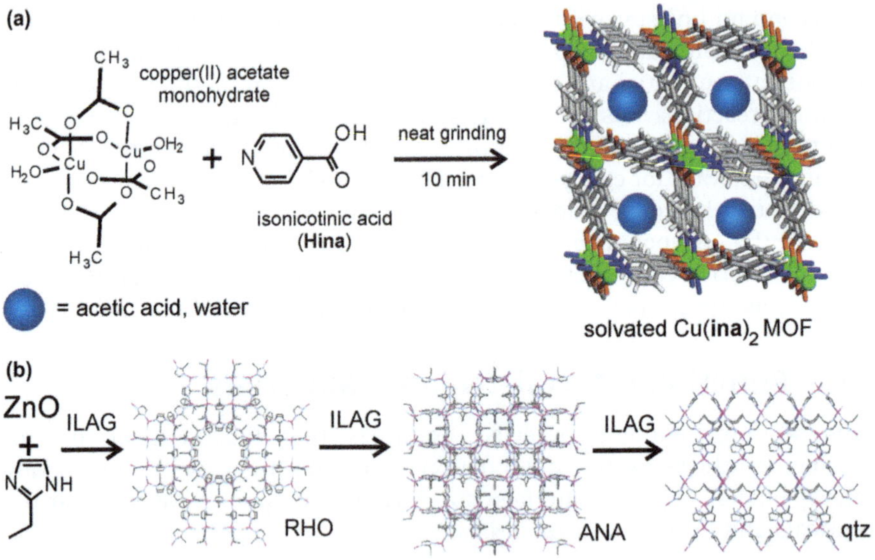

Figure 7.8 (a) The first reported mechanochemical synthesis of a porous MOF, by neat grinding of copper(II) acetate hydrate with isonicotinic acid (Hina).[45] Acetic acid and water produced in the reaction become included in the open MOF product and can be removed by thermal treatment. (b) Interconversion of porous and non-porous ZIFs in the ILAG mechanochemical reaction of ZnO with 2-ethylimidazole.[54]

architectures of intermediate porosity. Such transformation is reminiscent of the Ostwald's rule of stages in which the less stable frameworks, expected to be of low density, transform into more stable, increasingly dense structures. The role of the salt additive in ZIF mechanosynthesis by ILAG is different than that in the synthesis of pillared MOFs. While the synthesis of pillared MOFs using ILAG clearly reveals an anion-related effect, ZIF synthesis strongly depends on the use of weakly acidic ammonium salts.[54,79] The topological control in ILAG synthesis of ZIFs appears to arise from the ability of salt additives to control the rate of spontaneous interconversion of porous and non-porous structures, rather than from specific structure-templating effects.[79]

The described mechanochemical methodologies have obvious advantages over solvothermal approaches used for porous MOF synthesis in terms of time (minutes *versus* days), temperature (near room-temperature *versus* 100–150 °C) and reactant choice (metal oxides *versus* soluble salts). Further advantages are the absence of bulk solvent and a wide choice of catalytic liquid phases, including ethanol, which is one of the more preferred solvents from a green chemistry perspective. The only waste product of the oxide-based reaction is water. The functional nature of porous MOFs obtained mechanochemically was recently investigated by Yuan and co-workers, who found that the Brunauer–Emmett–Teller (BET) surface area of the well-known, commercially relevant MOF HKUST-1 obtained by neat milling was comparable to that of samples obtained by conventional routes.[78] A different study by Schlesinger and co-workers compared neat and liquid-assisted mechanosynthesis to room-temperature and reflux solution synthesis, solvothermal, microwave-assisted, sonochemical, as well as electrochemical syntheses.[80] The comparison revealed that BET surface area and specific pore volume of HKUST-1 samples are increased if the reaction is conducted by LAG, resulting in a material whose porous properties are comparable to those of products obtained by solvothermal or microwave-assisted reactions. Similar observations have been made for the mechanochemically made MOF-14.[81]

7.6.2 Pharmaceutical Derivatives and Metallodrugs

Mechanochemistry is becoming increasingly popular in pharmaceutical materials science for the synthesis of, as well as screening for, new solid forms of active pharmaceutical ingredients (APIs).[13,82] Whereas most such work has focused on making and screening for molecular materials, such as polymorphs, co-crystals, solvates and salts, there is a growing interest in mechanochemical derivatization of APIs using metal ions. Such use of coordination chemistry for API modification has mostly targeted the improvement in a specific property of a drug, such as solubility, with few studies directed towards making metallodrugs,[83] *i.e.* metal–organic materials whose pharmacological activity arises from biologically active metal ions.

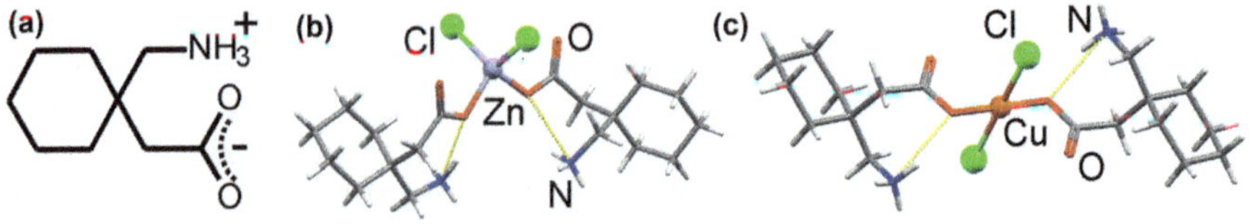

Figure 7.9 (a) Schematic representation of gabapentin. Molecular structures of mechanochemically prepared discrete complexes of gabapentin with: (b) zinc chloride and (c) copper(II) chloride.[84]

A simple procedure for the mechanochemical complexation of an API with a transition metal ion is the manual grinding of zinc chloride or copper(II) chloride with the neuroleptic drug gabapentin,[84] resulting in the formation of neutral, discrete complexes (Figure 7.9). Subsequent work by Quaresma and co-workers expanded the mechanochemical complexation chemistry of gabapentin towards the formation of extended coordination networks, by manual grinding of the drug with chlorides of manganese(II), yttrium, as well as several lanthanides.[85]

Complexation to transition metal or rare earth metal ions is of limited value for improving solid-state properties of APIs, due to the common toxicity of such ions. An attractive alternative, however, is complexation with less active and physiologically acceptable cations, such as Mg^{2+}, Ca^{2+}, Na^{+}, K^{+} and Li^{+}. Spontaneous solid-state formation of magnesium complexes with non-steroidal anti-inflammatory drugs (NSAIDs) was investigated by the Byrn group,[86] who observed the slow partial transformation of physical mixtures of magnesium oxide and the NSAID flufenamic acid into a magnesium flufenamate salt. This work served as an inspiration for Chow and co-workers to conduct the deliberate mechanosynthesis of magnesium complexes with NSAIDs ibuprofen and salicylic acid directly from MgO.[75] The obtained ibuprofen magnesium hydrate was found to be substantially more soluble than ibuprofen itself, explaining the increased activity of ibuprofen when formulated along with MgO as an antacid.[75]

Ionic co-crystals, consisting of ionic salt components as well as neutral molecules, are a range of metal–organic materials highly promising for applications in pharmaceutical materials development. It was established by the Braga group that ionic co-crystals can be readily obtained by mechanochemical complexation of alkaline metal or alkaline earth metal halides with APIs.[87] An example of such reactivity is mechanochemical grinding or kneading of the model API barbituric acid (ba) with a range of alkaline metal bromides and chlorides, leading to the formation of ionic co-crystals or ionic co-crystal hydrates with compositions $ba \cdot NaBr \cdot 2H_2O$, $ba \cdot KBr \cdot 2H_2O$, $ba \cdot RbBr \cdot 2H_2O$, $ba \cdot RbBr$, $ba \cdot CsBr \cdot 2H_2O$ and $ba \cdot CsI$.[88] These multicomponent materials demonstrated enhanced thermal stability compared to pure solid barbituric acid and, which is of particular importance in the context of crystal engineering of pharmaceutical materials, they demonstrated enhanced dissolution rates compared to pure ba. Mechanochemistry was subsequently applied for the formation of ionic co-crystals of APIs

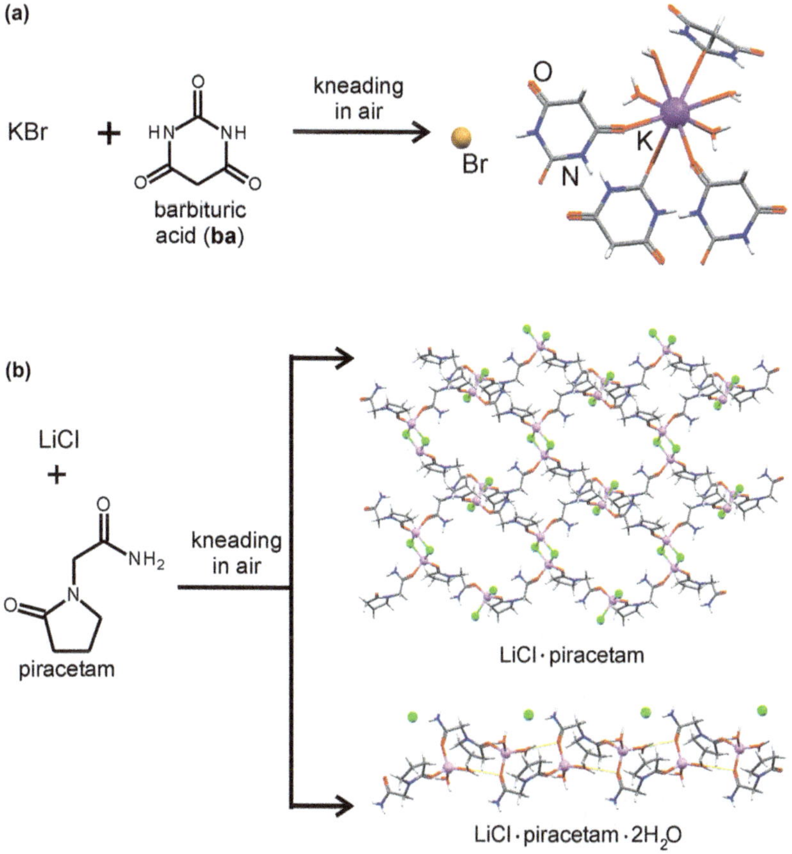

Figure 7.10 Examples of mechanochemical synthesis of ionic co-crystals: (a) by kneading of potassium bromide and barbituric acid with methanol[88] and (b) by kneading of lithium chloride with the API piracetam in the presence of methanol as the kneading liquid.[89,90]

piracetam, brivaracetam and seletracetam, as well as small organic molecules oxamide, nicotinamide (vitamin B6), malonamide and diacetamide (Figure 7.10).[89,90]

An example of using mechanochemistry to synthesize metallodrugs is the ILAG synthesis of bismuth subsalicylate,[56] the active component of the popular gastrointestinal drug Pepto-Bismol®. In addition to providing a clean and rapid route to the known API directly from the inexpensive metal oxide Bi_2O_3, variation of the reaction composition allowed the selective synthesis of two further bismuth salicylates. Quantitative yield and product purity facilitated the structural characterization of bismuth disalicylate monohydrate, an outstanding target from the 2002 Second Structure Determination by Powder Diffractometry Round Robin challenge (SDPDRR-2), from X-ray powder diffraction data.[91] The crystal structure revealed a non-porous 2D framework and corrected a previously proposed chemical composition involving hydroxide ligands. The structure of bismuth disalicylate hydrate is currently the only known crystal structure for a bismuth salicylate complex without auxiliary organic ligands.[92]

7.6.3 Luminescent Materials

One of the first investigations of the potential of mechanochemistry for the synthesis of luminescent metal–organic materials has been reported by Braga *et al.* who exploited neat grinding, as well as kneading, for the synthesis of coordination polymers involving luminophoric copper(I) iodide clusters.[93] The poorly soluble copper(I) iodide was readily transformed into a set of isostructural 2D coordination polymers by mechanochemical milling or kneading with either piperazine or dabco (Figure 7.11a). The luminescent properties of the prepared materials polymers arise from the presence of Cu_2I_2 dimeric units that are separated by organic ligands so as to form 2D frameworks.[94] The mechanochemical syntheses provided a clear benefit over conventional solution syntheses by allowing the direct use of copper(I) iodide precursor whose application in conventional solution-based synthesis is often complicated due to poor solubility. This work was subsequently extended towards the synthesis of discrete luminescent complexes by grinding of copper(I) iodide with the bidentate ligand diphenyl(2-pyridyl)phosphine (dpn) by LAG.[95] Ball milling of these two components in a 2 : 3 stoichiometric ratio and in the presence of a small amount of acetonitrile provided the discrete dimeric complex $Cu_2I_2(dpn)_3$, which exhibits a maximum of fluorescent emission at 525 nm, associated with a butterfly-shaped Cu_2I_2 unit. In contrast, milling in the presence of chloroform as the grinding

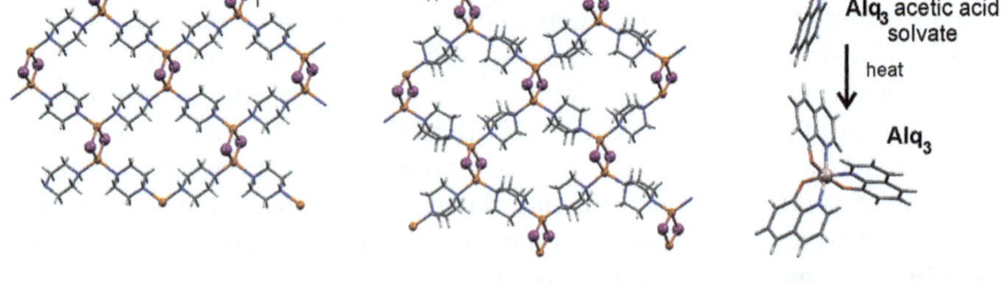

Figure 7.11 Mechanochemical synthesis of luminescent coordination polymers containing the Cu_2I_2 unit by milling of solid copper(I) iodide with piperazine or dabco;[93] (b) synthesis of the OLED material Alq_3 by a combination of mechanochemistry and thermal desolvation.[11]

liquid gave a mixture of two compounds with compositions $CuI(dpn)_{0.5}$ and $Cu_4I_4(dpn)_2 \cdot (CH_2Cl_2)_{0.5}$. The crystal structure of luminescent $CuI(dpn)_{0.5}$ was established from separately grown single crystals, and revealed the presence of 1D polymeric $(CuI)_\infty$ chains composed of edge-fused planar and butterfly-shaped Cu_2I_2. The detailed structure of $Cu_4I_4(dpn)_2 \cdot (CH_2Cl_2)_{0.5}$ is not known but its redshifted (*ca.* 575 nm) luminescence is most likely the result of a Cu_4I_4 core found in the isostructural $Cu_4I_4(dpn)_2$ obtained by thermal desolvation of $Cu_4I_4(dpn)_2 \cdot (CH_2Cl_2)_{0.5}$.[95] Consequently, mechanochemistry and associated control over reaction stoichiometry could be applied to explore, with the aid of structural characterization from XRPD data, the diversity of metal–organic subunits applicable for making fluorescent materials.

The, so far, only application of mechanochemistry for the synthesis of a commercially relevant luminescent material was reported by Ma and coworkers who described the synthesis of the important and archetypal OLED material (8-hydroxoquinolinate)$_3$aluminium (Alq_3) directly from basic aluminium acetate and the organic ligand 8-hydroxyquinoline (Figure 7.11b).[11] Besides being a pioneering (and successful) effort in the solvent-free manufacture of a commercial OLED material, this work also recognized the presence of an intermediate phase, characterized *via* structure solution from X-ray powder diffraction data, as a previously unknown acetic acid solvate of Alq_3, $Alq_3 \cdot AcOH$ (Figure 7.11b). Importantly, the solvate $Alq_3 \cdot AcOH$ is also luminescent, demonstrating the ability to discover and characterize novel potentially functional materials by a combination of mechanosynthesis and structure determination using X-ray powder diffraction. The authors also highlighted the ability to manufacture in facile fashion the OLED material Alq_3 on a scale of at least 50 g and, if high purity of the product is necessary, perform direct sublimation from the reaction mixture, in that way achieving an entirely solvent-free access to a technologically important metal–organic material.[11]

7.7 New Properties and Synthetic Opportunities arising from Mechanochemical Synthesis of Metal–Organic Materials

7.7.1 Structural Dynamics under Mechanochemical Conditions

A recently reported effect in MOF chemistry, which might become of considerable importance, is the highly dynamic nature of the frameworks under conditions of liquid-assisted grinding. For example, grinding of ZnO and terephthalic acid with water, DMF or methanol yields 1D, 2D or 3D forms of the zinc terephthalate polymer, respectively (Figure 7.12a).[52,71,95] Different forms were almost completely interconvertible by brief grinding with a suitable liquid and such fluxional behavior also permitted the

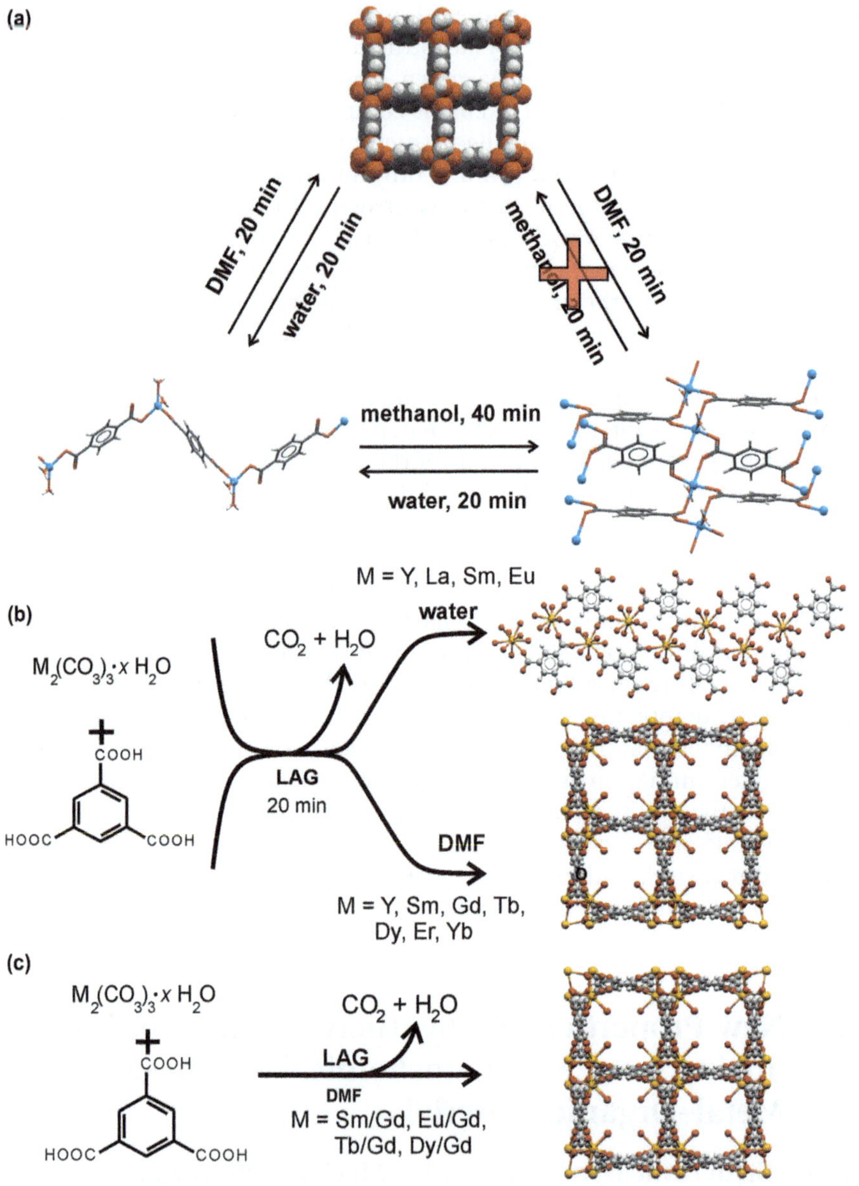

Figure 7.12 (a) Dynamic nature of coordination polymers and open MOFs under mechanochemical milling conditions: milling with a suitable liquid phase leads to the facile transformation of different open and close-packed zinc terephthalate frameworks materials, not observed in analogous slurrying experiments.[95] Mechanosynthesis of rare earth (lanthanide) ion MOF by LAG with trimesic acid leading to: (b) open or close-packed structures depending on the choice of the grinding liquid and (c) the formation of mixed-metal lanthanide MOFs.[97]

construction of open pillared structures by addition of bridging ligands such as dabco or bipy.[95] Such behavior is considerably different from solution-based chemistry, as the mechanochemically induced transformations do not readily occur by simple immersion of reactant materials in the bulk liquid.[95,96]

7.7.2 Synthesis of Solid Solutions

Although the isostructurality of components is often considered to be a very important factor in the synthesis of solid solutions, their formation *via* conventional solution-based methods is often hindered by the difference in solubility of individual components, which leads to the preferential crystallization of separate components. The potential advantage of mechanochemistry for the construction of solid solutions of isostructural coordination polymers has been demonstrated by James and co-workers who constructed close-packed and open-framework materials by LAG reactions of rare earth metal carbonates and trimesic acid.[97] In particular, LAG of each individual metal carbonate in the presence of water quantitatively yielded isostructural close-packed materials consisting of 1D coordination polymers.[1] In contrast, LAG with DMF resulted in the formation of isostructural lanthanide MOFs based on an open 3D architecture (Figure 7.12b). The isostructurality of the 3D MOF products suggested the possibility to form solid solutions. Indeed, grinding of different pairs of lanthanide metal carbonates (*e.g.* pairs Sm–Gd, Eu–Gd, Tb–Gd and Dy–Gd) with trimesic acid afforded mixed-lanthanide 3D open frameworks isostructural to the ones obtained using individual metal carbonate reactants (Figure 7.12c).[97]

The ability to mechanochemically form solid solutions of coordination polymers was exploited for the synthesis of metal–organic materials with controlled crystallographic lattice parameters and physical properties. Mechanochemical liquid-assisted grinding of anhydrous $CoBr_2$ with bipy results in the quantitative formation of the 2D sheet coordination polymer $Co(bipy)Br_2$.[98] The coordination polymer is isostructural to $Co(bipy)Cl_2$, which is obtained by analogous reaction of anhydrous $CoCl_2$ (Figure 7.4a). Correspondingly, LAG of different mixtures of anhydrous $CoCl_2$ and $CoBr_2$ yields solid solutions of the two coordination polymers, with overall composition $Co(bipy)Br_{2-x}Cl_x$. The solid solutions are homogeneous at length scales detectable by XRPD and UV/Vis spectroscopy, with the Br/Cl ratio determined by the relative amounts of $CoBr_2$ and $CoCl_2$ in the reaction mixture.[98] The coordination polymer solid solutions are isostructural to the single phases $Co(bipy)Cl_2$ and $Co(bipy)Br_2$, but with lattice parameters and UV/Vis absorption properties determined by the Br/Cl ratio. Specifically, the crystallographic lattice parameters and UV/Vis absorption of $Co(bipy)Br_{2-x}Cl_x$ were found to change in a continuous and linear fashion with x over the range 0–2.[98]

7.7.3 Binding of Carbon Dioxide upon Mechanochemical Treatment

Manual grinding of solid silver carbonate (Ag_2CO_3) with PPh_3 in air and in the presence of ethanol leads to the absorption of one equivalent of water (present in the form of air moisture) and one equivalent of CO_2 gas to yield bis(triphenylphosphino)silver(I) hydrogen-carbonate, which can again release CO_2 upon heating.[99] Monitoring of the reaction through solid-state infrared spectroscopy suggests a two-step mechanism involving the formation of the solid tetrakis(triphenylphosphino)silver(I) carbonate as the intermediate. Although this particular mechanochemical product is not an extended metal–organic structure, the ability to chemically bind carbon dioxide through a mechanochemical transformation of a metal–organic system is fascinating, especially when considering the growing importance of metal–organic materials for sequestration of waste CO_2.

7.8 Mechanistic Studies of Mechanochemical MOF Synthesis

7.8.1 Stepwise Analysis

The mechanochemical synthesis of coordination polymers and MOFs does not necessarily need to proceed in a single step. The simplest approach to detect crystalline reaction intermediates in mechanosynthesis is by step-by-step analysis, *i.e.* by periodically halting the mechanochemical milling and analyzing the composition of the reaction mixture. Such analysis revealed the stepwise transformation of ZIF topologies in ILAG synthesis of zinc 2-ethylimidazolate from ZnO, which was tentatively associated to relative thermodynamic stabilities of porous and non-porous frameworks.[54] Another example of a stepwise mechanochemical reaction mechanism is the synthesis of the 1D coordination polymer zinc fumarate tetrahydrate (Figure 7.13a).[52,71] Whereas milling together of ZnO, fumaric acid and three equivalents of water for *ca.* 45 min leads to the specific formation of the tetrahydrate phase, stopping the milling earlier reveals the presence of an intermediate pentahydrate phase. This stepwise process is explained by the rapid, liquid-assisted formation of the crystalline pentahydrate at early stages of the reaction in which water is abundant. The rapid formation of the pentahydrate phase, however, efficiently depletes the water in the reaction mixture and transforms the fast liquid-assisted reaction into a slower neat grinding process.[71] The resulting neat grinding process is speculated to proceed through an amorphous intermediate, deduced by the spontaneous formation of different coordination polymers by ageing of the partially reacted reaction mixture (Figure 7.13a). A similar stepwise mechanism was observed in the mechanochemical reaction of CuO and acetic acid, where the abundance of acetic acid early in the reaction leads to the formation of

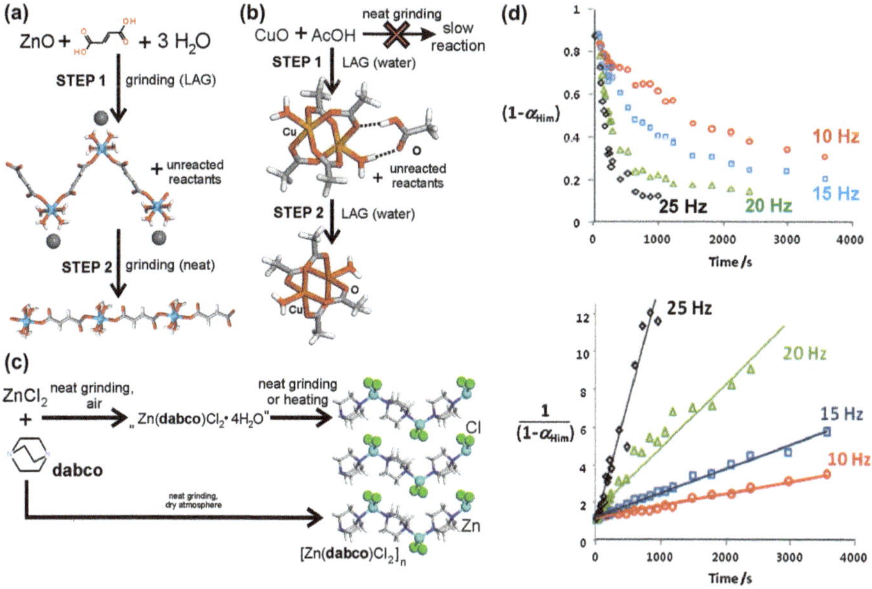

Figure 7.13 Stepwise reaction mechanism in the mechanochemical oxide-based LAG synthesis of: (a) zinc fumarate tetrahydrate coordination polymer and (b) copper(II) acetate monohydrate paddlewheel cluster. The stepwise nature of the reaction is associated with the large excess of liquid reactant present in the early stages of the reaction. (c) An example of how the surrounding atmosphere can affect the mechanism of mechanochemical formation of a coordination polymer: grinding $ZnCl_2$ and dabco in open air leads to the formation of the 1D $Zn(dabco)Cl_2$ *via* a hydrated intermediate. The same final product is obtained by grinding in a dry atmosphere, but without and observable crystalline intermediates. Results of stepwise Raman spectroscopy[100] monitoring of the LAG reaction of imidazole and ZnO: (top) fraction of unreacted imidazole plotted against time, at milling frequencies of 10, 15, 20 and 25 Hz; a pseudo second-order integrated plot of the kinetic data (bottom). Adapted from ref. 100.

the intermediate crystalline acetic acid solvate of the copper(II) acetate paddlewheel cluster (Figure 7.13b).[72]

An interesting example of a stepwise mechanism in mechanosynthesis is given by neat grinding reaction of anhydrous $ZnCl_2$ with the aliphatic ligand dabco.[66] The first reaction step is the formation of crystalline $Zn(dabco)Cl_2 \cdot 4H_2O$, which upon heating or further grinding transforms into the 1D zigzag polymer $[Zn(dabco)Cl_2]_n$. The latter was also obtained from solution and was characterized by single-crystal X-ray diffraction (Figure 7.13c). The formation of an intermediate was ascribed to the hygroscopic nature of dabco under grinding conditions. Consequently, this reaction clearly demonstrates how the surrounding atmosphere can influence the course of a mechanochemical reaction. The active role of moisture in the formation of the intermediate was also confirmed by conducting the mechanosynthesis in dry air and with carefully dried reactants. Under such conditions, $[Zn(dabco)Cl_2]_n$ formed directly, without observable intermediates.[66] That the sensitivity to moisture is largely related to dabco is supported by the observation that neat grinding of anhydrous $ZnCl_2$ and bipy to form of a similar zigzag 1D polymer proceeds in one step.

Whereas sample preparation for X-ray powder diffraction analysis often requires minutes, Raman spectroscopy is much more suited for rapid analysis of reaction mechanisms, as samples can be characterized within seconds without any particular preparation procedure. As a faster alternative to X-ray powder diffraction, the James group has applied Raman spectroscopy to monitor the mechanochemical LAG reaction of zinc oxide and imidazole in the presence of DMF to form the microporous framework known as ZIF-6.[100] This kinetic study revealed that product formation proceeds in a manner that highly resembles the kinetics of a second-order reaction in solution. These observations led the authors to propose a "pseudo-fluid" model of the reaction mixture under mechanochemical conditions, in which the reaction rate for this particular mechanochemical process is controlled largely by the number of reactive encounters involving surfaces of reactant imidazole and zinc oxide particles, and in which the role of milling is principally to stir the reaction mixture and expose fresh reactant surfaces by abrasion.[100] One of the most surprising aspects of such a pseudo-fluid reaction model, which was experimentally confirmed by conducting the reaction at different temperatures, is that the reaction rate is highly dependent only on the milling frequency (Figure 7.13d), but not temperature. That mechanochemical milling reactions by LAG and ILAG can indeed proceed in a manner that resembles established first- and second-order rate laws in solution was subsequently confirmed by the Halasz and Friščić groups by quantitative *in situ* and real time reaction monitoring using synchrotron X-ray diffraction.[101]

An impressive feat of rapid stepwise analysis of mechanochemical reactions has very recently been reported by Tröbs and Emmerling, who conducted X-ray powder diffraction, Raman spectroscopy and even rapid electron microscopy measurements on the three-component reaction of bismuth nitrate, terephthalic acid and imidazole.[102] This reaction, which yields a 3D MOF material, has successfully been monitored with a time resolution as low as 10 s.

7.8.2 *In Situ* and Real-time Studies

One of the major drawbacks of mechanochemical milling and grinding chemistry is the inability to carry out real-time observation of the transformation of reactants into products. As pointed out by Drebuschak and co-workers:[103]

> "It is a challenge to understand the processes taking place in a powder sample during its grinding in a mill, or compacting, since one can neither measure local temperature, pressure, shear stresses, nor follow the changes in the diffraction patterns or vibrational spectra *in situ*."

Indeed, *in situ* monitoring of mechanochemical reactions has been largely limited to temperature- and pressure-measurements on mechanically induced self-sustaining reactions, or on monitoring pressure changes in reactions adsorbing or releasing a gas. An example of the latter is the reactions kinetically analyzed by Urakaev and Boldyrev.[104] Very recently, however, the application of highly penetrating synchrotron X-ray radiation enabled an unprecedented technique for the real time *in situ* study of mechanochemical synthesis of porous metal–organic frameworks by powder X-ray diffraction.[15,17] The excellent penetrating power of X-rays with a short wavelength (0.1 Å) enabled recording of the powder diffraction patterns of reaction mixtures as they were undergoing milling, with a time resolution in seconds. Such real-time structural studies have painted an unexpectedly dynamic picture of mechanochemical milling reaction, involving highly labile intermediate phases whose lifetime is less than a minute, dependent on milling conditions and environment, and whose chemical and particle size evolution could be measured in real time and *in situ* (Figure 7.14a–c). For the mechanochemical synthesis of the porous framework material ZIF-8 *in situ* studies have enabled the quantification of product formation rates for reactions conducted by LAG and by ILAG.[17] Assuming Avrami–Erofe'ev kinetics, switching from LAG to ILAG by salt addition led to a *ca.* 15-fold increase in the rate constant for product formation.

The *in situ* monitoring technique readily captured the formation and interconversion of different ZIF topologies upon mechanochemical ILAG reaction of ZnO and 2-ethylimidazole, and plotting the time-dependent intensities of the characteristic X-ray reflections for all involved materials provided the first insight into the kinetics of ZIF mechanosynthesis and transformation (Figure 7.14d).[17] The real-time *in situ* monitoring of the mechanochemical milling reaction between imidazole (Him) and ZnO revealed a strong dependence of the reaction behavior on reaction conditions. Neat grinding of the two reactants was observed to proceed in a single step to provide the discrete coordination tetramer of composition $Zn_4(im)_8(Him)$. In contrast, ILAG with ethanol as the grinding liquid led to the rapid, one-step formation of the non-porous 3D zinc imidazolate framework with *zni* topology. Replacing ethanol with DMF as the grinding liquid led to formation of the open framework structure known as ZIF-4, containing included DMF guests. However, reducing the amount of DMF led to a stepwise process in which the initially formed ZIF-4 is subsequently, within minutes, replaced by a low porosity structure ZIF-6 (Figure 7.14d).[17]

The latest development in the synchrotron X-ray diffraction technique for monitoring mechanochemical reactions has been the introduction of an internal diffraction standard, in particular crystalline silicon.[101] The first studies using the internal standard technique allowed the quantitative analysis of mechanochemical reaction mechanisms using procedures of Rietveld refinement, which revealed not only the pseudo-fluid behavior first proposed by the James group, but also suggested the presence of a

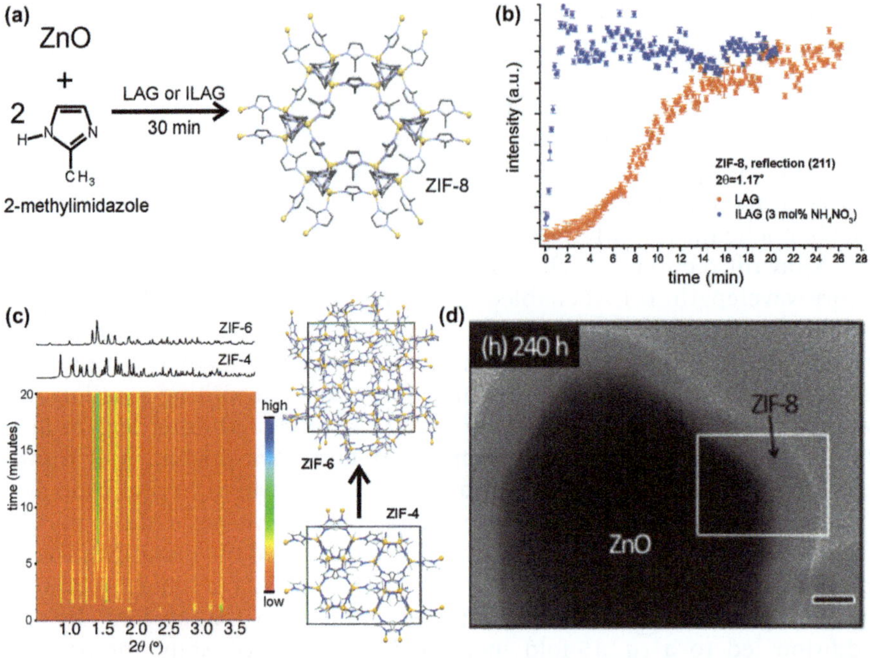

Figure 7.14 Mechanosynthesis of porous ZIFs investigated by time-resolved *in situ* powder X-ray diffractiony:[17] (a) quantitative synthesis of ZIF-8 by LAG or ILAG reaction of ZnO and 2-methylimidazole; (b) kinetic curves for the LAG and ILAG syntheses of ZIF-8, based on monitoring changes in intensity of the characteristic ZIF-8 (211) X-ray reflection during milling (adapted from ref. 17); (c) time-resolved X-ray powder diffractogram (left) and reaction scheme (right) for the mechanochemical reaction of ZnO and imidazole with a small amount of DMF, which first gives rise to the open ZIF-4 structure, which subsequently transforms into the low-porosity ZIF-6 (adapted from ref. 17). (d) Field emission scanning electron microscopy (FESEM) images of product particles after 240 h milling of ZnO and 2-methylimidazole, demonstrating the formation of core–shell structures composed of ZnO and ZIF-8.[103]
Reproduced from ref. 103.

considerable amount (*ca.* 30% of the reaction mixture) of an amorphous phase during mechanochemical LAG and ILAG reactions. While *in situ* and real time quantification of amorphous content is of particular importance for understanding mechanochemical reactions, the presence of large amounts of amorphous material during LAG reactions is in contrast to conclusions of previous *ex situ* studies, which suggested LAG processes involved very little amorphous phase or none at all. The most recent addition to the new toolkit of techniques of *in situ* and real time monitoring of mechanochemical reactions is Raman spectroscopy, enabled through the introduction of optically transparent poly(methyl methacrylate) (PMMA) reaction vessels. The first results of this new and readily accessible laboratory technique, obtained on the reaction system of $CdCl_2$ and cnge, confirm that

Raman spectroscopy enables monitoring of the reaction progress, including the transformations of different reaction intermediates and products.[16]

7.8.3 Electron Microscopy

There have been very few studies of the morphology of MOF particles obtained by mechanosynthesis. For example, scanning electron microscopy images of zinc fumarate pillared MOFs obtained by LAG from ZnO, fumaric acid and bipy reveal well developed rectangular crystals with approximate size of 200 nm.[1] A similar observation was also made for mechanochemically prepared microcrystals of zinc isonicotinate tetrahydrate, the monomeric precursor to an interpenetrated diamondoid zinc isonicotinate framework, which exhibited well-developed needle- and rod-like morphologies.[73] Overall, the several existing scanning electron microscopy (SEM) studies indicate that coordination polymers and MOFs prepared by LAG or ILAG will exhibit morphologically well-developed crystals with sizes in the range 100 nm–1 μm. The, so far, only systematic SEM study of particle development[105] during mechanochemical synthesis of MOFs has been conducted for slow (up to 240 h milling) neat grinding of ZnO and 2-methylimidazole to form ZIF-8. The study revealed that the transformation of ZnO particles into ZIF-8 can penetrate as far as 10 nm below the surface of the reactant oxide particle. Consequently, small particles of ZnO can be completely converted into ZIF-8. For larger ZnO particles, or their agglomerates, the mechanochemical reaction leads to the formation of particles consisting of a core of zinc oxide surrounded by a shell of ZIF-8 (Figure 7.14d).[105]

7.9 Outlook

The topics outlined in this brief overview of metal–organic mechanochemistry confirm the notion that this is a highly dynamic and rapidly developing area. Over a decade of intense research has provided sufficient information to begin establishing a systematic understanding of mechanochemical techniques and types of chemical reactivity most appropriate for the synthesis of metal–organic compounds. Several benefits arising from the application of mechanochemistry for the synthesis of functional metal–organic materials have been recognized and embraced by the research community, above all the ability to synthesize microporous MOFs rapidly and without solvent, as well as to activate normally inert mineral-like materials, such as metal oxides, into versatile precursors for MOF synthesis. Recent work indicates a growing interest for mechanochemistry as a means to prepare solid solutions[106] of coordination polymers or amorphous, non-crystalline forms of MOFs.[107,108] These areas are likely to become more visible in the further development of metal–organic mechanochemistry, as well as applications of mechanochemistry for organometallic synthesis,[109] which are still very rare. The understanding of reaction kinetics and mechanisms of mechanochemical reactions is still in its early stages and further *in situ* and real time studies will certainly lead to important developments in the future. The growing number of explorations of one-pot "all-at-once" multi-component reactions suggests that future research in metal–organic mechanochemical synthesis will become increasingly dedicated to finding routes to synthesizing complex products from the simplest possible reactants. In that light, the application of mechanochemistry to coordination bond formation has brought the research in metal–organic materials chemistry closer to the established traditions of synthetic organic synthesis: of designing less expensive, more efficient, more elegant and less environmentally harmful synthetic procedures.

Abbreviations

adc	acetylenedicarboxylate anion
AgOAc	silver(I) acetate
Alq_3	(8-hydroxoquinolinate)aluminium
ANA	analcime topology
API	active pharmaceutical ingredient
ba	barbituric acid
BET	Brunauer–Emmett–Teller
bipy	4,4′-dipyridyl
bpe	*trans*-1,2-bis(4-pyridyl)ethylene
CNacac	cyanoacetylacetonate ion
cnge	cyanoguanidine
Cp	cyclopentadienide anion
dabco	diazabicyclo[2.2.2]octane
dace	1,4-diaminocyclohexane
DMF	*N,N*-dimethylformamide
DMSO	dimethyl sulfoxide
dpn	diphenyl-(2-pyridyl)phosphine
etu	ethylenethiourea
FTIR	Fourier-transform infrared
H_2adc	acetylenedicarboxylic acid
H_2fum	fumaric acid
H_2salen	bis(salicylidene)ethylenediamine
HCNacac	cyanoacetylacetone
Him	imidazole
Hina	isonicotinic acid
ILAG	ion- and liquid-assisted grinding
im	imidazolate ion
ina	isonicotinate ion
LAG	liquid-assisted grinding
MOF	metal–organic framework
NMR	nuclear magnetic resonance
NSAID	non-steroidal anti-inflammatory drug
OAc	acetate ion

OLED	organic light-emitting diode
PMMA	poly(methyl methacrylate)
pn	1,3-bis(4-pyridyl)propane
qtz	quartz topology
RHO	zeolite ρ topology
SEM	scanning electron microscopy
TGA	thermogravimetric analysis
XRPD	X-ray powder diffraction
ZIF	zeolitic imidazolate framework

Acknowledgements

The author acknowledges the support of McGill University, NSERC Discovery Grant and FRQNT Nouveaux Chercheurs grant.

References

1. S. L. James, P. Collier, I. Parkin, G. Hyatt, D. Braga, L. Maini, W. Jones, C. Bölm, A. Krebs, J. Mack, D. Waddell, W. Shearouse, A. G. Orpen, C. J. Adams, T. Friščić, J. W. Steed and K. D. M. Harris, *Chem. Soc. Rev.*, 2012, **41**, 413–447.
2. V. Šepelák, A. Düvel, M. Wilkening, K.-D. Becker and P. Heitjans, *Chem. Soc. Rev.*, 2013, **42**, 7507–7520.
3. P. Baláž, M. Achimovičová, M. Baláž, P. Billik, Z. Cherkezova-Zheleva, J. Manuel Criado, F. Delogu, E. Dutková, E. Gaffet, F. J. Gotor, R. Kumar, I. Mitov, T. Rojac, M. Senna, A. Streletskii and K. Wieczorek-Ciurowa, *Chem. Soc. Rev.*, 2013, **42**, 7571–7637.
4. C. Suryanarayana, *Prog. Mater. Sci.*, 2001, **46**, 1–184.
5. G.-W. Wang, *Chem. Soc. Rev.*, 2013, **42**, 7668–7700.
6. A. Stolle, T. Szuppa, S. E. S. Leonhardt and B. Ondruschka, *Chem. Soc. Rev.*, 2011, **40**, 2317–2329.
7. S. K. Thabet, H. A. Tayim and M. U. Karkanawi, *Inorg. Nucl. Chem. Lett.*, 1972, **8**, 211–213.
8. A. Orita, L. Jiang, T. Nakano, N. Ma and J. Otera, *Chem. Commun.*, 2002, 1362–1363.
9. T. Friščić, *Chem. Soc. Rev.*, 2012, **41**, 3493–3510.
10. L. Maini, D. Braga, P. Mazzeo and B. Ventura, *Dalton Trans.*, 2012, 531–539.
11. X. Ma, G. K. Lim, K. D. M. Harris, D. C. Apperley, P. N. Horton, M. B. Hursthouse and S. L. James, *Cryst. Growth Des.*, 2012, **12**, 5769–5872.
12. M. D. Bala and N. J. Coville, *J. Organomet. Chem.*, 2007, **692**, 709–730.
13. A. Delori, T. Friščić and W. Jones, *CrystEngComm*, 2012, **14**, 2350–2362.
14. T. Friščić, I. Halasz, V. Štrukil, M. Eckert-Maksić and R. E. Dinnebier, *Croat. Chem. Acta*, 2012, **85**, 367–378.

15. I. Halaz, S. A. J. Kimber, P. J. Beldon, A. M. Belenguer, F. Adams, V. Honkimäki, R. C. Nihghtingale, R. E. Dinnebier and T. Friščić, *Nat. Protocol.*, 2013, **8**, 1718–1729.
16. D. Gracin, V. Štrukil, T. Friščić, I. Halasz and K. Užarević, *Angew. Chem., Int. Ed.*, 2014, **53**, 6193–6197.
17. T. Friščić, I. Halasz, P. J. Beldon, A. M. Belenguer, F. Adams, S. A. J. Kimber, V. Honkimäki and R. E. Dinnebier, *Nat. Chem.*, 2013, **5**, 66–73.
18. A. Lazuen-Garay, A. Pichon and S. L. James, *Chem. Soc. Rev.*, 2007, **36**, 846–855.
19. D. Braga, S. L. Giaffreda, F. Grepioni, A. Pettersen, L. Maini, M. Curzi and M. Polito, *Dalton Trans.*, 2006, 1249–1263.
20. V. Štrukil, L. Fábián, D. G. Reid, M. J. Duer, G. J. Jackson, M. Eckert-Maksić and T. Friščić, *Chem. Commun.*, 2010, **46**, 9191–9193.
21. G. A. Bowmaker, N. Chaichit, C. Pakawatchai, B. W. Skelton and A. H. White, *Dalton Trans.*, 2008, 2926–2928.
22. S. Karki, T. Friščić and W. Jones, *CrystEngComm*, 2009, **11**, 470–481.
23. J. Shen and R. Blachnik, *Thermochim. Acta*, 2003, **399**, 245–246.
24. R. Schmidt, A. Stolle and B. Ondruschka, *Green Chem.*, 2012, **14**, 1673–1679.
25. V. P. Balema, J. W. Wiench, M. Pruski and V. K. Pecharsky, *Chem. Commun.*, 2002, 1606–1607.
26. J. G. Hernández, N. A. J. Macdonald, C. Mottillo, I. S. Butler and T. Friščić, *Green Chem.*, 2014, **16**, 1087–1092.
27. S. R. Batten, N. R. Champness, X.-M. Chen, J. Garcia-Martinez, S. Kitagawa, L. Öhrström, M. O'Keeffe, M. P. Suh and J. Reedijk, *CrystEngComm*, 2012, **14**, 3001–3004.
28. L. J. Murray, M. Dincă and J. R. Long, *Chem. Soc. Rev.*, 2009, **38**, 1294–1314.
29. K. Sumida, D. L. Rogow, J. A. Mason, T. M. McDonald, E. D. Bloch, Z. R. Herm, T.-H. Bae and J. R. Long, *Chem. Rev.*, 2012, **112**, 724–781.
30. M. E. Foster, J. D. Azoulay, B. M. Wong and M. D. Allendorf, *Chem. Sci.*, 2014, **5**, 2081–2090.
31. Z. R. Herm, E. D. Bloch and J. R. Long, *Chem. Mater.*, 2014, **26**, 323–338.
32. L. E. Kreno, K. Leong, O. K. Farha, M. Allendorf, R. P. Van Duyne and J. T. Hupp, *Chem. Rev.*, 2012, **112**, 1105–1125.
33. B. Yilmaz, N. Trukhan and U. Müller, *Chin. J. Catal.*, 2012, **33**, 3–10.
34. A. Czaja, E. Leung, N. Trukhan and U. Müller, Industrial MOF synthesis, in *Metal-Organic Frameworks: Applications from Catalysis to Gas Storage*, ed. D. Farrusseng, Wiley-VCH Verlag GmbH, Weinheim, 2011.
35. T. Friščić and W. Jones, *Cryst. Growth Des.*, 2009, **9**, 1621–1637.
36. T. Friščić, *J. Mater. Chem.*, 2010, **20**, 7599–7605.
37. D. Braga, L. Maini and F. Grepioni, *Chem. Soc. Rev.*, 2013, **42**, 7638–7648.
38. A. V. Trask, N. Shan, W. D. S. Motherwell, W. Jones, S. Feng, R. B. H. Tan and K. J. Carpenter, *Chem. Commun.*, 2005, 880–882.

39. G. A. Bowmaker, *Chem. Commun.*, 2013, **47**, 334–348.
40. J. Yoshida, S.-i. Nishikiori and R. Kuroda, *Chem. Eur. J.*, 2008, **14**, 10570–10578.
41. J. Stojaković, B. S. Farris and L. R. MacGillivray, *Chem. Commun.*, 2012, **46**, 7958–7960.
42. Y. Fang, N. Salamé, S. Woo, D. S. Bohle, T. Friščić and L. A. Cuccia, *CrystEngComm*, 2014, **16**, 7180–7185.
43. A. Pichon and S. L. James, *CrystEngComm*, 2008, **10**, 1839–1847.
44. E. Boldyreva, *Chem. Soc. Rev.*, 2013, **42**, 7719–7738.
45. A. Pichon, A. Lazuen-Garay and S. L. James, *CrystEngComm*, 2006, **8**, 211–214.
46. D. Braga, M. Curzi, F. Grepioni and M. Polito, *Chem. Commun.*, 2005, 2915–2917.
47. C. J. Adams, H. M. Colquhoun, P.C. Crawford, M. Lusi and A. G. Orpen, *Angew. Chem., Int. Ed.*, 2007, **46**, 1124–1128.
48. K. L. Nguyen, T. Friščić, G. M. Day, L. F. Gladden and W. Jones, *Nat. Mater.*, 2007, **6**, 206–209.
49. T. Friščić, S. L. Childs, S. A. A. Rizvi and W. Jones, *CrystEngComm*, 2009, **11**, 418–426.
50. T. Friščić, A. V. Trask, W. Jones and W. D. S. Motherwell, *Angew. Chem., Int. Ed.*, 2006, **45**, 7546–7550.
51. D. Braga, M. Curzi, A. Johansson, M. Polito, K. Rubini and F. Grepioni, *Angew. Chem., Int. Ed.*, 2006, **45**, 142–146.
52. T. Friščić and L. Fábián, *CrystEngComm*, 2009, **11**, 743–745.
53. T. Friščić, D. G. Reid, I. Halasz, R. S. Stein, R. E. Dinnebier and M. J. Duer, *Angew. Chem., Int. Ed.*, 2010, **49**, 712–715.
54. P. J. Beldon, L. Fábián, R. S. Stein, A. Thirumurugan, A. K. Cheetham and T. Friščić, *Angew. Chem., Int. Ed.*, 2010, **49**, 9640–9643.
55. J.-P. Zhang, Y.-B. Zhang, J.-B. Lin and X.-M. Chen, *Chem. Rev.*, 2012, **112**, 1001–1033.
56. V. André, A. Hardeman, I. Halasz, R. S. Stein, G. J. Jackson, D. G. Reid, M. J. Duer, C. Curfs, M. T. Duarte and T. Friščić, *Angew. Chem., Int. Ed.*, 2011, **50**, 7858–7861.
57. D. Braga and F. Grepioni, *Angew. Chem., Int. Ed.*, 2004, **43**, 4002–4011.
58. K. D. M. Harris, Powder diffraction crystallography of molecular solids, in *Advanced X-Ray Crystallography*, ed. K. Rissanen, Springer-Verlag, Berlin, 2012, Topics in Current Chemistry, vol. 315 , pp. 133–177.
59. S. Karki, L. Fábián, T. Friščić and W. Jones, *Org. Lett.*, 2007, **9**, 3133–3136.
60. K. Fujii, A. Lazuen Garay, J. Hill, E. Sbircea, Z. Pan, M. Xu, D. C. Apperley, S. L. James and K. D. M. Harris, *Chem. Commun.*, 2010, **46**, 7572–7574.
61. S. A. Bourne, M. Kilkenny and L. R. Nassimbeni, *Dalton Trans.*, 2001, 1176–1179.
62. C. Hu and U. Englert, *Angew. Chem., Int. Ed.*, 2005, **44**, 2281–2283.
63. S. S. Manzini and N. J. Coville, *Inorg. Chem. Commun.*, 2004, **7**, 676–678.

64. A. Munyaneza, M. D. Bala and N. J. Coville, *S. Afr. J. Chem.*, 2009, **62**, 14–19.
65. W. J. Belcher, C. A. Longstaff, M. R. Neckenig and J. W. Steed, *Chem. Commun.*, 2002, 1602–1603.
66. D. Braga, S. L. Giaffreda, F. Grepioni and M. Polito, *CrystEngComm*, 2004, **6**, 458–462.
67. I. Stein and U. Ruschewitz, *Z. Anorg. Allg. Chem.*, 2010, **636**, 400–404.
68. C. J. Adams, M. A. Kurawa, M. Lusi and A. G. Orpen, *CrystEngComm*, 2008, **10**, 1790–1795.
69. C. J. Adams, M. A. Kurawa and A. G. Orpen, *Dalton Trans.*, 2010, 6974–6984.
70. J. Fernández-Bertrán, L. Castellanos-Serra, H. Yee-Madeira and E. Reguera, *J. Solid-State Chem.*, 1999, **147**, 561–564.
71. F. C. Strobridge, N. Judaš and T. Friščić, *CrystEngComm*, 2010, **12**, 2409–2418.
72. T. Friščić, I. Halasz, F. C. Strobridge, R. E. Dinnebier, R. S. Stein, L. Fábián and C. Curfs, *CrystEngComm*, 2011, **13**, 3125–3129.
73. P. Wang, G. Li, Y. Chen, S. Chen, S. L. James and W. Yuan, *CrystEngComm*, 2012, **14**, 1994–1997.
74. T. Friščić, E. Mešrović, D. Škalec Šamec, B. Kaitner and L. Fábián, *Chem. Eur. J.*, 2009, **15**, 12644–12652.
75. E. H. H. Chow, F. C. Strobridge and T. Friščić, *Chem. Commun.*, 2010, **46**, 6368–6370.
76. M. Ferguson, N. Giri, X. Huang, D. Apperley and S. L. James, *Green Chem.*, 2014, **16**, 1374–1382.
77. J. G. Hernández, I. S. Butler and T. Friščić, *Chem. Sci.*, 2014, **5**, 3576–3582.
78. W. Yuan, A. Lazuen Garay, A. Pichon, R. Clowes, C. D. Wood, A. I. Cooper and S. L. James, *CrystEngComm*, 2010, **12**, 4063–4065.
79. M. J. Cliffe, C. Mottillo, R. S. Stein, D.-K. Bučar and T. Friščić, *Chem. Sci.*, 2012, **3**, 2495–2500.
80. M. Schlesinger, S. Schulze, M. Hietschold and M. Mehring, *Microporous Mesoporous Mater.*, 2010, **132**, 121–127.
81. M. Klimakow, P. Klobes, A. F. Thunemann, K. Rademann and F. Emmerling, *Chem. Mater.*, 2010, **22**, 5216.
82. T. Friščić and W. Jones, *J. Pharm. Pharmacol.*, 2010, **62**, 1547–1669.
83. K. D. Mjos and C. Orvig, *Chem. Rev.*, 2014, **114**, 4540–4563.
84. D. Braga, F. Grepioni, L. Maini, R. Brescello and L. Cotarca, *CrystEngComm*, 2008, **10**, 469–471.
85. S. Quaresma, V. André, A. M. M. Antunes, L. Cunha-Silva and M. T. Duarte, *Cryst. Growth Des.*, 2013, **13**, 5007–5017.
86. X. Chen, J. G Stowell, K. R. Morris and S. R. Byrn, *J. Pharm. Biomed. Anal.*, 2010, **51**, 866–874.
87. D. Braga, F. Grepioni, G. I. Lampronti, L. Maini and A. Turrina, *Cryst. Growth Des.*, 2011, **11**, 5621–5627.
88. D. Braga, F. Grepioni, L. Maini, S. Prosperi, R. Gobetto and M. R. Chierotti, *Chem. Commun.*, 2010, **46**, 7715–7717.

89. D. Braga, F. Grepioni, L. Maini, D. Capucci, S. Nanna, J. Wouters, L. Aerts and L. Quéré, *Chem. Commun.*, 2012, **48**, 8219–8221.
90. J. Wouters, F. Grepioni, D. Braga, R. M. Kaminski, S. Rome, L. Aerts and L. Quéré, *CrystEngComm*, 2013, **15**, 8898–8902.
91. A. Le Bail and L. M. D. Cranswick, SDPD Round Robin 2002 results, *Commission Powder Diffraction IUCr Newsletter*, 2003, **29**, 31–34.
92. P. C. Andrews, G. B. Deacon, C. M. Forsyth, P. C. Junk, I. Kumar and M. Maguitre, *Angew. Chem., Int. Ed.*, 2006, **45**, 5638–5642.
93. D. Braga, F. Grepioni, L. Maini, P. P. Mazzeo and B. Ventura, *New J. Chem.*, 2011, **35**, 339–344.
94. L. Maini, P. P. Mazzeo, F. Farinella, V. Fattori, and D. Braga, *Faraday Discuss.*, 2014, DOI: 10.1039/C3FD00164D.
95. W. Yuan, T. Friščić, D. Apperley and S. L. James, *Angew. Chem., Int. Ed.*, 2010, **49**, 3916–3919.
96. T. Friščić, Towards mechanochemical synthesis of metal–organic frameworks (MOFs): from coordination polymers and lattice inclusion compounds to porous materials, in *Metal–Organic Frameworks: Design and Application*, ed. L. R. MacGillivray, John Wiley & Sons, Inc., Hoboken, New Jersey, 2010.
97. W. Yuan, J. O'Connor and S. L. James, *CrystEngComm*, 2010, **12**, 3515–3517.
98. C. J. Adams, M. F. Haddow, M. Lusi and A. G. Orpen, *Proc. Natl. Acad. Sci. USA*, 2010, **107**, 16033–16038.
99. G. A. Bowmaker, J. V. Effendy, P. C. Hanna, S. P. Healy, C. King, B. W. Pettinari, Skelton and A. H. White, *Dalton Trans.*, 2011, 7210–7218.
100. X. Ma, W. Yuan, S. E. J. Bell and S. L. James, *Chem. Commun.*, 2014, **50**, 1585–1587.
101. I. Halasz, T. Friščić, S. A. J. Kimber, K. Užarević, A. Puškarić, C. Mottillo, P. Julien, V. Štrukil, V. Honkimäki and R. E. Dinnebier, *Faraday Discuss.*, 2014, DOI: 10.1039/C4FD00013G.
102. L. Tröbs and F. Emmerling, *Faraday Discuss.*, 2014, DOI: 10.1039/C3FD00163F.
103. T. N. Drebuschak, A. A. Ogienko and E. V. Boldyreva, *CrystEngComm*, 2011, **13**, 4405–4410.
104. F. K. Urakaev and V. V. Boldyrev, *Powder Technol.*, 2000, **107**, 197–206.
105. S. Tanaka, K. Kida, T. Nagaoka, T. Ota and Y. Miyake, *Chem. Commun.*, 2013, 7884–7886.
106. M. Lusi, J. L. Atwood, L. R. MacGillivray and L. J. Barbour, *CrystEngComm*, 2011, **13**, 4311–4313.
107. S. Cao, T. D. Bennett, D. A. Keen, A. L. Goodwin and A. K. Cheetham, *Chem. Commun.*, 2012, **48**, 7805–7807.
108. T. D. Bennett, P. J. Saines, D. A. Keen, J.-C. Tan and A. K. Cheetham, *Chem. Eur. J.*, 2013, **19**, 7049–7055.
109. J. D. Egbert, A. M. Z. Slawin and S. P. Nolan, *Organometallics*, 2013, **32**, 2271–2274.

CHAPTER 8

A Chromatography-free Method for the Isolation of Products from Mechanochemical Reactions

JAMES MACK

University of Cincinnati, Department of Chemistry, 301 Clifton Court, Cincinnati OH, 45221-0172, USA
Email: james.mack@uc.edu

8.1 Introduction

From the beginning of time, chemistry has played a major role in the advancement of our society.[1] In the early periods of man, materials such as porcelain, copper and bronze were important chemical discoveries; later materials such as gun powder, soap and rubber further advanced our society. Although these and other chemical discoveries have had a positive impact, as time has passed questions arose as to the adverse effects of chemicals in the environment. In 1962, Rachel Carson's *Silent Spring* was one of the first books that directly challenged the chemical industry and the effect that chemical products has on the environment.[2] She claimed that dichlorodiphenyltrichloroethane (DDT) and other pesticides had a negative effect on the ecosystem. After many studies, it was shown that indeed DDT had an adverse effect on the ecosystem, which led to a ban of the use of DDT as a pesticide in 1972. Other incidents such as the Love canal tragedy and

RSC Green Chemistry No. 31
Ball Milling Towards Green Synthesis: Applications, Projects, Challenges
Edited by Brindaban Ranu and Achim Stolle

Published by the Royal Society of Chemistry, www.rsc.org

Figure 8.1 Victims of the Love canal tragedy protest (left). "Valley of the drums" waste site (right).

poor chemical waste management led to greater public awareness and more stringent government regulations, leading to the creation of governmental agencies such as the Environmental Protection Agency (EPA) (Figure 8.1).

Throughout the 1980s, society demanded to know more about what chemicals were being released into the environment. This outcry led to the passing of laws such as the Comprehensive Environmental Response, Compensation, and Liability Act (CERCLA)[3] and the Emergency Planning and Community Right-to-Know Act (EPCRA).[4] In 1990, the Pollution Prevention Act was the first law in the United States that sought to prevent waste rather than clean up the waste.[5] This new way of dealing with waste led chemists to consider how chemistry can be "greenified". Shortly after the Pollution Prevention Act, Anastas and Farris published *Benign by Design, Alternative Synthetic Design for Pollution Prevention*, which started to emphasize the principles of green chemistry.[6] A few years later Anastas and Warner published *Green Chemistry: Theory and Practice*, which became the foundation for the green chemistry movement.[7] Since then, green chemistry has been integrated into the fabric of the field of chemistry. Organizations such as the Green Chemistry Institute (GCI), the Interuniversity Consortium Chemistry for the Environment (INCA) and many others around the world lead the practices and support for the field. Journals such as *Green Chemistry*, *ChemSusChem* and *Sustainable Chemistry and Engineering* focus on the latest research in sustainable chemistry, which has increased significantly over the last decade.

Solvent-free chemical reactions seemed to be the perfect fit to reduce the total amount of waste generated. Solvents are typically the main component of chemical waste, especially in the pharmaceutical industry.[8] Some 80–90% of reaction mass and approximately 60% of energy use in the production of an active pharmaceutical ingredient (API) are attributed to solvents.[9] These solvents are used in reactions for API synthesis, providing a medium for reactions to take place, as well as separation and washing steps used to purify the API produced during the reaction. It is common practice in the pharmaceutical industry to incinerate solvent waste that is disposed.[9,10] One method that has demonstrated versatility in the field of solvent-free synthesis has been mechanochemistry. Mechanochemistry has been known for more than 100 years; however, given the attention to the reduction of chemical waste, there has been renewed focus on these novel conditions.[11–15] Mechanochemical reactions are typically performed in the absence of solvent and can give rise to higher yields and may lead to different products than what is observed in solution.

Although mechanochemistry has been utilized to limit solvent use during the reaction, often harsh solvents are needed to isolate the desired products. One popular method to isolate desired compounds after a mechanochemical reaction is column chromatography. Column chromatography is used to purify products, which can require a significant amount of solvent, more than would be used to conduct the reaction initially. As an example, Mack and co-workers demonstrated a significant rate enhancement of the Baylis–Hillman reaction under solvent-free mechanochemical conditions.[16] Although the reaction was conducted using solvent-free mechanochemical conditions, the product was isolated using column chromatography. Solvent needs for chromatography are dictated by polarity instead of toxicity. Solvents such as methylene chloride or hexane may be needed as a chromatographic solvent that cannot be easily replaced by a greener alternative. Recently, there have been reports of greener solvents used to conduct chromatography or the use of reversed-phase chromatography which utilizes greener solvents such as water or methanol as eluent.[17] Although greener methods for conducting chromatography are on the horizon, it would be best to remove the aspect of chromatography altogether. While other chromatography-free methods to isolate products from mechanochemical reactions, such as sublimation,[18] have been employed, this chapter will focus on the use of polymer supported resins for chromatography-free separations.

8.2 Polymer Supported Resins: Advantages and Disadvantages

Polymer supported resins were first reported by Bruce Merrifield in the 1963.[19] Merrifield resins made the ability to synthesize and isolate peptides significantly easier by conducting the reaction on a solid support. A Merrifield resin is made of polystyrene crosslinked with 1–2% divinyl benzene (Scheme 8.1).

By using the Merrifield resin, an amino acid unit can be anchored to the resin and various amino acids can be added to extend the peptide chain. Given the insolubility of the polymer resin, any unreacted amino acids can simply be filtered away, while the amino acids that were successfully added

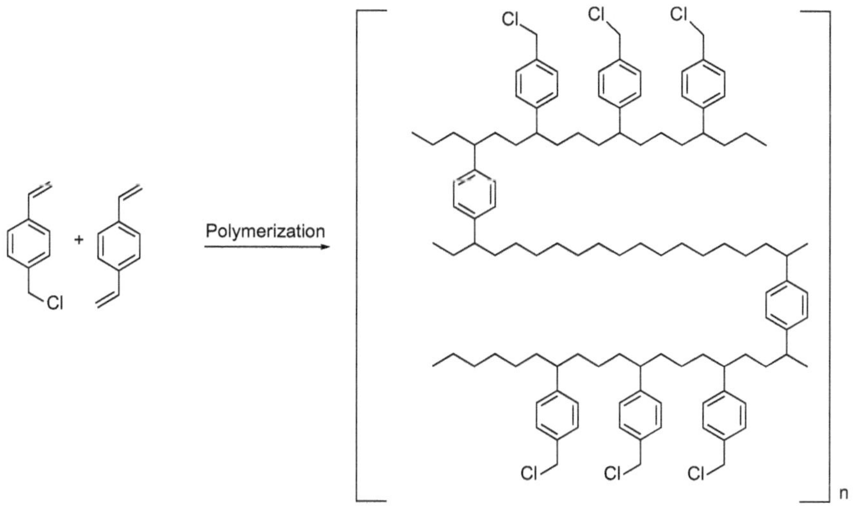

Scheme 8.1 Synthesis of polystyrene polymer resin cross-linked with 1,4-divinyl benzene.

to the polymer support will remain. Therefore, instead of conducting chromatography to isolate the peptide, you can simply wash away all of the undesired by-products, leaving behind the desired peptide attached to the polymer support. Subsequently, the desired peptide can be isolated by cleaving it from the polymer support (Scheme 8.2).

Use of this model of synthesis has been expanded to solid phase organic synthesis (SPOS), allowing for a wide variety of synthetic methods.[20–22] In principle this is a very clean and simple method for the isolation of compounds. Unfortunately, to use polymer supported resins, the resin needs to be activated in a solvent that allows the polymer to swell. Similar to the problems with column chromatography, the solvent choice to swell the polymer resin cannot be easily replaced by greener alternatives.

One of the major drawbacks of polymer supported resins is that only roughly 1% of the functionality resides on the outside of the polymer surface.[23] For the rest of the polymer supported reagents to be accessible to the reaction they must be placed in an organic solvent and swollen. The action of swelling allows the substrates to fit through the pores of the crosslinked polymer and interact with the internal reagents. Unfortunately, the act of swelling polymer supported reagents requires hazardous organic solvents such as benzene, toluene, or dichloromethane to name a few. Further, depending on the nature of the polymer resin, as well as the polarity of the reagents, the ability to swell the polymer can be unpredictable. As an example Hoff and co-workers used an analogue of the Corey–Bakshi–Shibata reagent attached to a poly-methacrylate polymer bead.[24] They discovered that THF and dichloromethane were suitable solvents to swell the polymer bead in order to conduct the reaction. However, similar ethereal solvents

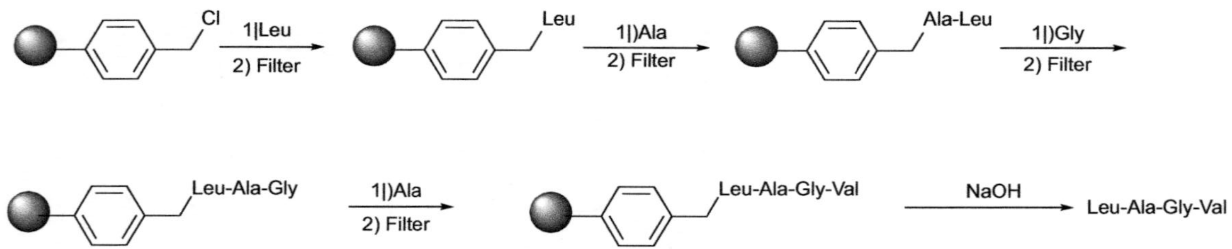

Scheme 8.2 Synthesis of tetrapeptide using a polystyrene cross-linked resin.
Reprinted (adapted) with permission from ref. 19. Copyright 1963 American Chemical Society.

such as 2-methyltetrahydrofuran or diethyl ether gave a low degree of swelling and were not suitable to conduct the reaction. There have been many theories developed to try to explain which factors strongly influence polymer swelling. Although many of these factors have been studied for decades, these subtle differences and interactions are still very much a mystery.[25–27]

Another drawback to polymer supported reagents is the long reaction times needed due to the rate of diffusion of the substrate inside of the polymer resin. Typically, reactions with polymer supported reagents can take anywhere from hours to days due to the slow rate of diffusion of the reagents inside the polymer surface. In certain reactions with polymer supported reagents, the rate of diffusion can be as slow as 2% of the non-supported reagent in solution. As an example Hu and co-workers demonstrated the use of a polymer supported chiral sulfonamide for the asymmetric reduction of ketones.[28] To attach (*S*)-diphenylprolinol, the desired organocatalyst in this case, to the polymer support they reacted (*S*)-diphenylprolinol with a polymer supported sulfonyl chloride. Even though they used an excess of the chiral catalyst roughly only 50% of the quantity of chiral reagent was attached to the polymer support after three days (Scheme 8.3). These long reaction times often make the use of the polymer supported reagents difficult and inefficient.

A third drawback of polymer supported resin in solution is the reactivity they have based on the size of the substrate; unlike simple solution phase experiments the rate of polymer supported reactions are dependent on the size of the substrate. Further, the amount of crosslinking also plays a critical role in the rate of the reaction, especially for larger substrates. Grubbs demonstrated through the use of hydrogenation experiments that the size of the substrate has a significant effect on the rate of a reaction as compared to the free reagent in solution.[29] When comparing the rate of hydrogenation of a simple small hydrocarbon like hexene to larger molecules like Δ^2-cholestene, he observed a significant difference between the rate in the free solution and polymer supported reagents (Table 8.1).

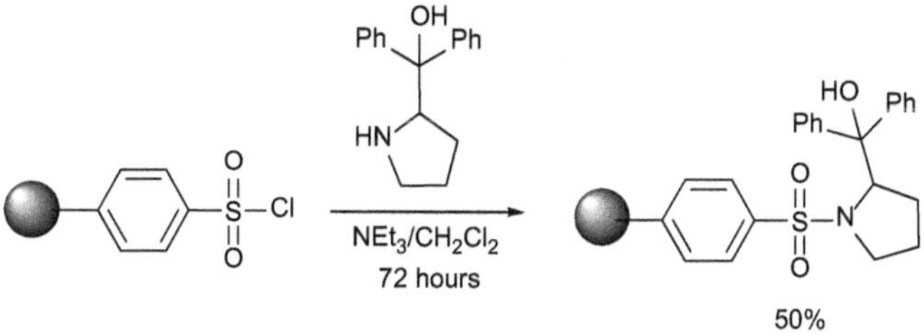

Scheme 8.3 Synthesis of polymer supported (*S*)-diphenylprolinol. Reprinted (adapted) with permission from ref. 28. Copyright 2001 American Chemical Society.

Table 8.1 Comparison of the rate of reduction of polymer supported catalyst to non-supported catalyst. Reprinted (adapted) with permission from ref. 29. Copyright 1971 American Chemical Society.

Olefin	$RhCl(Ph_3P)_3$ on polymer bead. Relative rate	Free $RhCl(Ph_3P)_3$ in solution
Cyclohexene	1	1
1-Hexene	2.55	1.4
Δ^2-Cholestane	1/32	1/1.4
Octadecene	1/2.06	1/1.4
Cyclooctene	1/2.54	1
Cyclododecene	1/4.45	1/1.5

8.3 Mechanochemical Activation of Polymer Resins

Given the difficulty of polymer supported resins in solution (*i.e.*, finding the right swelling solvent, long reaction times, and differences in reactivity depending on substrate) the use of polymer supported reagents in organic synthesis has dwindled, along with the ability to purify and isolate substrates without chromatography. It appears the very weakness of mechanochemical reactions (*i.e.*, the use of solvents to isolate products) is the very strength of polymer supported reactions. Further, the very weakness of polymer supported reagents (*i.e.*, the need for solvent to activate the polymer support) is the very strength of mechanochemical reactions. It is known that mechanical energy can have an effect on the size of polymer beads. Depending of the polymer bead, sometimes simply stirring the bead can have a noticeable impact. Therefore, instead of swelling the bead to allow the interaction of the bead with reagents the thought is to significantly reduce the particle size of the bead, whereby exposure of the surface will take place through mechanical activation. There are several benefits of using mechanochemistry to activate polymer beads over the traditional swelling mechanism. Using mechanochemistry would alleviate the use of harmful organic solvents in the process of swelling the polymer bead. The rate of the reaction would be significantly increased because the action of mechanochemical grinding forces reagents together, which leads to a significant increase in the reaction rate. The forcing of reagents together under mechanical action would also subdue the difference of reactivity of different substrates due to size.

There are many reports in the literature of the Wittig reaction using polymer supported resins with differing results. Simply to generate the phosphonium salt the reaction takes 48 h at 70 °C to complete. Once that has been satisfied there is an additional 4–24 h to conduct the deprotonation and addition steps needed to complete the reaction, depending on the amount of crosslinking used for the polymer support. The reaction does not

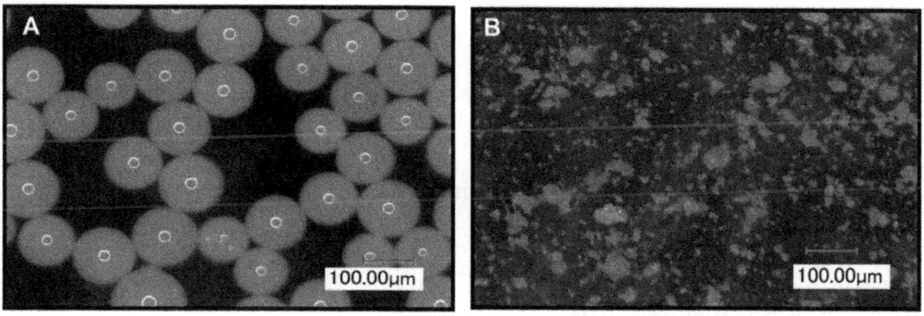

Figure 8.2 Pre-milled polymer resin (A); resin after 15 min of milling (B).[34]

take merely one solvent to swell the bead, but rather a combination of polar and non-polar solvents. A biphasic method using water/THF and phase transfer catalyst has been shown to be effective.[30–33]

In lieu of using a solution to conduct the polymer supported Wittig reaction, the use of polymer supported reagents under mechanochemical conditions was shown to be very effective.[34] Through mechanical grinding, the functional sites on the polymer bead could be accessed through increasing the surface area of the resin; more specifically, pulverizing the resin into smaller particle sizes (Figure 8.2). Before being ball milled, the functional resin particle size consistently measured between 75 and 125 μm.

Following ball milling, the resin measured approximately 1–5 μm. Traditionally, non-polar, organic solvents such as benzene or dichloromethane are needed to swell functional resins, allowing reagents to diffuse through the crosslinked pores of the resin. Although these solvents are a good choice for swelling the resin, they are not environmentally desirable. Polar solvents, such as ethyl acetate and ethanol, are much more desirable from an environmental perspective, but these solvents do not swell the functional resin. When comparing the Wittig reaction using solid support in an ethanol environment only 13% of the product stilbene was isolated over a two-day period. However, where pre-milled resin particles were used in the reaction 56% of the product stilbene was isolated in a two-hour period. When triphenylphosphine-functionalized polystyrene was used in the Wittig reaction under solvent-free mechanochemical conditions, the product stilbene was only recovered in 17% yield after 16 h of milling. This demonstrated, under mechanochemical conditions, that the unpredictable method of swelling is an unnecessary component of the reaction. However, for the polymer supported Wittig reaction, even through the use of mechanical activation, a polar environment is still necessary.

Liquid assisted grinding is a method whereby a small amount of liquid is used to aid in the grinding process. It was shown that by adding a small

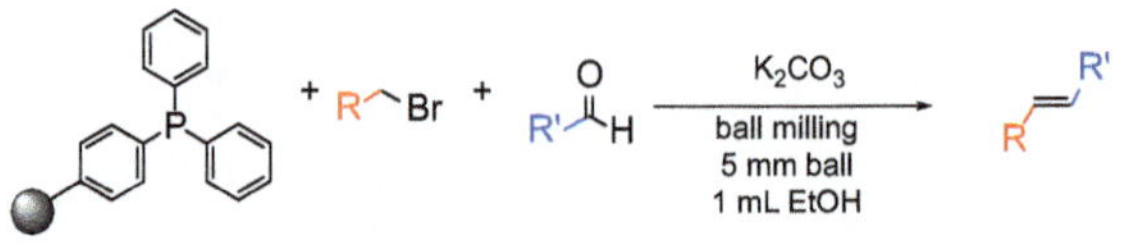

Scheme 8.4 Solvent-free Wittig reaction using polymer supported triphenylphosphine and liquid assisted grinding.

amount of ethanol to the milling conditions the reaction went to completion in less than 2 h (Scheme 8.4).

It was hypothesized that instead of swelling the resin particles the particles undergo conformational changes that depend on the polarity of the surrounding environment. The first step in the Wittig reaction is the attack of the triphenylphosphine on the halide, resulting in the formation of the phosphonium salt. The phosphonium salt needs to be accessible to the base and aldehyde for the subsequent Wittig reaction to take place. In a polar environment, the polar triphenylphosphonium bromide functions on the resin particles are exposed, whereby the non-polar polystyrene backbone is shielded (Figure 8.3). This presentation allows for greater interaction between the resin particles, base, and aldehyde, resulting in greater reactivity. Conversely, in a non-polar environment, the polystyrene backbone is exposed, shielding the polar triphenylphosphonium bromide functions. In this conformation, the base and aldehyde have decreased interaction with the triphenylphosphonium bromide, resulting in longer reaction times and reduced yields. Along with alleviating many of the problems with conducting the polymer supported reaction the desired products were still able to be isolated without the use of chromatography.

Another report demonstrating the power of ball milling and polymer supported reagents was given by Chen and co-workers.[35] The reaction of phenylacetylene with tetrakis(triphenylphosphine)palladium gave 1,4-diphenylbuta-1,3-diyne in high yields. When Bis(triphenylphosphine)palladium(II) dichloride was used as the catalyst, they observed 1,4-diphenylbut-1-en-3-yne as the primary product instead of the 1,4-diphenylbuta-1,3-diyne. These two catalysts showed similar reactivity under ball milling as in solution but higher yields and higher diastereoselectivity were achieved under ball milling. In an effort to develop chromatography-free conditions the authors used polymer supported palladium catalyst. Surprisingly, the polymer supported tetrakis(triphenylphosphine)palladium gave 1,4-diphenylbut-1-en-3-yne product as the major product. Furthermore, only the *trans* stereochemistry was observed where the *cis* product as well as the head to tail product is observed when the free catalyst was used in solution. To determine whether the polymer supported catalysts can give 1,4-diphenylbuta-1,3-diyne free triphenylphosphine was added to the polymer supported

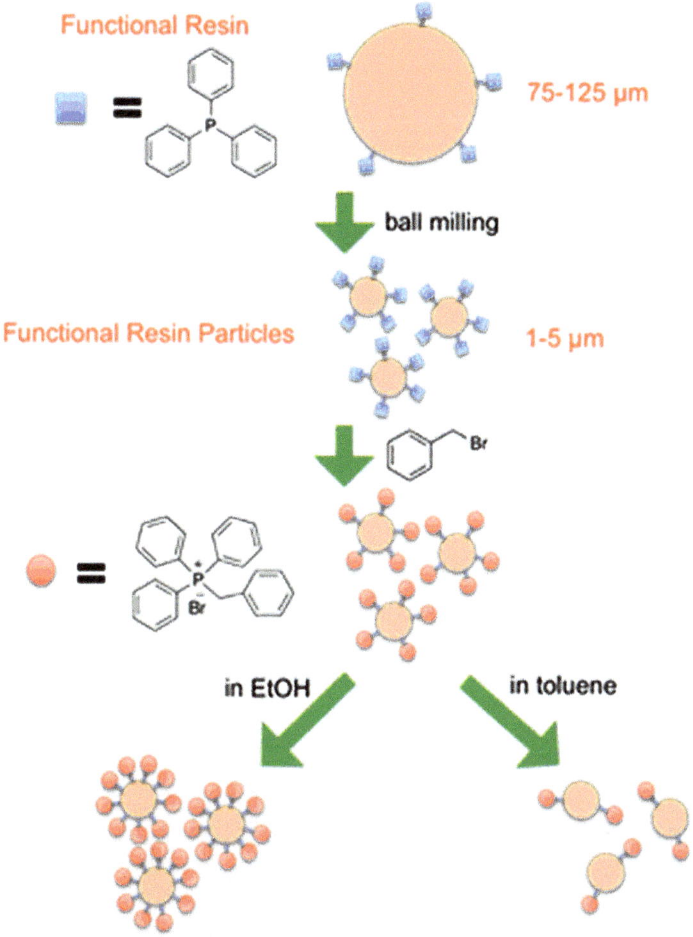

Figure 8.3 Accessibility of the polar functional group is highly dependent on environment polarity.

tetrakis(triphenylphosphine)palladium. After ball milling the polymer supported tetrakis(triphenylphosphine)palladium in the presence of excess triphenylphosphine, 1,4-diphenylbuta-1,3-diyne was formed as the major product (Scheme 8.5). This demonstrates the ability of the polymer supported catalyst under ball milling conditions to give rise to two different products where two completely different catalysts are needed in solution. Polymer supported catalysts have been known to be recyclable since you can easily recover the solid support by filtration. The authors found under mechanochemical conditions the polymer supported catalyst can be recycled but observed diminished yields after each run, citing that the palladium leaches out of the polymer over time.

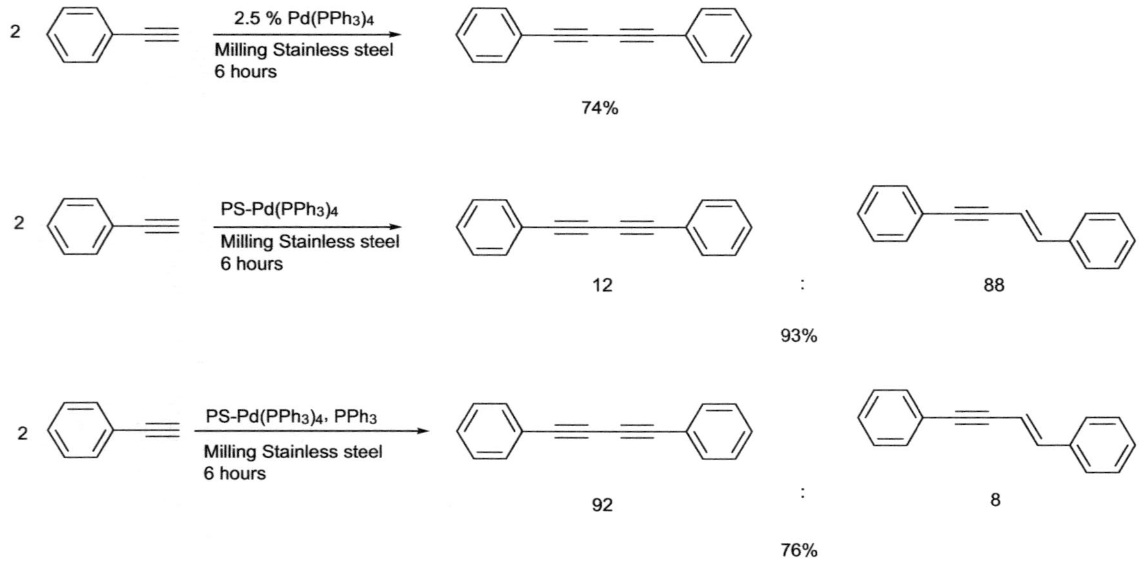

Scheme 8.5 Polymer supported Glaser coupling using mechanochemical conditions.

8.4 Conclusion

The development of environmentally benign methodology is an important aspect of the future of chemistry. When used together, ball milling and functional resins effectively carry out organic reactions, and provide an environmentally benign method for conducting chemical reactions and isolating products. In essence, ball milling significantly reduces organic solvent use during the reaction while the use of functional resins allows for the isolation of products in a chromatography-free manner. From the work presented in this chapter it was also demonstrated that polymer supported reagents under mechanochemical conditions can give products and selectivity not observed in solution. It is expected that further studies with ball milling and polymer supported reagents will lead to a significant advancement in the both the areas of mechanochemistry and polymer chemistry.

References

1. W. A. Schmidt, *J. Chem. Ed.*, 1928, 5, 1224.
2. R. Carson, *Silent Spring*, Houghton Mifflin, Riverside Press, Boston, Cambridge, Mass., 1962.
3. Comprehensive Environmental Response, Compensation, and Liability Act (CERCLA), U.S. Environmental Protection Agency, 2014. http://www.epa.gov/oecaagct/lcla.html. Accessed 7 July 2014.
4. Emergency Planning and Community Right-To-Know Act (EPCRA), U.S. Environmental Protection Agency. http://www.epa.gov/oecaagct/lcra.html. Accessed 8 July 2014.
5. Pollution Prevention Act of 1990, U.S. Environmental Protection Agency. http://www.epa.gov/p2/pubs/p2policy/act1990.htm. Accessed 7 July 2014.
6. P. T. Anastas and C. A. Farris (ed.) *Benign by Design: Alternative Synthetic Design for Pollution Prevention*, American Chemical Society, Washington, DC, 1994.
7. P. T. Anastas and J. C. Warner, *Green Chemistry: Theory and Practice*, Oxford University Press, New York, 1998.
8. P. G. Jessop, *Green Chem.*, 2011, **13**, 1391–1398.
9. C. S. Slater, M. J. Savelski, W. A. Carole and D. J. C. Constable, in *Green Chemistry in the Pharmaceutical Industry*, ed. P. J. Dunn, A. S. Wells and M. T. Williams, Wiley-VCH Verlag GmbH, Weinheim, 2010, pp. 49–82.
10. M. J. Raymond, C. S. Slater and M. J. Savelski, *Green Chem*, 2010, **12**, 1826–1834.
11. S. James, C. Adams, C. Bolm, D. Braga, P. Collier, T. Friscic, F. Grepioni, K. Harris, G. Hyett, W. Jones, A. Krebs, J. Mack, L. Maini, A. Orpen, I. Parkin, W. Shearouse, J. Steed and D. Waddell, *Chem. Soc. Rev.*, 2012, **41**, 413–447.
12. L. Takacs, *Acta Phys. Pol. A*, 2012, **121**, 711–714.
13. E. Boldyreva, *Chem. Soc. Rev.*, 2013, **42**, 7719–7738.
14. L. Takacs, *Chem. Soc. Rev.*, 2013, **42**, 7649–7659.

15. J. Kemsley, *Chem. Eng. News*, 2013, **91**, 35–36.
16. J. Mack and M. Shumba, *Green Chem.*, 2007, **9**, 328–330.
17. J. P. Taygerly, L. M. Miller, A. Yee and E. A. Peterson, *Green Chem.*, 2012, **14**, 3020–3025.
18. I. Huskic, I. Halasz, T. Friscic and H. Vancik, *Green Chem.*, 2012, **14**, 1597–1600.
19. R. B. Merrifield, *J. Am. Chem. Soc.*, 1963, **85**, 2149–2154.
20. A. Kirschning, W. Solodenko and K. Mennecke, *Chem. – Eur. J.*, 2006, **12**, 5972–5990.
21. M. Mentel and R. Breinbauer, *Eur. J. Org. Chem.*, 2007, 4283–4292.
22. S. A. Testero and E. G. Mata, *J. Comb. Chem.*, 2008, **10**, 487–497.
23. P. Hodge, *Chem. Soc. Rev.*, 1997, **26**, 417–424.
24. T. Thvedt, T. Kristensen, E. Sundby, T. Hansen and B. Hoff, *Tetrahedron: Asymmetry*, 2011, **22**, 2172–2178.
25. J. Heyda, A. Muzdalo and J. Dzubiella, *Macromolecules*, 2013, **46**, 1231–1238.
26. C. K. Lee, C. E. Diesendruck, E. Lu, A. N. Pickett, P. A. May, J. S. Moore and P. V. Braun, *Macromolecules*, 2014, **47**, 2690–2694.
27. P. J. Flory and J. Rehner, *J. Chem. Phys.*, 1943, **11**, 521–526.
28. J.-b. Hu, G. Zhao, G.-s. Yang and Z.-d. Ding, *J. Org. Chem.*, 2001, **66**, 303–304.
29. R. H. Grubbs and L. C. Kroll, *J. Am. Chem. Soc.*, 1971, **93**, 3062–3063.
30. S. V. McKinley and J. W. Rakshys, *J. Chem. Soc., Chem. Commun.*, 1972, 134–135.
31. M. Bernard and W. T. Ford, *J. Org. Chem.*, 1983, **48**, 326–332.
32. P. Leung, Y. Teng and P. Toy, *Org. Lett.*, 2010, **12**, 4996–4999.
33. P. Leung, Y. Teng and P. Toy, *Synlett*, 2010, 1997–2001.
34. W. Shearouse and J. Mack, *Green Chem*, 2012, **14**, 2771–2775.
35. L. Chen, B. E. Lemma, J. S. Rich and J. Mack, *Green Chem.*, 2014, **16**, 1101–1103.

CHAPTER 9

Ball-milling Promoted Chemical Bonding between Cellulose and Plastics

TAKAHIRO HIROTSU,*[a] TAKASHI ENDO[b] AND MYCO UMEMURA[c]

[a] Shikoku Collaboration Center, National Institute of Advanced Industrial Science and Technology, 2217-14 Hayashi-cho, Takamatsu 761-0395, Japan; [b] Biomass Refinery Research Center, National Institute of Advanced Industrial Science and Technology, 3-11-32 Kagamiyama, Higashi-hiroshima 739-0046, Japan; [c] Bioproduction Research Institute, National Institute of Advanced Industrial Science and Technology, 2-17-2-1 Tsukisamu-higashi, Toyohira Ward, Sapporo 062-8517, Japan
*Email: takahiro-hirotsu@aist.go.jp

9.1 Introduction

Cellulose has attracted growing interest as the most promising material to take the place of synthetic polymers to a considerable extent in the twenty-first century, because of its massive reproducibility and environmental benefits, as compared with the later polymers derived from fossil fuel resources.[1] While cellulose is photosynthesized in plants, algae, *etc.*, as much as two or three hundreds billion tons every year on the earth, it is a raw material that causes difficulties in chemical or engineering processes, owing to its insolubility in the usual solvents as well as having no thermoplasticity.

RSC Green Chemistry No. 31
Ball Milling Towards Green Synthesis: Applications, Projects, Challenges
Edited by Brindaban Ranu and Achim Stolle

Published by the Royal Society of Chemistry, www.rsc.org

Accordingly its applications have been limited to only a few areas such as paper and rayon.

Native cellulose is almost non-reactive with electrophiles or nucleophiles, because almost all the hydroxyl groups are involved in extensive networks of hydrogen bonds. Therefore, activation of hydroxyl groups in cellulose is considered to be a way to realize its extensive application. We have explored pulverization of fibers of cotton linter with high crystallinity as a native cellulose by ball milling, and found that the resultant amorphous powders exhibit unique characteristics with respect to specific interactions with water molecules added to them. Activated cellulose powders are so reactive that they can form ester bonds with maleic anhydride-grafted polyolefin (MPO)[2] in the solid state. This suggests strongly that novel cellulosic composites with thermoplasticity are possible by formation of an interphase with MPO chains bonded onto cellulose particles through ester bonds.

In this chapter, first, the molecular dynamics simulation of cello-oligosaccharides is described to understand their conformations as well as aggregation properties on the basis of intra- and intermolecular hydrogen bonding. Then, micro-pulverization of cotton linter by ball milling is characterized with respect to particle distribution, crystallinity, and degree of polymerization, and unique properties of the resultant powders are demonstrated on the basis of their specific interactions with water molecules. Ball milling is also effective for grafting of maleic anhydrides onto polyolefin like i-polypropylene. Finally, a description is given of the way in which ball milling of cellulose and polyolefin gives rise to novel composites characterized by an interphase between fine particles of cellulose and the polyolefin phase.

9.2 Intra- and Intermolecular Interactions of Cellulose

9.2.1 Conformation of Cello-oligosaccharides

Cellulose is a naturally occurring polymer consisting of D-glucopyranoses that are bonded with each other by $\beta(1 \rightarrow 4)$ glycosidic linkages. We can understand the major characteristics of cellulose by comparison with amylose, which is another polymer composed of D-glucopyranoses, but through $\alpha(1 \rightarrow 4)$ glycosidic linkage (Figure 9.1). Amylose is soluble in water, in contrast to cellulose. Umemura *et al.* carried out molecular dynamics simulation of malto- and cello-oligosaccharides from monomer to hexamer in two thousand TIP3P water molecules, and estimated the conformations of the isolated chains in water (Figure 9.2).[3] The conformations are designated with a pair of dihedral angles, Φ (H1–C1–O4′–C4′) and Ψ (C1–O4′–C4′–H4′). Estimated helical conformations with $(\Phi, \Psi) = (-30^\circ, -20^\circ)$ of malto-oligosaccharides are entirely consistent with those with $(\Phi, \Psi) = (-30^\circ, -20^\circ)$ determined from NMR spectroscopy;[4] for cello-oligosaccharides, their loose helical conformations with $(\Phi, \Psi) = (+40^\circ, +5^\circ)$ do not agree with those with

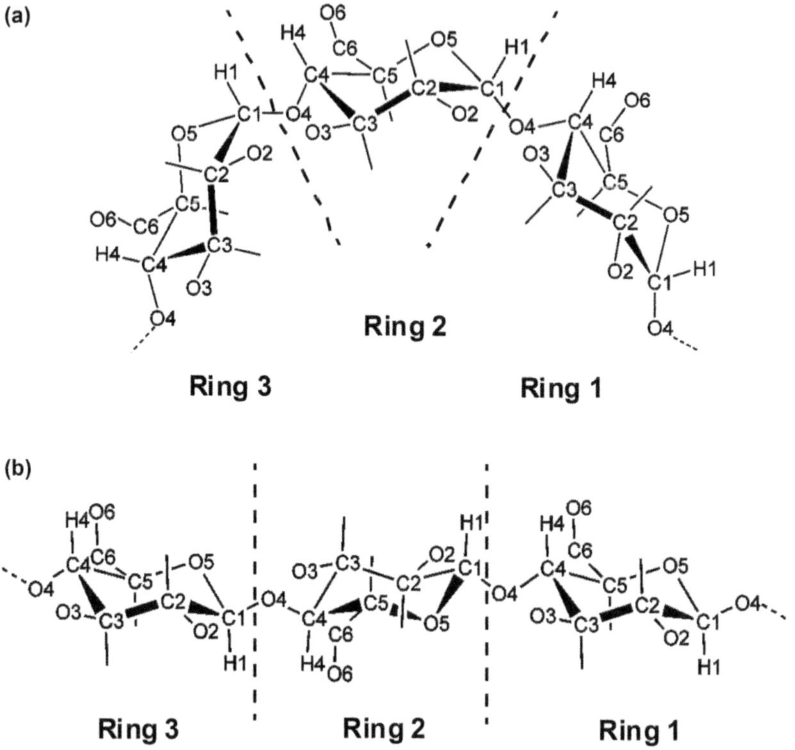

Figure 9.1 Molecular structures of malto- (a) and cello-oligosaccharides (b). Reprinted from ref. 3 with permission of Elsevier.

$(\Phi, \Psi) = (+45°, -38°)$[5] or $(+32°, -21°)$[6] determined experimentally, but still agree well with other calculation results with $(\Phi, \Psi) = (+44°, +2.9°)$.[6] For cello-oligosaccharides the relatively flat ribbon conformations obtained from experiments may reflect some interactions between the chains because such conformations are somewhat similar to that of crystalline cellulose.

Conformational flexibilities of the oligosaccharides are evaluated on the basis of the root-mean-square displacement (RMSD) per atom in each solute. Notably, for cello-oligosaccharides their standard deviations of the RMSDs are steady over the degree of polymerization (DP) of 2–6, while that for malto-oligosaccharides increases with DP, in particular when greater than 3. This indicates that malto-oligosaccharides fluctuate more with the DP, while cello-oligosaccharides keep rigidity. Distribution maps of the dihedral angle pair (Φ, Ψ) also reveal a difference in flexibility at the glycosidic linkages between both types of oligosaccharides. The angle pair distributes more widely in malto- than in cello-oligosaccharides, with an elliptical and a round shape, respectively, which is consistent with the results from the RMSDs. Every pair in malto- or cello-oligosaccharides has a maximal occupancy at a single point that does not differ among DPs, except for malto-hexamer (DP = 6) with plural maximal occupancies at another point in addition to it.

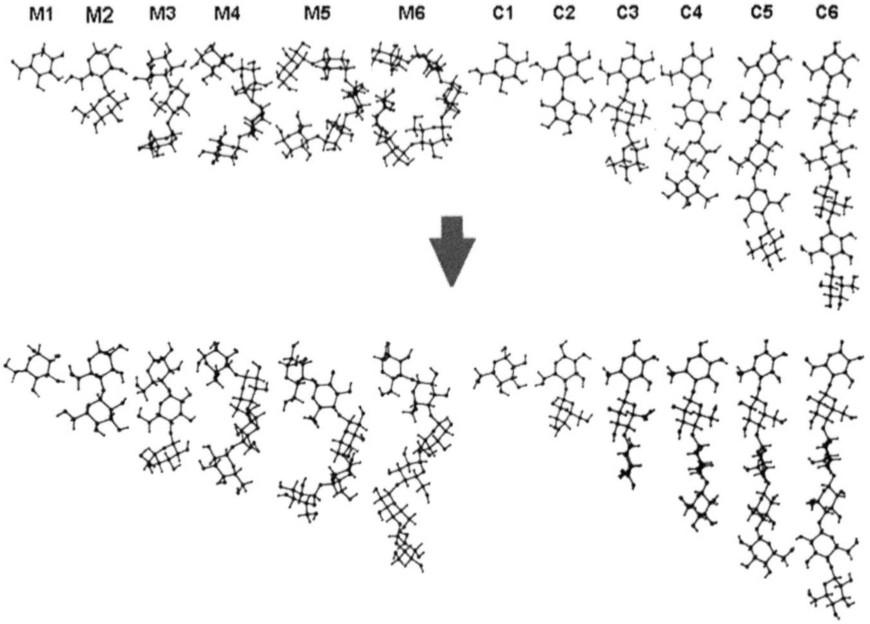

Figure 9.2 Average structures of each solute over the initial (above) and the last (below) 100 ps in the 1 ns calculation time. From left to right, malto (from M1 to M6) and cello-oligosaccharides (from C1 to C6) from monomer to hexamer, respectively.
Reprinted from ref. 3 with permission of Elsevier.

Intramolecular hydrogen bonding exerts a primary effect on stabilization of the conformations of malto- and cello-oligosaccharides. Two definitions of hydrogen bonds, (1) a distance between two oxygens of $\leq$0.35 nm and (2) an angle O–H$\cdots$O of 120–180°, make it possible to calculate the total number of intramolecular inter-residue hydrogen bonds for the conformations of malto- and cello-oligosaccharides (DP = 2–6). Inter-residue hydrogen bonds of interest occur more frequently with less fluctuation in cello- than in malto-oligosaccharides, in accordance with the results of conformational flexibility. Dominant hydrogen bonding between inter-residue O5–O3′, with an occupancy rate of 65%, causes the rigid conformation of cello-oligosaccharides at the glycosidic linkage. In contrast, malto-oligosaccharides have two modes of inter-residue hydrogen bonding of O2–O3′ and O6–O6′, with occupancy rates of 35% and 8.3%, respectively, in accordance with a helical conformation with a higher fluctuation on the side of O6 than O2 or O3.

The rigid conformation of cello-oligosaccharides is further assisted through the 1.3–1.5-times greater number of double hydrogen-bonded water molecules as compared to malto-oligosaccharides. The double hydrogen-bonded water molecules are located more around glycosidic linkages in cello- than in malto-oligosaccharides. Therefore, conformations of cello- and malto-oligosaccharides are determined primarily through different modes of intramolecular inter-residue hydrogen bonding, supported by relevant hydrations around glycosidic linkages with double hydrogen-bonded water molecules. Thus, cello-oligosaccharides have a relatively flat conformation with more rigidity, in contrast to malto-oligosaccharides, which have a more helical conformation with more flexibility. Considering a higher solubility of amylopectin primarily due to the flexibility of conformation at $\alpha(1\rightarrow6)$ linkage points,[7] cello-oligosaccharides with the rigid confirmation may exhibit self-aggregation.

9.2.2 Aggregations of Cello-oligosaccharides

Cello-oligosaccharides are less soluble in water than malto-oligosaccharides, and almost insoluble with DP ≥ 6, indicating a ready aggregation of cello-oligosaccharides through intermolecular hydrogen bonding. Umemura *et al.*[8] examined intermolecular interactions between cellotetraoses (DP = 4), cellopentaoses (DP = 5), or cellohexaoses (DP = 6) on the basis of molecular dynamics simulation of each double strand with 5000 TIP3P water molecules for 1 ns. Each double strand consists of a pair of cello-oligosaccharide chains initially in parallel, on the basis of the crystal structure of cellulose I. The double strands are denoted as **c4d, c5d,** and **c6d** for cellotetraose, cello-pentaose, and cellohexaose, respectively. The chain is numbered from the reducing glucose residue to the nonreducing glucose residue, like Ring 1 for the residue at the reducing end to Ring 6 for the residue at the nonreducing end in cellohexaose. Figure 9.3 shows the average structures of the three double strands in the initial and final 100 ps, compared with those of the three single strands, **c4s, c5s,** and **c6s,** in the last 100 ps, respectively, which are identical to relevant conformations in Figure 9.2. Close examination reveals that the conformations of chains in the double strands are closer to that in cellulose I than are those in the single strands. Notably, the double strands are separated for cellotetraose, in a half-aggregation state for cello-pentaose, and in a complete aggregation state for cellohexaose. From the time evolution of the distances between a pair of corresponding rings in the double strands, the aggregation states of **c5d** as well as **c6d** are maintained over 1 ns (apart from the initial 100 ps), while the chains of **c4d** separate from each other finally by more than 2.0 nm. In **c5d,** the distances are relatively unstable, particularly at Ring 1, with fluctuations of 0.48–1.0 nm. Of these three double strands, **c6d** exhibits the most stable state of aggregation with the smallest distances around 0.48 nm (except Ring 6). A specific separation from Ring 1 in **c5d** remains unresolved, and requires further considerations on the simulation. Nevertheless, the simulation results show clearly that the aggregation of cello-oligosaccharides in water depends on the degree of polymerization, with a critical point around DP = 5, which is very consistent with experiments.

The diffusivity of an isolated solute in water depends on interactions of the solute with water molecules around it. The self-diffusion coefficient *D*

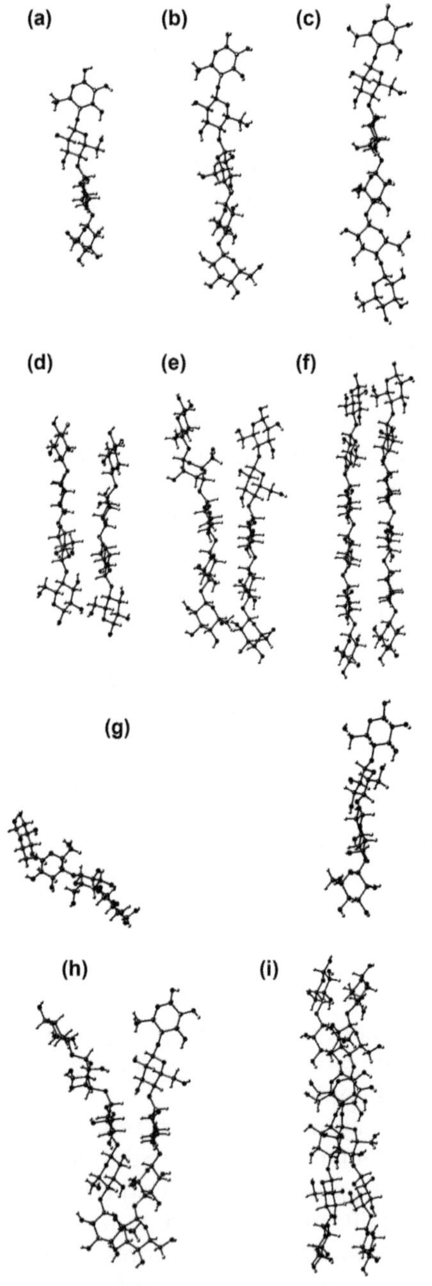

Figure 9.3 Average structures of solutes through the initial or last 100 ps in 1 ns calculated time: **c4s** (a), **c5s** (b), and **c6s** in the last 100 ps (c), and **c4d** (d and g), **c5d** (e and h), and **c6d** in the initial and last 100 ps (f and i), respectively.
Reprinted from ref. 8 with permission of American Chemical Society.

can be calculated from the Einstein relation; the calculated values are 6.87×10^{-10} (**c4s**), 6.35×10^{-10} (**c5s**), and 5.26×10^{-10} m^2 s^{-1} (**c6s**) for the single strands, and 6.29×10^{-10} (**c4d**), 5.31×10^{-10} (**c5d**), and 3.68×10^{-10} m^2 s^{-1} (**c6d**) for the double strands,[8] decreasing with the DP. Assuming that a cello-oligosaccharide is a flat rectangular parallelepiped $\{4\times8\times(6\times\mathrm{DP}+0.3)\ \mathrm{nm}^3\}$ and its double strand is a stack of two parallelepipeds $\{8\times8\times(6\times\mathrm{DP}+0.3)\ \mathrm{nm}^3\}$, the DP value is approximately inversely proportional to the surface area in the single and double strands. This proves that the self-diffusion of each strand is determined primarily by interactions of hydroxyl groups of the strand with water molecules. It is expected that such interactions exert an influence on the fluctuation of conformation of the single or double strand.

Figure 9.4 describes the root-mean-square displacement (RMSD) per atom in each system, to examine the positional fluctuation of an atom from its average position during an interval. Except for the **c4d**, all the systems stabilize in 100 ps. The atomic positions of the double strands are stabilized, with fewer fluctuations, by *ca.* 0.2 nm than those of the single strands, probably due to intermolecular interactions in the former strands. In addition, fluctuation within a 30–40 ps interval is observed in each system with fine fluctuations, except for another unique *ca.* 100 ps interval of fluctuation in **c5d** owing to a closing–separating movement at the reducing end. Fluctuations at intervals of 5 and 40 ps both increase with DP in the single strand, as expected from increasing interactions of the chain with water molecules with DP. It is, however, noted that the fluctuations of the double strand decrease with DP, and are in good agreement with those of the single strand when DP = 5. Thus, the fluctuations of the single and the double strands go into reverse around DP = 5. This suggests that the fluctuation in the double strands is suppressed by stronger intermolecular interaction between the chains with increasing DP.

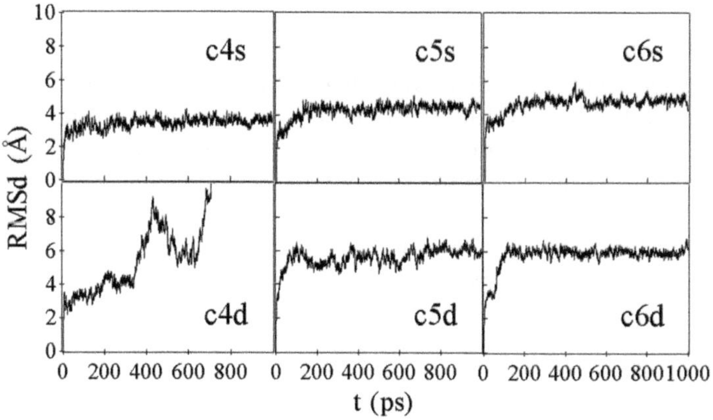

Figure 9.4 Root-mean square distances (RMSD) per atom in solutes after RMSD fittings in each system.
Reprinted from ref. 8 with permission of American Chemical Society.

Hydrogen bonding is considered as a primary interaction even between a pair of oligosaccharide chains as well as within each chain, in the double strand, with the same definition of a hydrogen bond described above but with an extended region of the angle O–H···O of 90–180°. First, the intramolecular inter-residue hydrogen bonding in the double strands is also characterized mostly by the mode of O3–O5′; the averaged occupancies are 81% for **c4d**, 66% for **c5d**, and 88% for **c6d**, compared with those for the single strands of 78–80%. The value of **c4d** is almost equal to that of **c4s**, reflecting the separated state of the chains in **c4d**. A significant decrease in the occupancy for **c5d** is probably due to the bending of a chain between Ring 2 and Ring 3 (Figure 9.3). The double strand **c6d** exhibits the greatest value of 88% among the six systems, as a result of the formation of a nearly complete aggregation state (Figure 9.3). This means that a flat ribbon surface in cellulose germinates already in cello-hexamer. It is also recalled that such a O3–O5′ mode of intramolecular inter-residue hydrogen bonding possesses crucial significance for the firm structure of cellulose.[9] Secondary inter-residue hydrogen bonds are O3–O6′ and O2–O6′ modes, with a greater occupancy for the former than for the latter, differing from native cellulose, which predominantly has the O2–O6 mode due to the *tg* conformation[†] of O6. Exchange between the *tg* and the *gt* conformation[†] at O6 is probably due to frequent attacks on O6 atoms by water molecules around them still in the double strands.

Intermolecular hydrogen-bonding can occur between the chains in the double strands. The average occupancy of the hydrogen bonds increases with DP, 0.64 (**c4d**), 2.93 (**c5d**), and 4.43% (**c6d**), which is consistent with their aggregation state. Notably, almost all of these hydrogen bonds are formed with O6: 59.0 (**c4d**), 84.1 (**c5d**), and 97.2% (**c6d**). Among the intermolecular hydrogen bonds with O6, the rate for a pair of O3 and O6′ increases the most significantly with DP, for example 10.7% (**c4d**) to 39.0% (**c6d**), in line with the fact that native cellulose possesses dominant intermolecular hydrogen bonds between O3 and O6′.[10]

Molecular dynamics simulation of the double strands of cello-oligosaccharide with DP = 4–6 describes clearly their aggregation states with a critical point at DP = 5. While the double strand of cello-pentamers is in a half-aggregated state, the double strand of cello-hexamers exhibits a complete aggregation state stabilized through intermolecular hydrogen bonds between O3 and O6′ with an almost flat ribbon surface by intramolecular inter-residue hydrogen bonds of O3–O5′. Even the double strand of cello-hexamers represents fairly well the nature of native cellulose.

[†]The *tg* and *gt* conformations are related with O6 in a D-glucopyranose. The *tg* conformation denotes that O6 is in a *trans* and a *gauche* position with respect to O5 and O4, respectively, around the C5–C6 axis, while for the *gt* conformation O6 is in a *gauche* and a *trans* position for O5 and O4 in the same manner.

9.2.3 Polymorphism of Cellulose

Cellulose exhibits polymorphism, as expected from the possibilities of a variety of networks of hydrogen bonding between the chains. So far, four major types of polymorphs have been confirmed and denoted as cellulose I, II, III, and IV. Cellulose I is the polymorph of native cellulose, and occurs in plants and algae. In plants, polymer chains of cellulose, bio-synthesized, are self-assembled to form cellulose microfibrils 3–5 nm in width and height and several μm in length, which have been considered to have a parallel-packed structure of 6×6 polymer chains of cellulose as shown schematically in Figure 9.5. In addition, tunitin, which is a constituent forming the tunica of Ascidiacea, and bacterial cellulose also exhibit cellulose I. Studies from NMR spectroscopy[11] and electron diffraction[12] have clarified that native cellulose is not simply a single form of cellulose I but strictly a composite of two distinct crystalline forms, I_α and I_β. The I_α form of cellulose is dominant in bacterial and algal cellulose, while the other form is dominant in woods and Ascidiacea.[10] The crystalline phase I_α undergoes irreversible transformation into the phase I_β,[13] and the latter is considered to be more stable than the former.[14]

Cellulose II is the phase of regenerated cellulose. It is well known that rayon is a regenerated cellulose produced from wood pulp or cotton linter. Native cellulose is transformed into the cellulose II phase by treatment with *ca.* 20 wt% NaOH solution, *i.e.*, mercerization. Such transformation is also possible by dissolution in a solution of carbon disulfide or Cu(II)-ammonia complex followed by regeneration in dilute sulfuric acid solution. These extreme treatments cause a drastic change in morphology of native cellulose,

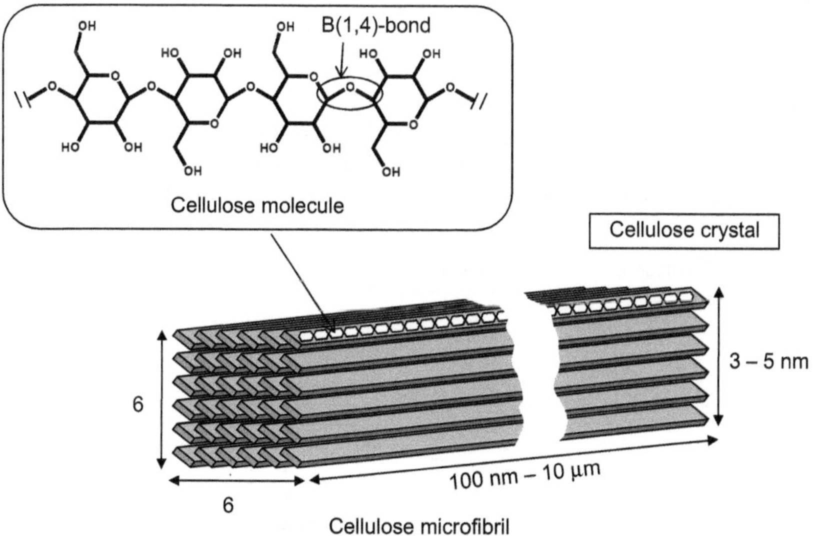

Figure 9.5 Schematic drawing of cellulose microfibril.

due to a crucial change in alignment of the chains from a parallel type to an anti-parallel one, as described below. The transformation from the phase I into the phase II is irreversible, because of the much greater stability of the latter phase than the former. Another phase, phase III, is generated by treatment of native cellulose with liquid NH_3, without an extreme change in morphology. Phases I and II are transformed into III_I and III_{II}, respectively. The structure of phase IV remains unclear. Figure 9.6 describes a simplified relation between the polymorphs of cellulose with respect to phase transformation and alignment of the chains; further understanding of cellulose requires crystallographic characterization, in particular of cellulose I and II.

The ordered packing of cellulose chains has been steadily revealed by intensive studies for nearly 100 years, and some important results are concisely described here. Table 9.1 lists crystallographic data of cellulose I_β[10] and II.[15] With the *c* axis as the unique axis, the space group of $P2_1$ requires two types of twofold screw axes parallel to the *c* axis: one through the origin and the other through the center point on the *ab* plane of a unit cell.[16] Accordingly, the symmetry operations generate two independent chains at the origin and the center of the cell, both parallel to the *c* axis; and origin and center chains can form two relevant sheets parallel to the *bc* plane, independent of each other. The directions of the reducing ends of the origin chains are arranged to be consistent with that of the *c* axis.

For cellulose I_β, center chains are aligned in a sheet with the same directions as that of the origin chains,[10] with a shift of *ca.* 1/4*c* as compared with the origin chains, so as to avoid repulsive interactions between the origin and center sheets. Cellulose chains in an origin and a center sheet are not equivalent, having slightly different values of Φ and Ψ (Table 9.1), with both holding a flat ribbon shape. In addition, all the hydroxymethyl groups are in a *tg* conformation, and each can readily form a hydrogen bond with the hydroxyl group at the second place, O2–H2, in the adjacent residue of the same chain. According to Langan *et al.*,[10] interestingly, the hydrogen atoms linked to O2 and O6 take more than one location, in contrast to that linked to O3, which occupies a well-defined localized position, causing two distinct networks of hydrogen bonding in each sheet: I_p and I_d for the principal and disordered positions of both H2 and H6, respectively.

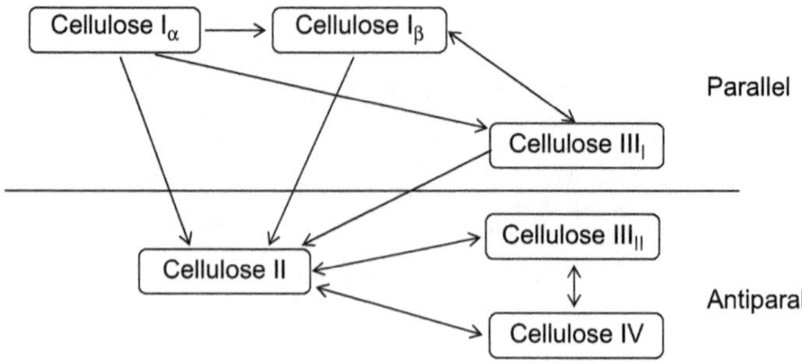

Figure 9.6 Cellulose polymorphs and their transformations.

Table 9.1 Crystallographic data of major cellulose polymorphs. Reproduced from refs 10 and 15 with permission of American Chemical Society.

	Cellulose I_β[10]	Cellulose II[15]
Chemical formula	$C_{12}H_{20}O_{10}$	$C_{12}H_{20}O_{10}$
Space group	Monoclinic, $P2_1$	Monoclinic, $P2_1$
a (nm)	0.7784(8)	0.801
b (nm)	0.8201(8)	0.904
c (nm)	1.0380 (10)	1.036
γ (°)	96.5	117.1
V (nm^3)	0.6583(11)	0.6678
Z	2	2
Density (g cm^{-3})	1.64	1.61
Chain alignment	Parallel	Anti-parallel
Conformation:		
Φ^a (°)	−98.5 (o), −88.7 (c)	−95.4 (o), −91.3(c)
Ψ^b (°)	−142.3 (o), −147.1 (c)	92.3 (o), 89.4 (c)
χ^c (°)	170 (o), 158 (c)	— (o), — (c)
χ'^c (°)	−70 (o), −83 (c)	−172.4 (o), 148.4 (c)
	tg (o and c)	*gt* (o and c)

[a]The dihedral angle to describe the relative orientation of adjacent glycosyl residues in the same chain, defined by O5–C1–O1–C4. Note that the definition is different from that in Section 9.2.1.
[b]Another dihedral angle, defined by C1–O1–C4–C5 and C1–O1–C4–C3 for cellulose I_β and II, respectively. Note that the definition is different from that in Section 9.2.1.
[c]Dihedral angles with regard to the conformation of hydroxyl methyl group, which are defined by χ(O5–C5–C6–O6) and χ'(C4–C5–C6–O6).

Notably, for I_p, H2 is located principally in the direction to O6, with co-operative orientation of H6 toward O3′ in a neighboring chain in the same sheet, whereas for I_d another orientation of H6 to O2 causes H2 to occupy a position off the line of O2–O6. Accordingly, for I_p, origin chains are hydrogen-bonded with each other in a sheet, through intermolecular hydrogen bonds O6–H6···O3′ with the aid of intramolecular hydrogen bonds O2–H2···O6/O1 and O3′–H3···O5′, while center chains are bonded through intermolecular hydrogen bonds O6–H6···O3′/O2′ supported by intramolecular hydrogen bonds O2–H2···O6 and O3′–H3′···O5′ (Figure 9.7). These results prove that chains are aligned in both sheets through intermolecular hydrogen bonds of principally O6–H6···O3′ and partly O2–H2···O6′, holding a flat ribbon structure through intramolecular hydrogen bonds of dominantly O3–H3···O5. Notably, there are almost no hydrogen bonds between sheets, which are held together only by hydrophobic interactions.[10]

On comparing cellulose II with cellulose I_β, the significant difference is the orientation of center chains opposite to that of origin chains. In addition, the hydroxymethyl groups in both chains are in a *gt* conformation. Thus, both origin and center chains are aligned in each sheet in a zigzag fashion when projected on the *ab* plane, with a relative displacement along the *c* axis of *ca.* 1/4*c*, allowing a network of hydrogen bonding not only between origin chains or center chains but also between origin and center

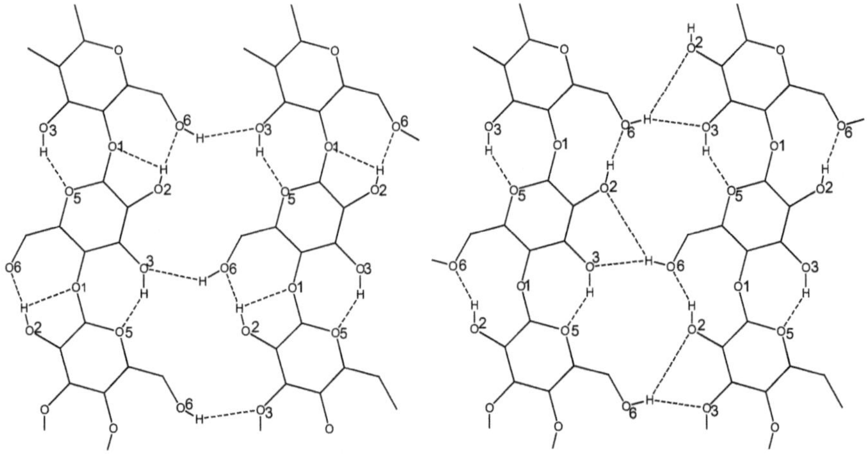

Figure 9.7 Schematic representation of the principal network of hydrogen bonds (I_P) in the origin (left) and center (right) sheets of cellulose Iβ. Hydrogen bonds are represented by dotted lines. Only the oxygen atoms involved in hydrogen-bonding have been labeled for clarity.
Reproduced from ref. 10 with permission of American Chemical Society.

chains. Langan *et al.*[15] have proved on the basis of neutron diffraction that a systematic intramolecular three-center hydrogen bond is formed in both chains with a three-center mode of O3–H3···O5/O6, that intermolecular hydrogen bonds are formed between origin chains with a mode of O2–H2···O6′ and between center chains with a mode of O6–H6···O2′, and that intermolecular hydrogen bonds are formed even between origin and center chains with a mode of O2c–H2c···O2o as well as with a four-center mode of O6o–H6o···O6c/O5c/O3c, where suffixes "c" and "o" denote "center" and "origin", respectively. Such a network of hydrogen bonding between chains in a sheet as well as in neighboring sheets substantiates well the highest stability of cellulose II among the polymorphs.

Native cellulose is, as clarified above, an assembly of cellulose molecules aligned orderly in sheets, based on a beautiful network of hydrogen bonding with all the hydroxyl groups. Such hydrogen bonding is the fundamental cause of mechanical as well as aggregation properties of cellulose as described later. This also explains the extremely low reactivity of cellulose because all the active sites are completely involved in a network of hydrogen bonding.

9.3 Activation of Cellulose by Ball Milling

9.3.1 Micro-pulverization of Native Cellulose

Ball milling is considered to be a mechanical method to pulverize and activate native cellulose in a solid state, because it is expected that reactive hydroxyl groups can be liberated by collapse of the network of hydrogen bonding in it through collision impacts. It is generally accepted that cellulose consists of crystalline as well as amorphous regions, and water molecules are adsorbed preferentially in the amorphous regions. Pulverization of native cellulose is significantly dependent on its water content.[17] Figure 9.8 shows size distributions of cellulose powders prepared by ball milling of fibers of cotton linter CF11 (cellulose I) with a water content of *ca.* 7.0 wt% or smaller than 0.1 wt%, the former corresponding to an air-dried sample and the latter to a vacuum-dried one. Vacuum-dried cellulose exists as smaller powders (with particles most frequently <10 μm for 42% of the powders), in a milling time of 60 min, beyond which larger particles grow gradually. As compared with vacuum-dried cellulose,

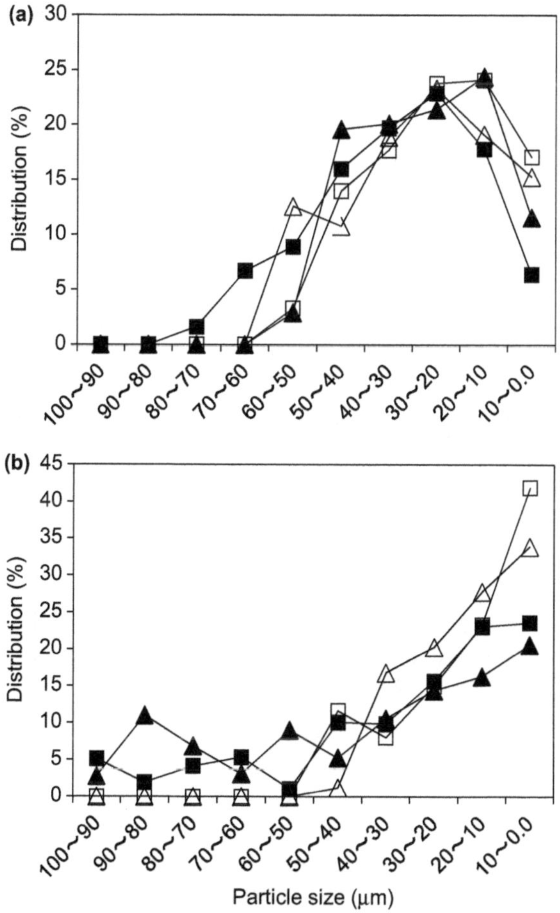

Figure 9.8 Particle distributions of cellulose powders obtained by ball milling of air-dried (a) and vacuum-dried fibers of cotton linter (CF11) (b) for 20 (△), 60 (□), 180 (▲), and 300 min (■) at room temperature.
Reprinted from ref. 17 with permission of the Society of Polymer Science, Japan.

air-dried cellulose yields greater particles with sizes mostly distributed between 10 and 50 μm, due to more extensive aggregation of primary particles with an increase in milling time, as seen from scanning electron microscopy. Such aggregates cannot be dispersed into primary particles even by an ultrasonic treatment, but can with dimethyl sulfoxide, which causes cellulose to swell by cleavage of hydrogen bonds between molecular chains in it.[18] This suggests strongly that primary particles of cellulose with a certain amount of water aggregate extensively by formation of hydrogen bonds between them, with the aid of water molecules.

The crystalline property of cotton linter CF11 is affected profoundly by ball milling, depending on its water content. Cotton linter is originally in a crystalline phase of cellulose I, with a crystallinity index (*CI*)[19] of 93% for an air-dried state, decreasing to 85% when vacuum-dried. The *CI* of cellulose decreases more rapidly down to zero with milling time when vacuum-dried than when air-dried (Figure 9.9). With vacuum-dried cellulose, removal of water molecules induces hydrogen bonds to be rearranged directly between intra- and inter-chains of cellulose in amorphous regions with no water molecules, accumulating strains there. With mechanical impacts on the amorphous regions, such strains cause extensive collapse of an ordered state in crystalline regions. In contrast, for air-dried cellulose, water molecules adsorbed in amorphous regions reduce strains owing to more stable hydrogen bonds through their involvement between intra- and interchains, relaxing mechanical impacts as a plasticizer.

The degree of polymerization (DP) of cotton linter hardly varies upon ball milling and is *ca.* 200, regardless of its dry state. Ball milling of

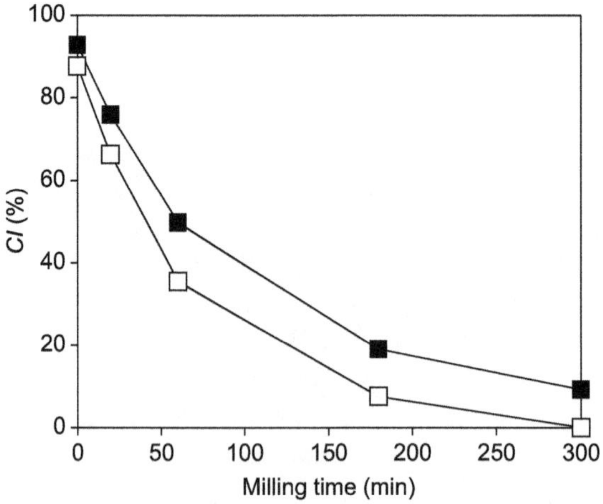

Figure 9.9 Crystallinity indexes (*CI*) of cellulose powders obtained by ball milling air-dried (■) and vacuum-dried CF11 samples (□) at room temperature. Reprinted from ref. 17 with permission of the Society of Polymer Science, Japan.

wood pulp with DP of *ca.* 790 causes a decrease in DP down to *ca.* 200. These results indicate that ball milling exhibits a comparable effect on native cellulose as acid-hydrolysis[20] in regard to a leveling-off of DP.

On the basis of the above results, ball milling of native cellulose with a content of water smaller than 0.1 wt% realizes amorphous particles with the most frequent sizes being smaller than 10 μm. With cellulose containing *ca.* 7 wt% water, which is relevant to an air-dried state, water molecules cause intensive aggregation of primary particles of cellulose with a reduced crystalline degree, because of acceleration of hydrogen bonds between primary particles through the aid of them. Accordingly, addition of 20 wt% acetone to vacuum-dried cellulose causes much smaller particles with a distribution degree of particle sizes smaller than 10 μm for 70%, compared to 42% without addition of acetone.[17] Much smaller particles of cellulose are prepared by ball milling of cellulose with poly(ethylene glycol) (PEG) or poly(vinyl alcohol) (PVA).[21] Both polymers, with high affinity to cellulose, form films coating the particles of cellulose during ball milling, so that the primary particles are isolated from each other without hydrogen bonds between them. Suspension of the resultant composite in water liberates fine particles of cellulose because of ready resolution of the polymer in water. Such particles of cellulose are almost all distributed with sizes smaller than 10 μm, in contrast to original cellulose (Figure 9.10).

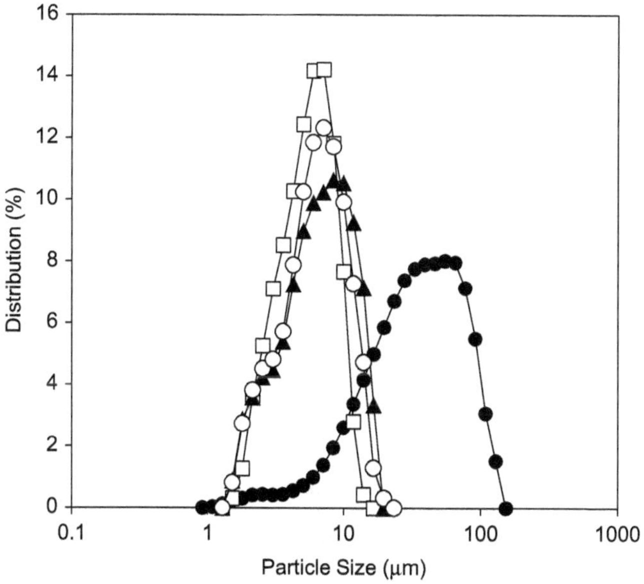

Figure 9.10 Particle distributions of cellulose powders obtained by ball milling of a mixture of CF11 and PEG or PVA: Original CF11 (●) and the powders derived with 4 (▲) and 15 wt% PEG (□), and 15 wt% PVA (○).

9.3.2 Characteristics of Micro-pulverized Cellulose

Ball milling of native cellulose, with a water content of less than 0.1 wt%, produces fine powders of amorphous cellulose. Resultant cellulose powders are so reactive that they exhibit unique properties due to the collapse of a network of hydrogen bonding in a phase of cellulose I. Here two characteristics are described: one is a crystalline transformation into cellulose II,[22] the other promoted aggregation of cellulose particles.[23]

Amorphous cellulose powders, which are prepared by ball milling in advance, are further ball-milled with a certain amount of water added, for a fixed time; the product is denoted by MC_w^t, where the subscript and superscript are relevant to the amount of water added (wt%) and the milling time of cellulose with water added (h). Figure 9.11 shows X-ray diffraction patterns of products as well as original cellulose. The products with ≤ 30 wt% water content are powdery, while those with 50 wt% are somewhat viscous. Original cellulose is in phase I, with signals at 14.8°, 16.8°, and 22.6° assigned to the reflections (1,−1,0), (1,1,0), and (2,0,0), respectively. Notably, MC_{30}^{2h} and MC_{50}^{2h} both have new signals at 12.1°, 19.8°, and 22.0°,

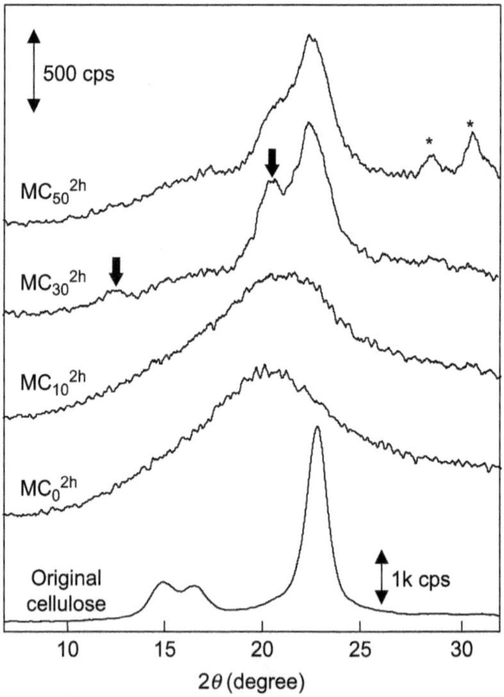

Figure 9.11 X-Ray diffraction patterns of original CF11 (cellulose I) and samples milled for 2 h with controlled amounts of water added (0, 10, 30, and 50 wt%, from bottom to top). The peaks marked with (*) are attributed to contaminants from the milling pot and balls loaded in it. Reprinted from ref. 22 with permission of Springer Science and Business Media.

characteristic of cellulose II,[19] while MC_0^{2h} and MC_{10}^{2h} remain amorphous. The signals due to cellulose II are stronger for MC_{30}^{2h} than for MC_{50}^{2h}. Further, for the samples MC_{30}^{t}, the signals show increasing intensities with the milling time (t). These results indicate clearly that addition of greater than 30 wt% water to the amorphous cellulose induces a crystalline transformation into cellulose II, accelerated by ball milling. Of most interest, a diffraction pattern of cellulose II even appears in MC_{30}^{0}, which is prepared only by addition of 30 wt% water to the amorphous cellulose powders. This is a spontaneous transformation into cellulose II. It is the first demonstration that the transformation from a phase of cellulose I into a cellulose II phase even occurs in the solid state.

Table 9.2 lists spin–lattice relaxation times (T_1^{1H}) of some samples, based on a proton inversion–recovery sequence with CP/MAS ^{13}C-NMR spectroscopy, in order to understand the mobility of cellulose chains and their interactions with water molecules. The amorphous cellulose MC_0 has a smaller T_1^{1H} than original cellulose (cellulose I). Addition of 30 or 50 wt% water decreases significantly T_1^{1H}; ball milling causes further a decrease in T_1^{1H} for MC_{30}^{2h} but, conversely, a slight increase for MC_{50}^{2h}. The smaller T_1^{1H} values correspond well to the greater degrees of crystalline transformation into cellulose II described above. This kind of relaxation is principally due to interactions of water molecules with cellulose chains in amorphous or disordered regions.[24] Accordingly, on addition of 30 wt% rather than 50 wt% water, a specific kind of interaction between cellulose chains and water molecules probably greatly increases the mobility of cellulose chains, particularly in disordered or amorphous parts, so that these interacting chains can self-assemble to transform into a phase of cellulose II.

As described above, cellulose chains are parallel in a phase of cellulose I but anti-parallel in a phase of cellulose II. It is reasonable that amorphous cellulose, prepared by ball milling of native cellulose, is assumed to consist of microfibrils with parallel-arrayed cellulose chains. This leads to the logical deduction that the transformation into cellulose II is based on interactions between oppositely oriented cellulose chains belonging to different microfibrils.

Table 9.2 ^{1}H spin–lattice relaxation times (T_1^{1H}) of cellulose–water systems determined by CP/MAS ^{13}C-NMR. Reprinted from ref. 22 with permission of Springer Science and Business Media.

Sample	T_1^{1H} (s)
Original cellulose CF11 (cellulose I)	5.77
MC_0	4.53
MC_{30}^{0h}	1.22
MC_{30}^{2h}	1.11
MC_{50}^{0h}	1.33
MC_{50}^{2h}	1.43
MC_{30}^{0h} (D_2O)	3.20
MC_{30}^{2h} (D_2O)	3.04

Such interactions are promoted most by addition of a specific amount (30 wt%) of water to amorphous cellulose. However, no mechanism has been resolved to explain this unique crystalline transformation into cellulose II.

Another characteristic of amorphous cellulose prepared by ball milling of native cellulose is related to promoted aggregation between powders by hot-pressing. Hot-pressing of amorphous cellulose powders with a controlled amount of water added (W_c) of *ca.* 3–6 wt% yields a transparent plate, if molding temperature and pressure are arranged according to the amount of water. For example, the addition of 3.8 wt% water yields a most transparent plate at 150 °C and 200 MPa; a small increased addition of 5.8 wt% water at a decreased temperature of 120 °C and the same pressure has the same effect. With W_c smaller than 3 wt%, a resultant plate remains white and fragile, while it becomes slightly brown with $W_c = ca.$ 6 wt% and contracts with cracks when W_c is greater than 10 wt%. With W_c smaller than 15 wt%, cellulose exhibits recrystallization to cellulose I during hot-pressing. Figure 9.12 shows crystallinity indexes (*CI*)[19] of the surface part of cellulose plates at 200 MPa against molding temperature. With $W_c = 0$ wt% the *CI* hardly changes in the range 25–180 °C, indicating that cellulose in the plate is still amorphous. With $W_c = 3.8$ wt%, which gives a most transparent plate at 150 °C and 200 MPa, the *CI* increases with milling temperature, particularly at greater than 120 °C, suggesting that formation of the transparent plate is accompanied by recrystallization. For $W_c = 14.8$ wt%, which often results in plates with cracks, recrystallization occurs even at the lower temperature, and the *CI* levels off at temperatures greater than 150 °C, probably due to pyrolysis of cellulose molecules, considering the marked decrease in the DP at the same temperature.[23] Note that the averaged *CI* of the whole plate, which is prepared with $W_c = 3.8$ wt% at 150 °C and 200 MPa, is 24%, which is smaller than the 33% shown in Figure 9.12.

If amorphous powders of cellulose are hydrothermally treated at 120 °C for 10 min, the resultant powders do not yield a transparent plate under any hot-press conditions. The *CI* value of the powders is *ca.* 40%, suggesting that a network of hydrogen bonds is formed principally within each powder, but not between powders. Therefore, a mechanism for producing a transparent plate by hot-pressing is based on formation of hydrogen bonds between cellulose powders during recrystallization to cellulose I. Water molecules adsorbed in cellulose are bound primarily to hydroxyl groups of cellulose chains in amorphous regions; breaking hydrogen bonds originally formed there, and consequently activating them, elevates the mobility of cellulose chains. Accordingly, formation of a new network of hydrogen bonds accompanied by removal of water molecules under the hot-pressing treatment causes rearrangement and recrystallization of cellulose molecules particularly near the surface of a plate more than in the inner region where water cannot be fully removed.

The two characteristics of amorphous cellulose demonstrate that native cellulose can be activated by ball milling. As described below, such amorphous cellulose can react even with maleic anhydride in the solid state.

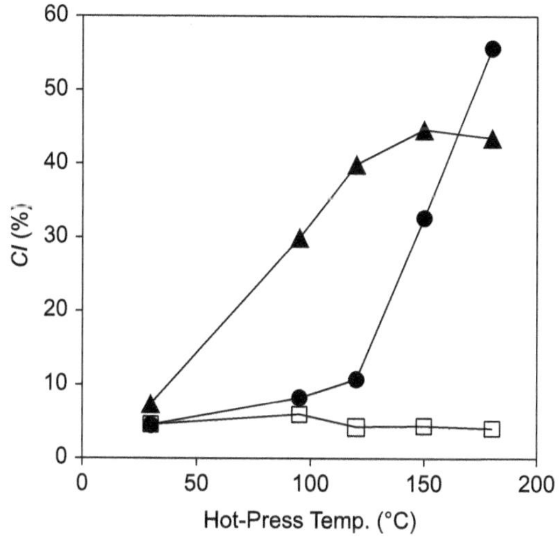

Figure 9.12 Effects of water content and temperature on the *CI* of the surface of the plate: $W_c = 0$ (□), 3.8 (●), and 14.4 wt% (▲).
Reprinted from ref. 23 with permission of the Society of Polymer Science, Japan.

9.4 Grafting of Maleic Anhydride onto Polyolefins by Ball Milling

Maleic anhydride (MA)-grafted polyolefins are a most attractive and extensively employed compatibilizer for the synthesis of polymer-blends or composites of polar materials with non-polar ones. MA-grafted polyolefins are synthesized mainly in solution phase as well as in molten phase,[2] and recently we have succeeded in mechanochemical grafting of MA monomers onto polyolefin.[25,26]

9.4.1 Free-radical Grafting of Maleic Anhydride onto Polyolefins

In the past several decades, grafting of MA onto polyolefin substrates has been studied most intensively in the functionalization of polyolefins.[27] However, our fundamental understanding of the grafting mechanism has been quite limited because of a lack of knowledge of the microstructures of grafts as well as side reactions. Free-radical grafting of MA onto polyolefin has been recognized to start with the formation of macroradicals by hydrogen abstraction of primary radicals on secondary (for polyethylene, PE) or tertiary carbons (for polypropylene, PP) in the polymer chains, followed by grafting of MA monomers on the macroradicals, although for PP the grafting of MA monomers is complicated by β-scission of the macroradicals, which generates two fragments (Scheme 9.1): one with the terminal isopropenyl

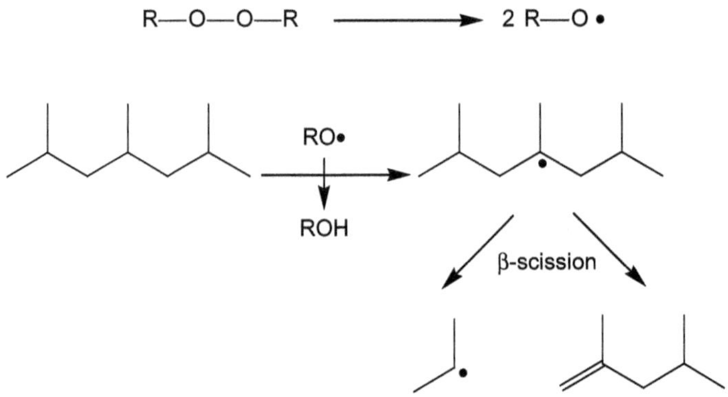

Scheme 9.1 β-Scission of a macroradical initially formed from PP.

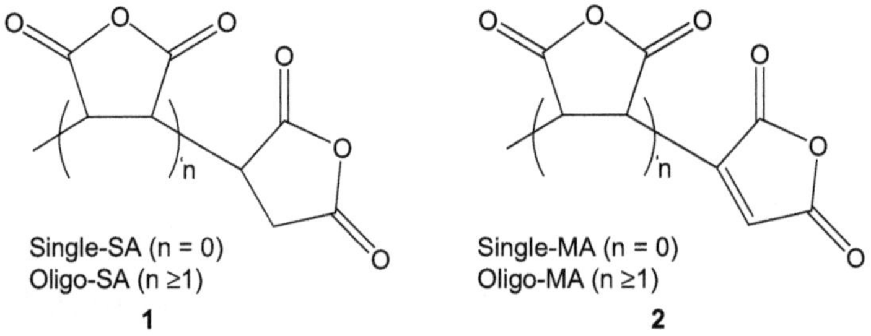

Figure 9.13 Graft structures in PE, determined by ^{13}C NMR.

group, the other with a secondary carbon radical on the terminal carbon, which probably abstracts a hydrogen to yield a tertiary carbon radical on a PP chain or might cause grafting of MA. Usually, the overall grafting reaction is terminated by recombination of macroradicals for PE to cause an increase in molecular weight and by combination of fragmentary radicals for PP to yield a product with a smaller molecular weight than the original PP.

Heinen *et al.* have determined precisely the structure of species grafted onto polyolefins on the basis of ^{13}C NMR spectroscopy.[28] For PE, they observed single- and oligo-succinic anhydride rings (single-SA and oligo-SA, **1**, Figure 9.13),[28] which are formed by hydrogen abstraction of the relevant graft radicals (oligo-R) from PE consisting of secondary carbons. Besides this species, we have observed a single MA ring and oligo-succinic anhydride rings terminated with a MA ring (single-MA and oligo-MA, **2**), with di-*tert*-butyl peroxide (DTBP) as an initiator, on the basis of ^{13}C NMR spectroscopy.[29] Notably, the single- and oligo-MA grafts are fewer than the single- and oligo-SA grafts. Furthermore, for a molar ratio of initial MA/DTBP smaller than 1, single-MA grafts are formed predominantly, while longer oligo-MA grafts are more probable with the molar ratio exceeding 1. The single- or oligo-MA graft can be formed through disproportionation between a corresponding radical (oligo-R) and a macroradical on a secondary carbon in PE or another oligo-R. The combination between secondary carbon radicals on PE substrates terminates the overall free-radical grafting, causing an increase in molecular weight of the resultant graft polymer.

For PP, the graft mechanism of MA is more complicated and still controversial[27] because of β-scission of polypropylene substrate (Scheme 9.1). The grafting probably occurs on tertiary carbon radicals rather than on secondary carbon radicals at the chain end of fragments due to β-scission (Scheme 9.1). Heinen *et al.*[28] determined, on the basis of ^{13}C NMR spectroscopy, two species of the graft: one is the *single* succinic anhydride ring on a tertiary carbon of PP substrate (**3**) and the other is the itaconic anhydride terminal (**4**) (Figure 9.14). The species **3** can be formed through addition of an MA monomer to a tertiary carbon radical of a PP chain followed by a ready intra- or intermolecular hydrogen transfer from a PP chain with methine carbons, which is quite different from PE chains with only methylene carbons, because the hydrogen on a tertiary carbon is abstracted more readily than that on a secondary carbon.[30] The formation of these species has been also confirmed by ^{13}C NMR spectroscopy of several industrial MA-grafted PP products.[31] A mechanism for generating the species **4** is that the single succinic anhydride radical on a tertiary carbon in a PP chain undergoes intramolecular hydrogen abstraction *via* a six-membered transition state followed by β-scission, and finally by disproportionation (Scheme 9.2);[27] this is different from the original mechanism proposed by Heinen *et al.*[28] The mechanism in Scheme 9.2 expects that the formation of the itaconic anhydride terminals **4** is accompanied by a decrease in molecular weight of the backbone polymer. This agrees well with the inverse relationship between the molecular weight and the content of the itaconic anhydride terminals **4**, which was observed from NMR examination of industrial products of MA grafted PP with different graft contents of MA.[31]

Roover *et al.*[32] proposed the formation of single or oligo-grafts on the secondary carbon terminals, from IR spectroscopy. Although with IR spectroscopy it is fairly difficult to determine exactly the structure of such grafts, it is possible to evaluate qualitatively the grafting of MA.

Grafting of maleic anhydrides on poly(ethylene-*co*-propylene) can be understood on the basis of the graft structures on polyethylene and polypropylene described above, depending on the mode of sequence of ethylene

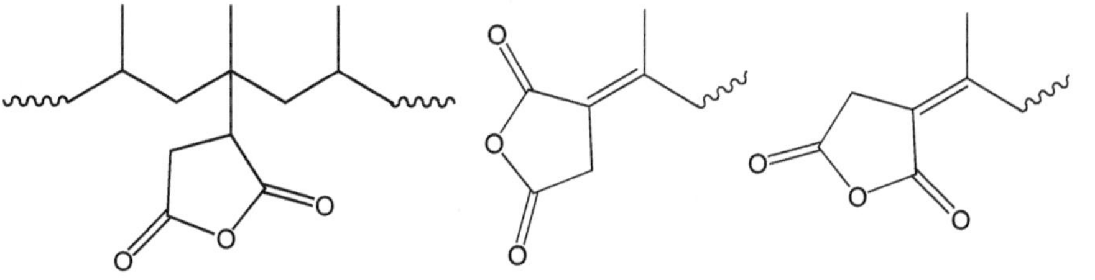

3 **4**

Figure 9.14 Graft structures in PP, determined by ^{13}C NMR.

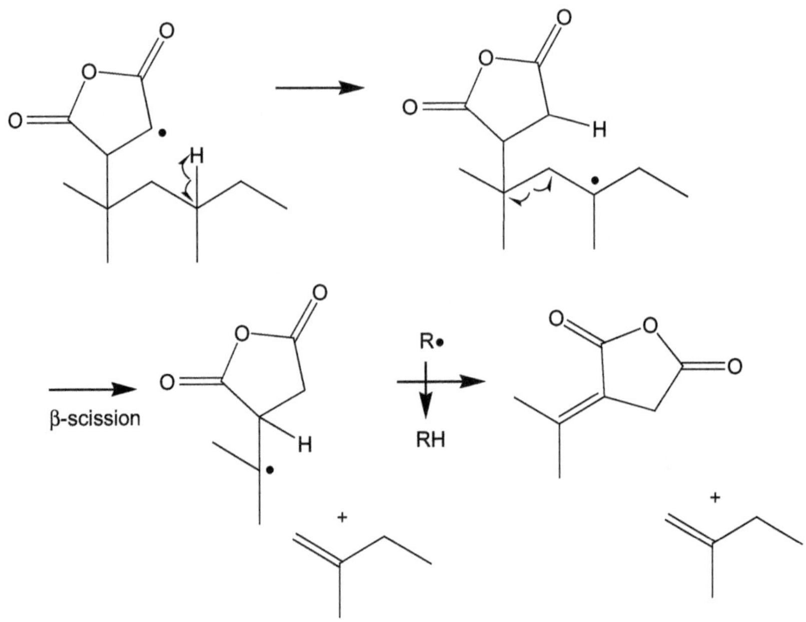

Scheme 9.2 Mechanism for formation of the graft **4**.
Reproduced from ref. 27 with permission of Elsevier.

and propylene units in the polymer.[28] Notably, only single succinic anhydride rings (**5**) on tertiary carbons are observed in poly(ethylene-*alt*-propylene). For poly(ethylene-*stat*-propylene) with a molar ratio of ethylene/propylene ≥1, the grafts are assigned to the structures of **1** and **5**. On the other hand, poly(ethylene-*stat*-propylene) with a molar ratio of 1 : 3 has more variable grafts, and the assigned grafts are the structures **3** and **4** as well as **1** and **5**, with the former two being observed for i-PP.

9.4.2 Mechanochemical Grafting of Maleic Anhydride onto Polypropylene

Free-radical grafting of maleic anhydride (MA) onto polyolefin starts with formation of macroradicals through hydrogen abstraction by primary radicals. The resultant macroradicals, which are mostly secondary and tertiary carbon radicals for PE and PP respectively, cause not only reaction with MA monomers resulting in their grafting but also side reactions. The main side reaction is a recombination for PE, while for PP β-scission generates fragments with the secondary carbon radical and the terminal isopropenyl group. Efficient grafting of MA monomers onto polyolefin requires suppression of such side reactions causing changes in the molecular chain structure from that of the original polymer. Increasing temperature is favorable not only for formation of macroradicals from PP but also for β-scission of the macroradicals leading to degradation of the PP. Grafting of

MA onto PP in melt state[33] is usually operated for 10–20 min but at high temperature, ranging from its melting temperature to 200 °C, accompanied inevitably by the degradation of the polymer. In contrast, mechanochemical grafting of MA by ball-milling can be performed at lower temperatures, though longer times are needed.[25,26]

Mechanochemical grafting of MA onto i-PP[25,26] has been performed by use of a Planetary Ball Mill (Pulverisette 5, Fritsch, Germany) equipped with four jars, each of which is 500 cm^3 in volume and loaded with 25 balls of 15 mmϕ diameter. Usually, 50 g of PP is loaded in each jar with given amounts of MA as well as benzoyl peroxide (BPO) as an initiator, and is milled at a rotational speed of 300 rpm for 8 h, with a cyclic mode of milling (50 min) followed by a pause (10 min). The temperature inside the jars is 75–80 °C during the milling steps, which is much lower than that in the melt-mixing. Figure 9.15 shows that the grafting of MA onto i-PP proceeds abruptly during the reaction time of 1–2 h to almost finish after 4 h; the melt flow index (MFI) increases significantly from the beginning to 1 h, and then quickly decreases until 2 h, after which it almost maintains a plateau (Figure 9.16). These results indicate that β-scission occurs at the beginning, generating fragments with smaller molecular weights, until a certain amount of macroradicals with tertiary carbon radicals is attained, beyond which grafting of MA on the macroradicals occurs competitively with combination of fragment radicals (secondary carbon radicals) resulting from β-scission of the macroradicals. It is, therefore, expected that the grafting of MA onto i-PP depends strongly on the initially added amounts of MA as well as BPO. The amounts of MA and BPO hereafter are given in the unit of wt% that is based on the weight of i-PP used.

The graft yield is plotted against the initial amount of MA under a constant amount of BPO of 1.0 wt% in Figure 9.17. The graft yield is *ca.* 90%, which is very high, in the range of MA ≤ 1.0 wt%, beyond which it decreases

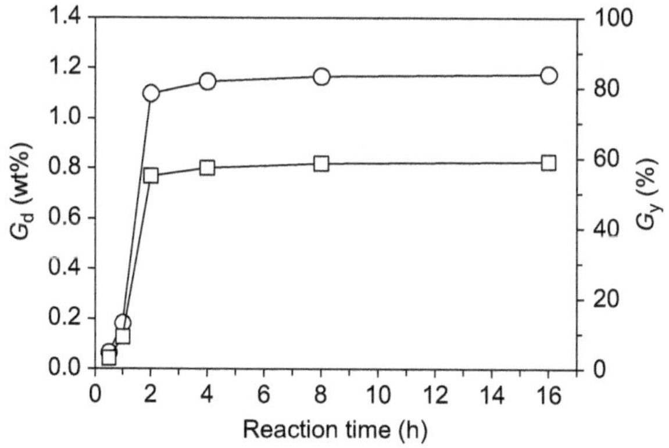

Figure 9.15 Changes of graft degree (○) and yield of MA onto i-PP (□) with reaction time under a weight ratio of i-PP/MA/BPO = 100/2/1.
Reprinted from ref. 26 with permission of John Wiley & Sons.

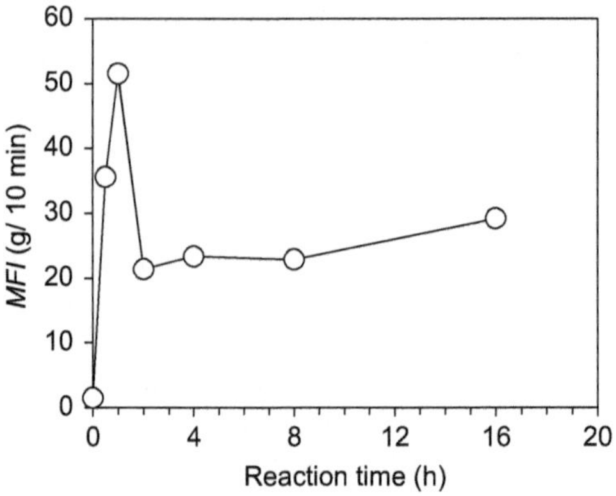

Figure 9.16 Change of MFI of the graft product with reaction time under a weight ratio of i-PP/MA/BPO = 100/2/1.
Reprinted from ref. 26 with permission of John Wiley & Sons.

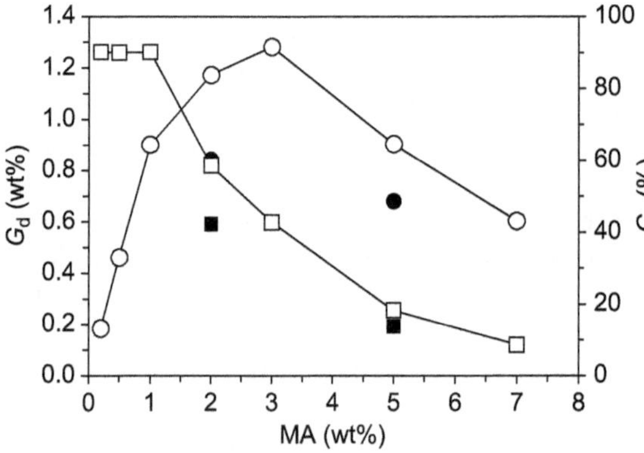

Figure 9.17 Effects of MA content on graft degree (○) and yield (□) of the graft product with BPO = 1.0 wt%. For comparison, the degree (●) and yield (■) from melt mixing (190 °C, 75 rpm, and 15 min) are also shown.
Reprinted from ref. 26 with permission of John Wiley & Sons.

with increasing MA amount. Such a decrease in the graft yield, which is also observed in the melt grafting,[33] has been attributed to formation of a two-phase system of i-PP in the melt state and MA in the liquid state, which causes a greater loss of BPO with an increasing content of the MA phase.[34] Accordingly, the graft degree exhibits a maximal value of 1.3 wt% at an MA amount of 3.0 wt%.

Figure 9.18 presents the dependence of the graft yield on the initial amount of BPO for an MA amount of 2.0 wt%. The graft yield increases with increasing amount of BPO up to 1.0 wt%, attaining a plateau of 70% for ≥1.5 wt% BPO. A similar relation is also observed with a greater amount of MA (5.0 wt%), even though the smaller rate for BPO up to 2 wt% results in a smaller maximal graft yield of 50% for ≥3.0 wt% BPO.

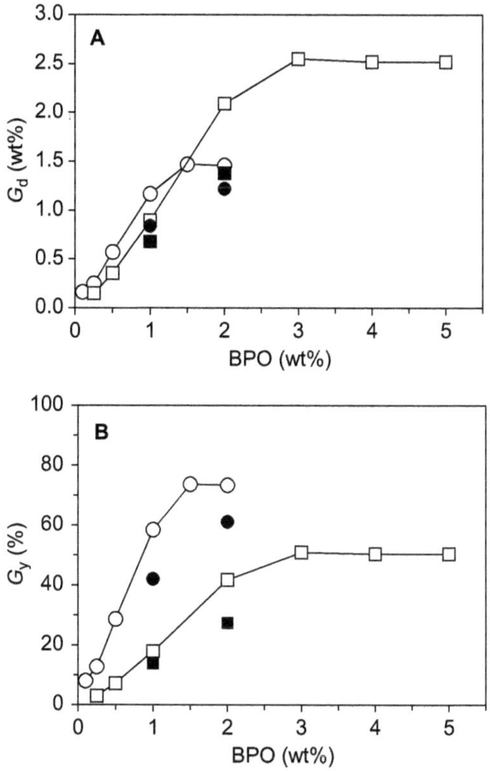

Figure 9.18 Effects of BPO content on graft degree (A) and yield (B) of the graft product with MA = 2.0 (○) and 5.0 wt% (□). The results shown by (●) and (■) are from melt mixing (190 °C, 75 rpm, and 15 min) with MA = 2.0 and 5.0 wt%, respectively.
Reproduced from ref. 26 with permission of John Wiley & Sons.

These results on ball-milling grafting of MA onto i-PP appears to indicate that the graft yield is the greatest with MA ≤ 1 wt%, beyond which it rather reduces with increasing amount of MA, but, conversely, with decreasing amount of BPO. This leads us to consider the relation of the graft yield with the weight ratio of BPO/MA, a new parameter. All the results given in Figures 9.17 and 9.18 are re-plotted in this mode in Figure 9.19. Notably, the graft yield is expressed by a linear relation with the BPO/MA ratio except for extreme cases that do not satisfy the linearity, regardless of the individual content of MA or BPO. Of much interest, this linearity holds over the greater extent of the BPO/MA with decreasing the MA amount; thus the graft yield reaches 50%, 73%, and 90% at the upper limit of the BPO/MA weight ratio of 0.40, 0.61, and 0.81, with the initial amounts of MA of 5.0, 2.0, and ≤1.0 wt%, respectively. Furthermore, if such linearity is not satisfied, the graft products exhibit extremely low MFI values, in some cases smaller than the original i-PP, with the MA amounts of 2.0 and 5.0 wt% (Figure 9.20). This indicates that the free radical grafting of MA is forced to terminate and

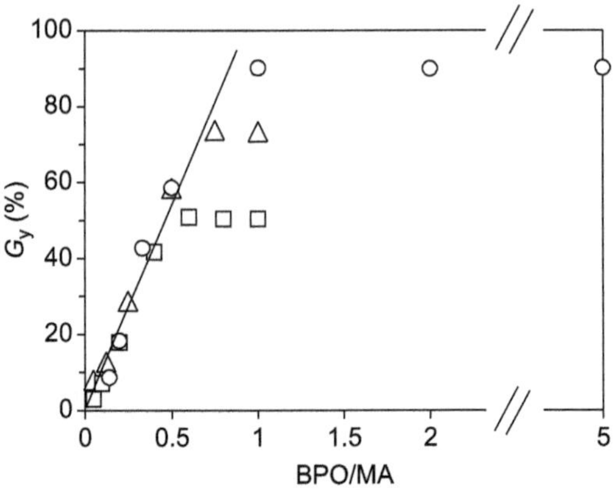

Figure 9.19 Linear relationship between the graft yield and the weight ratio of BPO/MA with data in Figures 9.17 (○) and 9.18 (MA = 2.0 (△) and 5.0 wt% (□)). Reproduced from ref. 26 with permission of John Wiley & Sons.

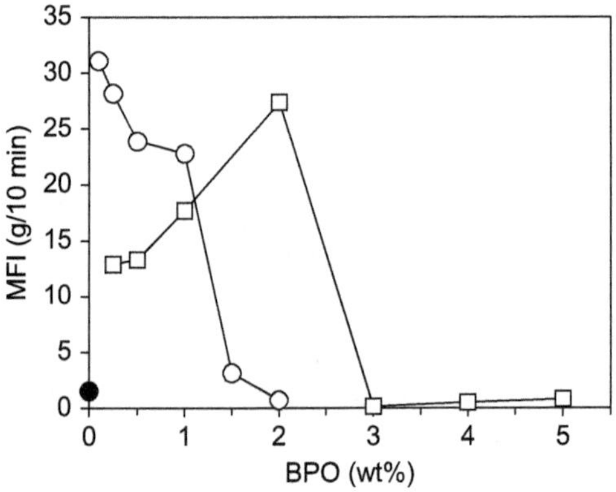

Figure 9.20 Effects of BPO content on MFI of the graft product with MA amounts of 2.0 (○) and 5.0 wt% (□), compared with i-PP without ball milling (●). Reproduced from ref. 26 with permission of John Wiley & Sons.

results in the smaller graft yield when the amount of macroradicals attains a critical value depending on the initial amount of MA > 1.0 wt%, within which it proceeds to a nearly perfect level of 90%, satisfactorily competing with side reactions such as combination.

On the basis of the above results, ball-milling grafting of MA onto i-PP is determined primarily by the weight ratio of BPO/MA, a new parameter. Thus, the graft efficiency, *i.e.*, the graft yield of MA by ball-milling, exhibits linearity with the BOP/MA ratio, the range of which is dependent on the initial amount of MA > 1.0 wt%, regardless of the formation of a two-phase component of i-PP/MA. The linear efficiency is broken mainly due to termination by dominant competing reactions like the combination of macroradicals, when the BPO/MA ratio exceeds a critical value for MA ≥2.0 wt%

(Figures 9.19 and 9.20). The data depicted in Figures 9.19 and 9.20 are also valuable from an engineering viewpoint, because they make it possible to choose the grafting conditions when synthesizing an MA-graft product with desired graft degree and MFI.

9.5 Novel Composites of Cellulose and Plastics by Ball Milling

9.5.1 Compatibilization of Cellulose with Poly(ethylene glycol)

Cellulose exhibits essentially no thermoplasticity and moreover insolubility in usual solvents. Accordingly, mechanochemical compounding of cellulose with thermoplastic polymers is considered to be a useful method to yield novel thermoplastic composites with cellulose. First, we describe a composite of native cellulose with poly(ethylene glycol) (PEG; $M_W = 2.0 \times 10^6$) as a polymer with high affinity to cellulose, using highly crystalline cotton linter as a pure cellulose for basic studies. As described briefly in preparing fine particles of cellulose in Section 9.3.1, ball milling of a mixture of native cellulose and PEG yields a new composite through hydrogen bonds between cellulose particles and PEG molecules surrounding them.[35] Interactions between cellulose and PEG have been confirmed by differential scanning calorimetry (DSC), which is useful in determining the temperature and enthalpy of a phase transition of PEG. The transition temperature (onset) and enthalpy of PEG in the composite, which is smaller than those of genuine PEG, decrease with decreasing PEG content (Figure 9.21), suggesting stronger restrictions of PEG molecules by cellulose with a decreasing content of PEG in the composite. Such phenomena do not occur in a simple mixture of native cellulose and PEG without ball milling. This is also confirmed by DSC analysis of the constituents of cellulose and PEG separated by washing with water a composite with a PEG content of up to 20 wt%, where the separated cellulose and PEG are hereafter named washed cellulose and extracted PEG. Thus, the washed cellulose shows no phase transition in the region of temperature concerned, while the extracted PEG exhibits almost the same melting temperature and enthalpy as genuine PEG shown in Figure 9.21. Further, IR spectra of the composites reveal almost simple superposition of spectra due to amorphous cellulose and the extracted PEGs without any new signals, except for a broad OH signal that shifts to smaller wavenumbers with an increase in PEG content. These results prove that the PEG and cellulose constituents in the composite undergo no chemical change, except for interactions of cellulose and PEG as well as a decreasing molecular weight of PEG (*e.g.*, $M_W = 1.5 \times 10^5$ for the extracted PEG from the composite).

The cellulose phase in a composite has a slightly increasing crystallinity index[19] (*CI*) due to cellulose I, ranging from 0 to 11%, nearly amorphous,

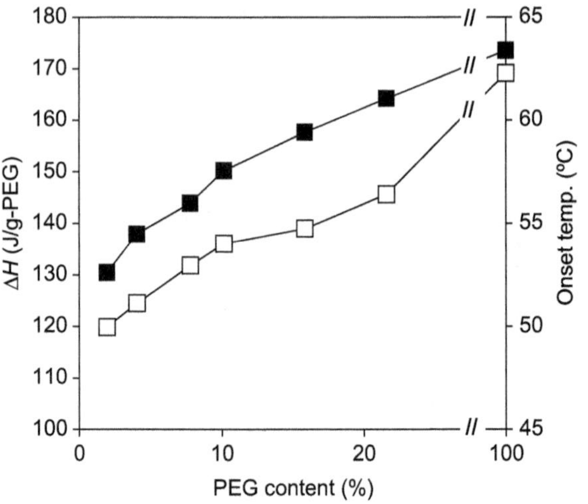

Figure 9.21 Dependences of the onset melting temperature (□) and enthalpy (Δ*H*) (■) of the composite of cellulose and PEG on the PEG content. Reprinted from ref. 35 with permission of the Chemical Society of Japan.

with an increase in PEG content of 0–22 wt%. Strikingly, the *CI* value of the washed cellulose increases to 30–50% according to an increase in the PEG content, suggesting intensive recrystallization of cellulose into a phase of cellulose I by extracting PEG molecules from the composite. Notably, amorphous cellulose prepared by ball milling only native cellulose remains amorphous even if it is washed with water. These results indicate that, when a mixture of native cellulose and PEG is ball-milled, the mechanical stress primarily makes cellulose amorphous and allows extensive insertion of parts of PEG molecules between microfibrils or even between cellulose chains in microfibrils, with other parts of the PEG molecules involved in the PEG phase. Accordingly formation of hydrogen bonds between cellulose and PEG molecules causes compatibilization between cellulose and PEG phases, and the resultant composite can be classified as a polymer alloy.

The composite is completely different from a corresponding blend prepared from a solution of cellulose and PEG, because in the blend cellulose and PEG molecules are considered to be miscible with each other.[36] The cellulose phase in the composite exhibits recrystallization to cellulose I on extraction of PEG, while removal of PEG from the blend makes cellulose chains assemble to form a phase of cellulose II, because of its higher stability than cellulose I. Therefore, the composite of cellulose and PEG is characterized not only by high dispersion of fine particles of cellulose in the PEG matrix, due to strong interactions between cellulose and PEG molecules, but also by a parallel orientation of cellulose chains in each primary phase of cellulose. The composite exhibits a thermoplastic property owing to the high fluidity of fine particles of cellulose in it, and accordingly yields various molds by usual hot-pressing. A similar composite is obtained by ball milling of cellulose and poly(ε-caprolactone), another polymer with high affinity for cellulose, and is readily hot-pressed to yield various molds with water-resistance, differing from the composite with PEG.

9.5.2 Composite of Cellulose and Maleated Polyolefin

Synthesis of composites consisting of hydrophilic cellulose and hydrophobic polyolefin usually requires a compatibilizer such as maleated polyolefin, in marked contrast to the composite of cellulose with PEG described above. Melt mixing of fibrous crystalline cellulose (FC) with maleic anhydride-grafted polyethylene (MPE) at 200 °C for 10 min yields strong adhesion of FC to MPE in an FC–MPE composite, while there is no adhesion of FC to unmaleated PE (UPE) in an FC–UPE composite.[37] The strong adhesion between FC and MPE phases is obtained even on using MPE with a slight MA content of 0.25–0.5 wt%. Such interfacial structures of the FC–MPE and the FC–UPE composites cause quite different behaviors of crystallization as well as melting.[38] Accordingly, the FC–MPE composite exhibits an increase in tensile strength and Young's modulus in the range of FC content ≤60 wt%, suggesting improved transfer of the high tensile strength and elasticity of FC to the MPE phase, in marked contrast to the FC–UPE composite, which exhibits decreases in both tensile properties. Furthermore, the FC–MPE composite exhibits a greater storage modulus than the FC–UPE composite, particularly in the FC content of 60 wt%, strongly suggesting the more improved dispersion of FC fibers in the MPE matrix.[37] Nevertheless, the FC–MPE composite still exhibits a significant decline in toughness[‡] with an increasing FC content, owing to a remarkable decrease in elongation.[37] Similar results have been observed also on FC–MPP and FC–UPP composites obtained by melt-mixing, where MPP and UPP denote maleated and unmaleated polypropylene, respectively.[39,40]

The marked decrease in toughness and fluidity of cellulosic composites with an increase in cellulose content makes it difficult to develop novel composites with a greater content of cellulose. This suggests strongly the necessity of the formation of an interphase between FC and PE or PP phases, with PE or PP chains bonded onto cellulose particles. The formation of ester bonds between OH groups of FC and MA grafts of MPE or MPP was not confirmed for the FC–MPE or the FC–MPP composites obtained by melt-mixing.[37,39]

[‡]Toughness is directly proportional to the multiplication of tensile strength and elongation.

9.5.2.1 Formation of an Interphase between Cellulose and Polyethylene Phases by Ball Milling

The crystallinity of FC is retained even through melt-mixing FC and MPE[41] and, accordingly, there are almost no OH groups free from hydrogen bonding on the surface of the FC particles that are probably capable of forming ester bonds with MA groups of MPE. As described before, ball milling is an effective method to activate crystalline cellulose, that is, to change the crystalline state of cellulose into an amorphous state so that a great number of free OH groups can be formed on the surface of the resultant fine particles of amorphous cellulose. Therefore, our interest is directed to mechanochemical synthesis of composites of fibrous crystalline cellulose (FC) with MPE by ball milling.[41]

Mechanochemical compounding of FC and MPE has been performed in a Planetary Ball Mill (Pulverisette 5, Fritsch, Germany) equipped with four jars, each 500 cm^3 in volume and loaded with 25 balls (diameter: 1.5 cm). Milling of FC (13.5 g) and MPE (MA graft content: *ca.* 0.8 wt%; 31.5 g) in each jar at 250 rpm with a cyclic mode of 50 min milling followed by a 10 min pause, for 24 h, yields a mixture with a FC content of 30 wt% (MCMix); part of the MCMix was then further melt-mixed in an R60 mixer (Toyo Seiki; Japan) at 180 °C for 30 min to obtain a mechanochemically synthesized composite (MCCom). For comparison, another composite (UMCom) has been prepared with the same composition but only by melt-mixing under the same conditions as described above.

The extent of MPE bonded onto cellulose in a product can be estimated by completely removing free MPE in the product by solvent extraction. The residue fractions of MPE are 3.1% (UMCom), 19.3% (MCMix), and 22.3% (MCCom), based on the total weight of MPE in each material. As the weight loss of genuine MPE is 100% by the same treatment, this demonstrates clearly that ball milling induces extensive bonding of MPE molecules onto cellulose even in a solid state without any solvent, in marked contrast to the case of UMCom by usual melt-mixing.

IR spectra of the UMCom, MCMix, and MCCom are shown in Figure 9.22, compared with those of MPE. The MPE has characteristic bands at 1792 cm^{-1} and at 1713 cm^{-1} due to saturated succinic anhydride groups[32] and their hydrolysis derivatives, respectively, as the latter band disappears and the former one increases in intensity after drying. Such behavior is observed on the MCMix before and after drying (Figure 9.22d and e), suggesting free MPE still remaining in the mixture after ball milling. Notably, the MCCom and dried MCMix samples exhibit characteristic signals at 1733–1735 cm^{-1} as well as at 1717–1719 cm^{-1}, in marked contrast to the UCCom sample, which shows no signals in this region. The signal at 1733–1735 cm^{-1} is assigned to ester groups between OH groups of cellulose and MA groups of MPE. The formation of such ester bonds liberates a carboxylic acid in each relevant MA graft, and this is consistent with the appearance of the signal at 1717–1719 cm^{-1}. These results indicate clearly that ball milling causes

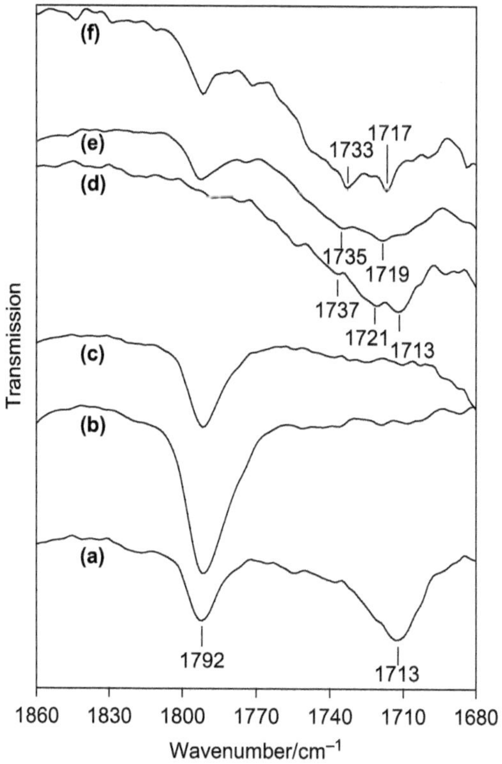

Figure 9.22 IR spectra of MPE (a), dried MPE (b), UMCom (c), MCMix (d), dried MCMix (e), and MCCom (f).
Reprinted from ref. 41.

extensive formation of ester bonds between FC and MPE, which is quite different from melt-mixing. It is remarkable that such mechanochemical esterification proceeds even between high-crystalline cellulose and MPE in a solid state by ball milling, as confirmed by the results of the MCMix sample. Ball milling of FC changes drastically its crystalline state into an amorphous one, as clearly shown in Figure 9.23, and this indicates that crystalline FC is so activated as to generate a great number of free OH groups on the surfaces of fine particles of amorphous cellulose. Of special interest, the ball-milled FC, even if melt-mixed with MPP, can generate ester bonds with MPP.[42]

From the aforementioned results, we can describe a mechanism for the mechanochemical compounding of crystalline cellulose with MPE by ball milling: (1) activation of cellulose to yield free OH groups and (2) reaction of resultant free OH groups with MA grafts of MPE to generate ester bonds. The activation of cellulose is essential for the compounding of crystalline cellulose with MPE through ester bonds. Accordingly, such esterification can proceed extensively by ball milling cellulose and MPE or even by ball milling only cellulose in advance to melt-mixing with MPE, only if cellulose particles have a number of free OH groups on the surfaces.

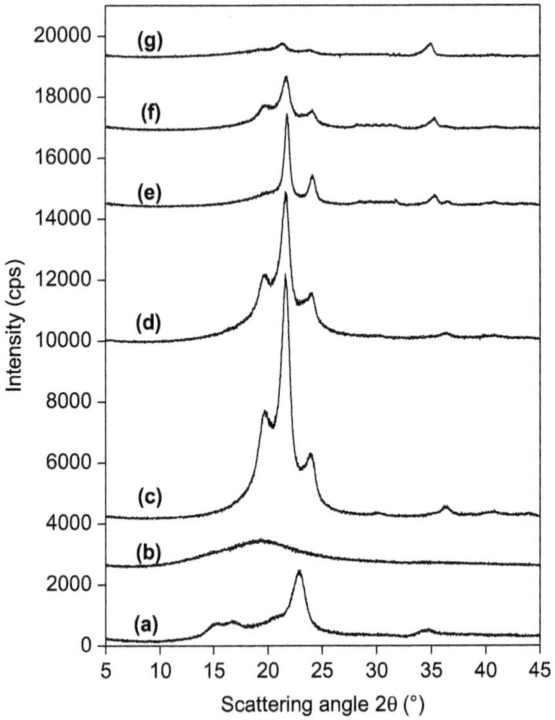

Figure 9.23 X-Ray diffractograms of original cellulose (a), milled cellulose (b), MPE (c), UMCom (d), MCMix (e), MCCom (f), and the sample after removing non-bonded MPE from the MCCom with xylene (g).
Reprinted from ref. 41.

Thermal and mechanical properties of the MCCom are shown in Table 9.3, in comparison with those of the UMCom. The MCCom exhibits a lower melting enthalpy and a lower thermal expansion coefficient than UMCom, which is consistent with the lower crystallinity of MPE in the MCCom (Figure 9.23). Note that the MCCom exhibits almost the same tensile strength as, but a much greater elongation than, the UMCom. Further, the Izod impact strength of the MCCom is surprisingly enhanced by 173%. This indicates that the mechanochemical method can greatly improve the toughness and ductility of the MCCom, probably owing to formation of an interphase between cellulose particles and the MPE matrix.

9.5.2.2 Characteristics of an Interphase Formed in a Mechanochemical Composite of Cellulose and Polypropylene

Let us consider in detail such an interphase in a mechanochemical composite of FC (30 wt%) and MPP (70 wt%).[43] For this purpose, by removing thoroughly free MPP from the relevant products UMCom, MCMix, and

Table 9.3 Thermal and mechanical properties of the UMCom and the MCCom. Reprinted from ref. 41.

	Thermal properties		Mechanical properties		
	Melting enthalpy[a] (J g^{-1})	Linear expansion coefficient[b] (mm mm^{-1} °C^{-1})	Tensile strength[c] (MPa)	Tensile elongation[c] (%)	Izod impact strength[d] (kJ m^{-2})
MPE	139.6	36.2×10^{-5}	19.6	>500	No peak
UMCom	97.0	27.6×10^{-5}	23.9	32	9.3
MCCom	87.2	23.4×10^{-5}	24.1	73	25.4

[a]Values determined from differential scanning calorimetry (DSC) at a heating rate of 5 °C min^{-1} with a Perkin-Elmer Pyris 1 DSC apparatus.
[b]Values determined from thermal mechanical analysis (TMA) at a heating rate of 2 °C min^{-1} with a Perkin-Elmer DMA 7e.
[c]Values determined with a Shimazu AG-100A tensile tester at a strain speed of 50 mm min^{-1}, according to the Japan Standard Society, JIS K7113-1995, Testing method for tensile properties of plastics, 1995.
[d]Values determined with a Yasuda No. 258 impact tester with a hammer of 5.5 J, according to the Japan Standard Society, JIS K7110-1984, Determination of Izod impact strength, 1984.

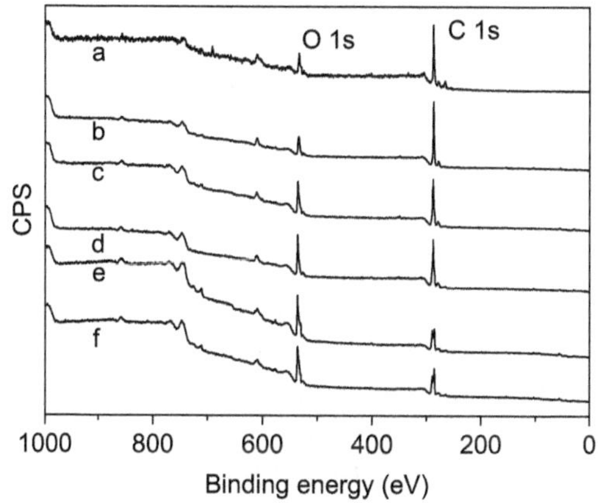

Figure 9.24 XPS spectra of MPP (a), rMCCom (b), rMCMix (c), rUMCom (d), milled CF11 (e), and original CF11 (f).
Reprinted from ref. 43 with permission of John Wiley & Sons.

MCCom with xylene in a Soxhlet extractor, the resultant residues are denoted as rUMCom, rMCMix, and rMCCom, respectively. The contents of MPP bound onto FC are 0.9 wt% for the rUMCom, 4.7 wt% for the rMCMix, and 10.2 wt% for the rMCCom on the basis of the total weight of the MPE phase. The rMCMix and the rMCCom certainly possess ester bonds, according to IR spectroscopy, in contrast to the rUMCom, and quite similarly to the cases of the systems of FC with MPE as mentioned earlier.

X-Ray photo-electronic spectrometry (XPS) is very useful for characterization of the surface of a material; XPS spectra of these residue samples are shown in Figure 9.24. Two characteristic signals due to C(1s) and O(1s) are observed at 285 and 532 eV, respectively, owing to C–C bonds of the main chain of MPP and O–C=O of MA grafts of MPP. Notably, these residue materials have different ratios of peak areas due to C(1s) and O(1s), ratios that are greater than that of FC. Thus the order of the ratios C/O is FC (0.76 and 0.70 for original and after milling) < rUMCom (1.12) < rMCMix (1.60) < MPP (1.78) < rMCCom (2.05). It is remarkable that the rMCMix and the rMCCom have greater C/O ratios than the rUMCom, with the rMCCom exhibiting an even greater C/O ratio than does MPP. The greater C/O ratios of the residue samples than of CF11 means that, in the residue samples, cellulose particles are covered by MAPP chains, because MAPP has the greater C/O ratio than FC. Furthermore, the order of the C/O ratios for the residues is consistent with that of the weight increases mentioned above. These results suggest strongly esterification of MPP on the surface of FC. The greater C/O ratio of the rUMCom than FC indicates a slight occurrence of esterification even in the UMCom, although this is not been confirmed by IR spectrometry.[39]

The XPS spectrometry reveals that the surfaces of cellulose particles are covered MPP chains, owing to the formation of ester bonds between OH groups of cellulose and MA groups of MPP. The greater C/O ratio of the rMCCom than MPP indicates that the PP chain side of an MPP chain bound to a cellulose particle stretches out from the surface of the particle to outside, as the C/O ratio of MPP reflects just the average constitution of a MPP chain. Therefore, the structure of the MCMix or the MCCom can be described with a core–shell model: a cellulose particle as a core and MPP chains bound on the surface composing a shell, which is very effective in allowing the bound MPP chains to interact with the hydrophobic PP phase.

Interactions between MPP chains bound on cellulose particles can be characterized by differential scanning calorimetry (DSC), by comparison with that of neat MPP chains. Table 9.4 shows several behaviors of crystallization as well as melting of MPP bonded onto cellulose particles in the rUMCom, the rMCMix, and the rMCCom, where ΔH_c and ΔH_m are the

Table 9.4 DSC characteristics of MPP chains bound to cellulose in the rUMCom, the rMCMix, and the rMCCom. Reprinted from ref. 43 with permission of John Wiley & Sons.

	Crystallization		Melting	
Sample	Temperature[a] (°C)	$\Delta H_c{}^b$ (J g^{-1})	Temperature[a] (°C)	$\Delta H_m{}^b$ (J g^{-1})
MPP				
Melt mixing	120	−92.4	158	93.8
Milled/melt-mixing	123	−90.6	159	91.7
rUMCom	120	−34.2	150	17.1
rMCMix	122	−37.0	150	24.1
rMCCom	124	−45.3	148	30.0

[a]Value corresponds to the center of each DSC peak.
[b]Value calculated on the basis of the weight of bound MPP for the rUMCom, the rMCMix, and the rMCCom.

enthalpies for crystallization and melting of the MPP phase in the residue samples, respectively. The values of $-\Delta H_c$ and ΔH_m are much smaller for the residue samples than for neat MPP; but, of much interest, these values increase in the order of rUMCom < rMCMix < r MCCom. This means clearly that MPP chains bonded onto the particles of cellulose can crystallize, even if the degree of crystallization is much smaller than that of neat MPP and that the degree increases with the number of MPP chains bound onto the cellulose particles. The MPP chains bound onto the cellulose particles can interact not only within themselves but also with each other. Therefore, the MPP chains bonded onto the cellulose particles in the MCCom or the MCMix can form an interphase that causes tight interfacial adhesion between hydrophilic cellulose and hydrophobic PP phases.

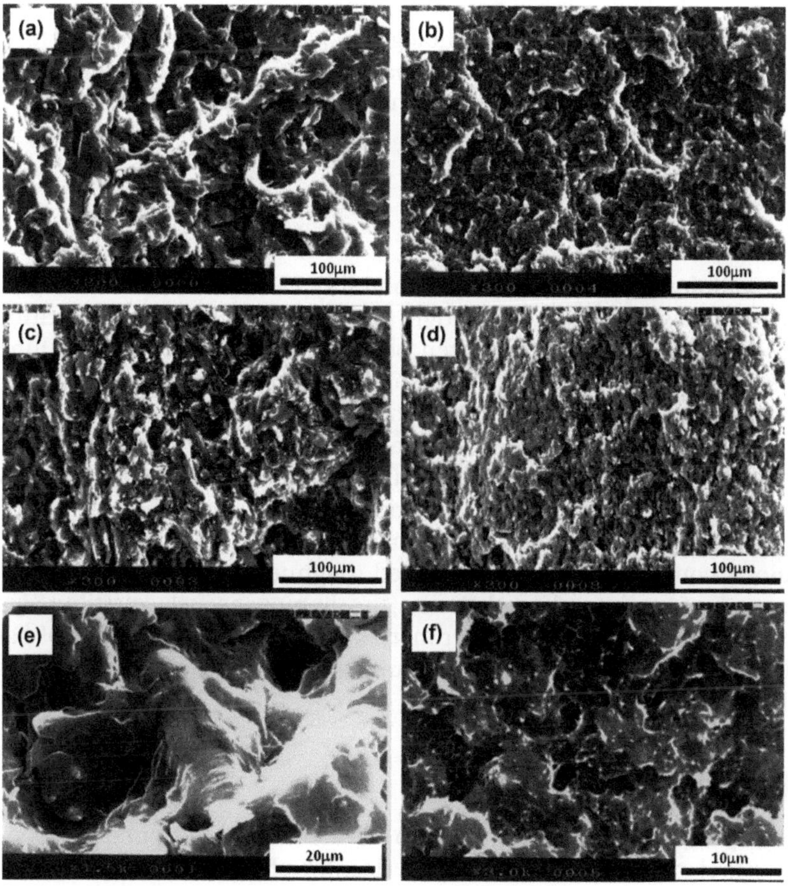

Figure 9.25 SEM microphotographs of fractured surfaces of UMCom ((a) ×300; (e) ×1500), MCCom ((b) ×300; (f) ×3000), UMCom60 ((c) ×300), and MCCom60 ((d) ×300).
Reprinted from ref. 43 with permission of John Wiley & Sons.

Morphological characteristics of the MCCom and the UMCom are revealed from observations of their fractured surfaces by scanning electron microscopy (SEM) (Figure 9.25). Even more improved distribution of cellulose particles are observed in the MCCom than in the UMCom, with both cellulose contents of 30 and 60 wt%. Magnified micrographs of these composites reveal a marked difference in interfacial adhesion between cellulose particles and the MPP matrix; for the MCCom, micro-particles of cellulose are distributed uniformly and tightly in the matrix, while for the UMCom the surface of a particle of fibrous cellulose is only partly covered with MPP, yielding a much rougher interface. This suggests strongly a much more effective transfer of the elasticity of cellulose into the MPP matrix in the MCCom than in UMCom. Accordingly, the MCCom exhibits greater tensile strength than does the UMCom for a cellulose content of $\leq$70 wt%, with the Young's modulus being greater than that of the UMCom at a content of 70 wt%, below which both are almost equal. The tensile properties of the MCCom reflect well its morphology, *i.e.*, the homogeneous distribution as well as tight adhesion of micro-particles of cellulose in and with the MPP matrix, both owing to the nature and structure of the interphase described above.

9.6 Concluding Remarks

Native cellulose exhibits not only insolubility in the usual solvents but also no thermoplasticity, due to the tight packing of cellulose chains through intra- and intermolecular hydrogen bonding all over them. Ball milling of fibers of native cellulose in the solid state produces fine powders of amorphous cellulose with parallel chains. The amorphous powders exhibit unique properties, particularly emphasized by a transformation into cellulose II on addition of a specific amount of water. The amorphous powders are so reactive that they can effectively form ester bonds with maleic anhydride-grafted polyethylene or polypropylene. Consequently, resultant cellulosic composites possess improved toughness and ductility due to the formation of an interphase between the homogeneously distributed particles of cellulose and the polymer matrix. Therefore, the activation of native cellulose by ball milling will be an important base for its extensive applications.

Acknowledgements

T. H. expresses his sincere thanks to all members, who pursued cooperatively novel research on the characterization of biomaterials for their extensive applications, of his research group in the former Shikoku National Industrial Research Institute (SNIRI) and then National Institute of Advanced Industrial Science and Technology (AIST) during the years 1996–2004. T. H. greatly appreciates Dr Y. Kimura, a former Director General of

SNIRI, for his encouragement and patience, and Dr T. Miyazaki, the former Director of the Institute for Marine Resources and Environment, AIST, for his considerate support.

References

1. S. Kobayashi, *Encyclopedia of Cellulose* (Japanese), ed. The Cellulose Society of Japan, Asakura Publishing Co., Ltd., Tokyo, 2000, ch. 1, pp. 1–2.
2. G.-H. Hu, J.-J. Flat and M. Lambla, *Reactive Modifiers for Polymers*, ed. S. Al-Malaika, Blackie Academic & Professional, London, 1997, ch. 1, pp. 1–83.
3. M. Umemura, Y. Yuguchi and T. Hirotsu, *J. Mol. Struct: Theochem*, 2005, **730**, 1.
4. H. Sugiyama, T. Nitta, M. Horii, K. Motohashi, J. Sakai, T. Usui, K. Hisamichi and J. Ishiyama, *Carbohydr. Res.*, 2000, **325**, 177.
5. H. Sugiyama, K. Hisamichi, T. Usui, K. Sakai and J.-i. Ishiyama, *J. Mol. Struct.*, 2000, **556**, 173.
6. N. W. H. Cheetham, P. Dasgupta and G. E. Ball, *Carbohydr. Res.*, 2003, **338**, 955.
7. R. B. Best, G. E. Jackson and K. J. Naidoo, *J. Phys. Chem. B*, 2002, **106**, 5091.
8. M. Umemura, Y. Yuguchi and T. Hirotsu, *J. Phys. Chem. A*, 2004, **108**, 7063.
9. K. Tashiro and M. Kobayashi, *Polymer*, 1991, **32**, 1516.
10. Y. Nishiyama, P. Langan and H. Chanzy, *J. Am. Chem. Soc.*, 2002, **124**, 9074.
11. R. Atalla and D. L. Vanderhart, *Science*, 1984, **223**, 283.
12. J. Sugiyama, R. Vuong and H. Chanzy, *Macromolecules*, 1991, **24**, 4148.
13. H. Yamamoto, F. Horii and H. Odani, *Macromolecules*, 1989, **22**, 4130.
14. M. Wada, *Macromolecules*, 2001, **34**, 3271.
15. P. Langan, Y. Nishiyama and H. Chanzy, *J. Am. Chem. Soc.*, 1999, **121**, 9940.
16. T. Hahn (ed.), *International Tables for Crystallography*, D. Reidel Publishing Company, Dordrecht, 2nd revised edn, 1987, vol. A (Space-Group Symmetry), pp. 112–113.
17. T. Endo, R. Kitagawa, T. Hirotsu and J. Hosokawa, *Kobunshi Ronbunshu* (Japanese), 1999, **56**, 166.
18. K. Chitumbo, W. Brown and A. De Ruvo, *J. Polym. Sci.: Polym. Symp.*, 1974, **47**, 261.
19. A. Isogai and M. Usuda, *Sen-i Gakkaishi*, 1990, **46**, 324.
20. J.-M. Lauriol, J. Comtat, P. Froment, F. Pla and A. Robert, *Holzforschung*, 1987, **41**, 165.
21. T. Endo, T. Hirotsu and J. Hosokawa, *Jpn Pat.* JP2979135, 1999.
22. M. Ago, T. Endo and T. Hirotsu, *Cellulose*, 2004, **11**, 163.

23. T. Endo, F. Zhang, R. Kitagawa, T. Hirotsu and J. Hosokawa, *Polym. J.*, 2000, **32**, 182.
24. V. J. Mcbrierty and K. J. Packer, *Nuclear Magnetic Resonance in Solid Polymers*, Cambridge University Press, Cambridge, 1993, ch. 3, pp. 51–81.
25. W. Qiu, T. Endo and T. Hirotsu, *Eur. Polym. J.*, 2005, **41**, 1979.
26. W. Qiu and T. Hirotsu, *Macromol. Chem. Phys.*, 2005, **206**, 2470.
27. G. Moad, *Prog. Polym. Sci.*, 1999, **24**, 81.
28. W. Heinen, C. H. Rosenmöller, C. B. Wenzel, H. J. M. de Groot, J. Lugtenburg and M. van Dunin, *Macromolecules*, 1996, **29**, 1151.
29. L. Yang, F. Zhang, T. Endo and T. Hirotsu, *Macromolecules*, 2003, **36**, 4709.
30. G.-H. Hu, J.-J. Flat and M. Lambla, *Reactive Modifiers for Polymers*, ed. S. Al-Malaika, Blackie Academic & Professional, London, 1997, ch. 1, pp. 10–13.
31. K. Miyauchi and K. Saito, *Bunseki Kagaku* (Japanese), 2006, **55**, 547.
32. B. de Roover, M. Sclavons, V. Carlier, J. Devaux, R. Legras and A. Momtaz, *J. Polym. Sci. A.: Polym. Chem.*, 1995, **33**, 829.
33. W. Qiu, T. Endo and T. Hirotsu, *J. Appl. Polym. Sci.*, 2006, **102**, 3830.
34. A. H. Hogt, J. Meijer and J. Jelenič, *Reactive Modifiers for Polymers*, ed. S. Al-Malaika, Blackie Academic & Professional, London, 1997, ch. 2, pp. 112–116.
35. T. Endo, R. Kitagawa, F. Zhang, T. Hirotsu and J. Hosokawa, *Chem. Lett.*, 1999, 1155.
36. T. Kondo, C. Sawatari, R. St. J. Manley and D. G. Gray, , *Macromolecules*, 1994, **27**, 210.
37. F. Zhang, T. Endo, W. Qiu, L. Yang and T. Hirotsu, *J. Appl. Polym. Sci.*, 2002, **84**, 1971.
38. F. Zhang, W. Qiu, L. Yang, T. Endo and T. Hirotsu, *J. Appl. Polym. Sci.*, 2003, **89**, 3292.
39. W. Qiu, T. Endo and T. Hirotsu, *J. Appl. Polym. Sci.*, 2006, **102**, 3830.
40. W. Qiu, F. Zhang, T. Endo and T. Hirotsu, *J. Appl. Polym. Sci.*, 2003, **87**, 337.
41. F. Zhang, W. Qiu, L. Yang, T. Endo and T. Hirotsu, *J. Mater. Chem.*, 2002, **12**, 24.
42. W. Qiu, F. Zhang, T. Endo and T. Hirotsu, *J. Appl. Polym. Sci.*, 2004, **91**, 1703.
43. W. Qiu, T. Endo and T. Hirotsu, *J. Appl. Polym. Sci.*, 2004, **94**, 1326.

CHAPTER 10

Technical Implications of Organic Syntheses in Ball Mills

ACHIM STOLLE

Institute for Technical Chemistry and Environmental Chemistry, Friedrich-Schiller University Jena, Lessingstr. 12, D-07743 Jena, Germany
Email: Achim.Stolle@uni-jena.de

10.1 Introduction

Synthesis chemists, regardless from which field of chemical synthesis, search for new methods to optimize task-specific synthetic procedures.[1,2] The reasons for this are as numerous as the types of chemical reactions that exist and are determined by mainly external forces. On the one hand, sociological or ecological pressure can drive a synthesis chemist to look for alternative reaction paths employing, for example, less hazardous chemicals or economic reasons might drive an industrial researcher to develop a new or advanced process or catalyst to circumvent processes patented by competing companies. Thus, social, economic, and ecological but also legislative, regulative, or political reasons are limiting factors catalyzing the development of new and advanced reaction protocols, process techniques, or reaction types. Facing these factors many researchers seek for protocols to:

i. increase product yield (by increasing yield and/or selectivity),
ii. speed-up the reaction rate and therefore increase the space–time yield,
iii. decrease investment costs by application of methods that are as simple as possible,

RSC Green Chemistry No. 31
Ball Milling Towards Green Synthesis: Applications, Projects, Challenges
Edited by Brindaban Ranu and Achim Stolle

Published by the Royal Society of Chemistry, www.rsc.org

iv. save reagents, solvents, additives, or catalysts to reduce operating expenses,
v. increase the energy efficiency of a process,
vi. employ new or alternative reagents (*e.g.* renewable materials).

Catchphrases like *green chemistry*, *sustainable chemistry*, *novel process windows*, or *process intensification* summarize the endeavours to deal with the limiting factors.

For sure, the simplest solution for process intensification is the possibility to optimize an existing process or a reaction by variation or process parameters like pressure, temperature, concentration profile or pH. These methods allow us to use existing technologies and apparatus, reducing the investment costs for implementation of the optimized reaction conditions.

If one of the aims is the employment of alternative solvents or reagents (*e.g.* catalysts), substantial modification of the process itself is necessary, including up-stream processes and reactor type.[1–3] Hence, the balance sheet for an assessment has to be enlarged by incorporating the respective process steps. A change of material properties (*e.g.* state of aggregation, heat capacity) often forces process engineers to change the reactor concept, increasing the effort with respect to process development and investment costs.

In connection with the last mentioned fact, a switch of reactor technology might be applicable, influencing either the process concept (continuous *versus* batch) or the technology itself. Integration of alternative synthesis concepts like micro-reaction technology,[4] high-pressure synthesis with supercritical solvents,[5] microwave-assisted synthesis,[6] or the application of ball mills for materials[7–9] or organic syntheses[10] leads to novel fields of application and products that are difficult to synthesize by other routes. The implementation of those concepts can result in omission of complete reaction steps or sequences and contributes to reduction of the ecological footprint and energy intensity.[11,12]

10.1.1 Mechanochemistry

The energy that is released due to the movement and collision of grinding bodies does not lead to particle refinement only, but can also provide enough energy to initiate chemical reactions in the bulk material. This field of chemistry is defined as *tribochemistry* or *mechanochemistry*.[7] Heinicke provided the following definition:

> "Tribochemistry is the branch of chemistry dealing with the chemical and physic-chemical changes of matter due to the influence of mechanical energy."[7,13]

However, there has been discussion of what *mechanical energy* is or, specifically, if solely mechanical energy drives the chemical reactions or if thermal energy released by frictional forces contributes to the reaction progress.[7,8,14,15] This led Fernández-Bertrán to slightly change the definition, specifying the type of energy that initiates the chemical reaction and also defining the types of reagents that are comminuted in ball mills ("Tribochemistry are reactions generated by friction during the milling of solid reagents").[16] However, most authors stick to the original definition of Heinicke, also accepting that local temperature increases in the collision zone can activate the reactions. This consideration led to IUPAC's definition of a *mechanochemical reaction*:

> "[A mechano-chemical reaction] is a chemical reaction that is induced by the direct absorption of mechanical energy."[17]

It has to be mentioned that this definition is based on considerations with respect to polymer disintegration by mechanical processes (tensile load), but it is generally accepted for mechanochemical reactions of small organic molecules[15] and furthermore for organic reactions carried out in ball mills.[10,18] During the action of the grinding bodies in the ball mill various processes can take place beside solely particle refinement:[16]

i. comminution of particles to very small size (down to nm-scale),[8,19–21]
ii. generation of large new surfaces,[9,21]
iii. formation of dislocation and point defects (micro-strains),[21]
iv. phase transformation in polymorphic materials,[22–24]
v. chemical reactions.[8,14–16,18,24,25]

10.1.2 Using Mortar and Pestle for Organic Syntheses

With respect to the definition of mechanochemistry the question of *how the energy is distributed to the system* is unanswered. The most obvious way and simplest possibility of initiating a mechanochemical reaction is the application of mortar and pestle. Although this method still finds widespread application in solvent-free organic synthesis,[26–28] its use comes along with several disadvantages considering process safety and stability.[10] Comminuting solids in an open environment by hand can cause several hazards (dust, explosive heat evolution) for the operator if no special precautions are implemented. Furthermore, the energy entry strongly depends on the physiognomy of the operator and significant differences in yield can occur if the operator changes.[29] Additionally, manually operated grinding devices tend to provide the energy not continuously due to fatigue of the operator over long reaction times. Thus, the accomplishment of reactions by different persons may lead to different results (Figure 10.1), except for reactions with a low activation barrier or a high thermodynamic driving force. From a chemical engineering point of view ball mills should be applied since those offer the possibility to control the reaction time, energy density, and other parameters.[9,10]

10.1.3 Ball Mills

The application of ball mills in the field of organic synthesis contributes to *process intensification* of an existing process or reaction. On the other hand,

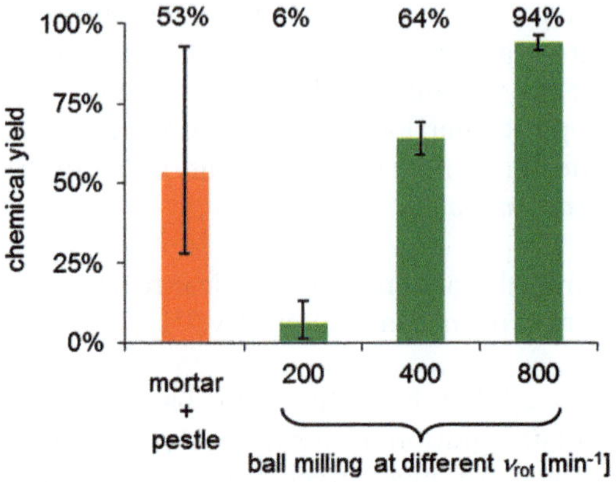

Figure 10.1 Comparison of the results for a Suzuki-cross coupling carried out in the solid state with mortar and pestle (results from nine persons of different age and gender) and in a planetary ball mill at different rotation frequencies.[29]

implementation of this philosophy might lead to new or advanced reaction procedures and or products. In general, ball mills are machines designed for a specific purpose – particle refinement.[8,9,19,30,31] Owing to continuous grinding of particles with grinding bodies their size is reduced, resulting in fine particles with a defined size distribution.[19,20] Reduction of the particle size increases the specific surface area and also the surface concentration of defects, resulting in higher reactivity or improved processing of the particles.[21,31]

The application of ball mills is one of the various possibilities by which to increase the energy density of a system, *i.e.* increase the kinetic energy and thus the bulk temperature. The energy provided by the propulsion unit of the ball mill causes movement of grinding bodies (also known as milling balls) in the milling chamber (also known as milling beaker, milling vessel) resulting in frictional forces between the grinding bodies, the grinding bodies with the milling chambers wall, and the feed material in-between grinding bodies. Those frictional forces cause wear stresses of the feed material and lead to particle refinement.[8] The wear mechanisms depend on the type of mill and include impact, friction, or shear.

10.1.4 Scope of this Chapter

This chapter aims to provide an overview of the influence of various parameters concerning the application of ball mills for organic syntheses. The classification of parameters outlined below will focus on those with technical and/or engineering relevance. For an overview of the diversity regarding the chemical aspects in terms of different reaction protocols the reader is referred to the other chapters of this book.

10.2 Ball Mills as Reactors in Organic Synthesis

The prerequisite of efficient mixing and particle refinement for a successful mechanochemical reaction has been pointed out above. A technical solution that combines these two tasks in one apparatus are ball mills, whose application as reactor in organic synthesis has attracted attention within the last decade, as manifested in numerous review articles dedicated to the field in general.[10,27,32–37] With respect to *process intensification*, the utilization of ball mills seemed a promising way to improve the material and energy intensity of a chemical reaction. Beside the combination of mixing, particle refinement and energy entry in one apparatus, the most obvious advantage is the omission of solvents during the reaction. Therefore, reactions carried out in ball mills are generally considered as solvent-free.[38] However, the terms *solvent-free* or *solvent-less* are not defined by IUPAC, but definitions disclosed by Rothenberg *et al.* or Welton account for processes that are carried out in the absence of solvents in general.[38,39]

With regard to reactions in ball mills and the fact that solids interact with each other, *i.e.* a heterogeneous disperse reaction system is present, they can be classified as solvent-free.[26,28,38] In some cases the addition of small amounts of solvents (liquid-assisted grinding = LAG)[40] or the processing of liquids is reported in ball mills. In any event, compared to the amount of solid materials present in most cases the reaction mixtures appeared as free flowing powders that are not agglomerated.

Generally, processes in ball mills are characterized by different parameters, regardless of the type of process, *i.e.* a mechanochemical or a mechanically-assisted reaction or particle comminution, and the type of ball mill. In contrast to particle refinement solely, the balance volume for reactions in ball mills has to be enlarged taking into account chemical parameters of the specific reactions.[9,10,41] The large number of influencing variables (Scheme 10.1) can be categorized into:

i. chemical parameters,
ii. technological parameters,
iii. process parameters.

Chemical parameters summarize all those variables that are directly linked to the chemical transformations taking place in the chamber of the ball mill, which include (i) the type of chemical reaction, (ii) the presence of catalysts

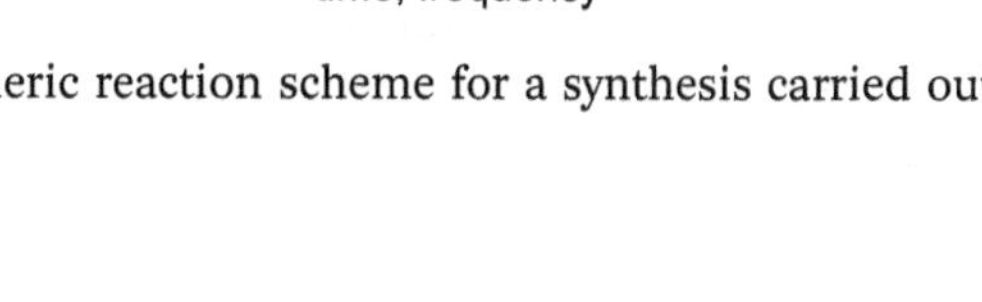

Scheme 10.1 Generic reaction scheme for a synthesis carried out in a ball mill.

or additives, (iii) the reagent ratio, and (iv) the presence of low amounts of liquid(s) in the case of liquid-assisted grinding, LAG.[40,42–46]

With respect to *technological parameters*, the importance of (i) the type of ball mill, (ii) the milling material, (iii) the number and (iv) size of milling balls as well as (v) the filling degree of the milling beakers for the accomplishment of chemical reactions should be considered. This set of parameters describes the technology that is used for comminution.

Process parameters are useful for controlling the energy entry during the process. The main values in this class of parameters are (i) operating frequency (rpm), (ii) reaction time ($t_{reaction}$), and (iii) temperature (T). Whereas rpm is directly correlated to the kinetic energy of the grinding media and, thus, to the energy that can be dissipated as heat to the mill charge, $t_{reaction}$ and T determine the progression of a chemical reaction and are linked to the reaction kinetics.

10.3 Technological Parameters

This first set of parameters describes the technology that is used for the comminution, such as type of ball mill, grinding material, number and size of the milling balls as well as the filling degree of the milling beakers.

10.3.1 Type of Ball Mill

Particle refinement technology includes various types of ball mills for applications in industry and at laboratory scale.[9,30,31,47] From the large number of technical possibilities for the realization of mechanical action of grinding bodies to a mill feed, only a manageable number has been applied so far for organic synthesis on the laboratory scale.[10,48] The main types of ball mills used so far for synthetic purposes and their working principles are summarized in Figure 10.2.[49] Figure 10.3 provides a general overview of the existing technologies applicable for mechanochemical reactions of solids on laboratory to (pilot) plant scale.[50] Beside differences in the operating principle they differ in the maximum capacity of the milling vessel, $V_{MV,max}$, ranging from 0.010 L for planetary (PBM), mixer (MBM) or vibration ball mills (VBM) to several 100 L for horizontal rolling ball mills.

10.3.1.1 Planetary Ball Mills

Planetary ball mills (PBMs, Figure 10.3B) are a special form of drum mills used, for example, in the ore refining industry at large-scale levels (see for instance Figure 10.3D). A PBM is a device by which on a central disk (sun disk) disks of smaller diameter are mounted containing the holding mechanism(s) for the milling beakers, whereby the number of vessels is 1, 2 or 4. The disks rotate in opposite directions and, therefore, the movement of the milling beakers is similar to the rotation of a planet around their central celestial body in the sun system.[49,51,52] The grinding bodies inside the

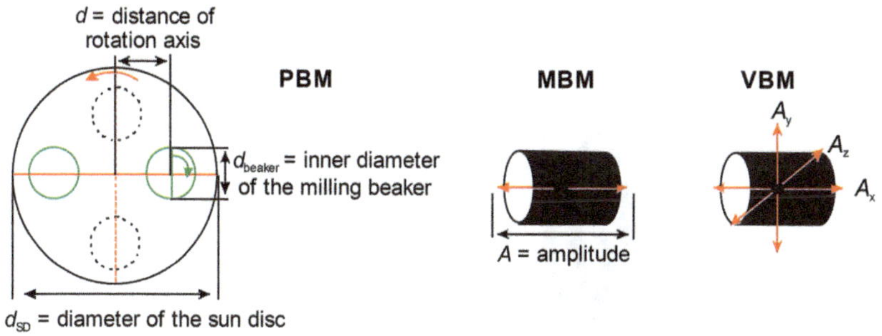

Figure 10.2 Working principle of different ball mills used for organic synthesis. PBM: planetary ball mill; MBM: mixer ball mill; VBM: vibration ball mill.

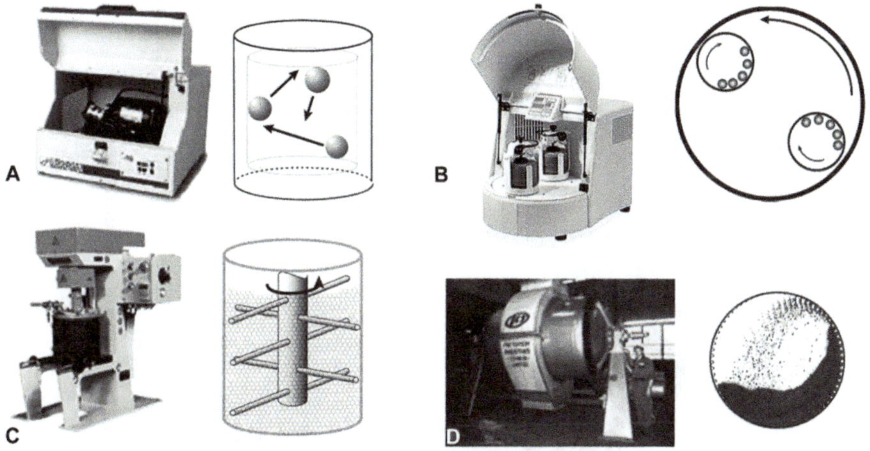

Figure 10.3 Set of milling technologies currently used at the laboratory and (pilot) plant scale for the performance of mechano-chemical reactions and the processing of biomass. (A) SPEX mixer mill ($V_{MV,max}$ = 0.065 L). (B) Planetary ball mill ($V_{MV,max}$ = 0.5 L). (C) Attritor (V_{MV} = 4 L). (D) Rolling ball mill ($V_{MV,max}$ = up to m^3-scale).
Reproduced from ref. 50 with permission of the Royal Society of Chemistry.

milling beakers follow typical trajectories that recently have been monitored by video analysis and simulated with the help of the discrete element method (DEM; Figure 10.4).[53,54] Despite improvements it is still complicated to calculate the specific trajectories of the milling balls inside the beakers due to overlap of the two opposite rotation directions of the sun and the planetary disks.[53,55] It is generally accepted that the milling balls are accelerated and follow the rotation of the milling beaker for approximately one quarter of its perimeter, then they are moving through the volume of the milling beaker and collide with the opposite wall. The point at which the grinding bodies leave their centrifugal trajectory (*release point*; with respect

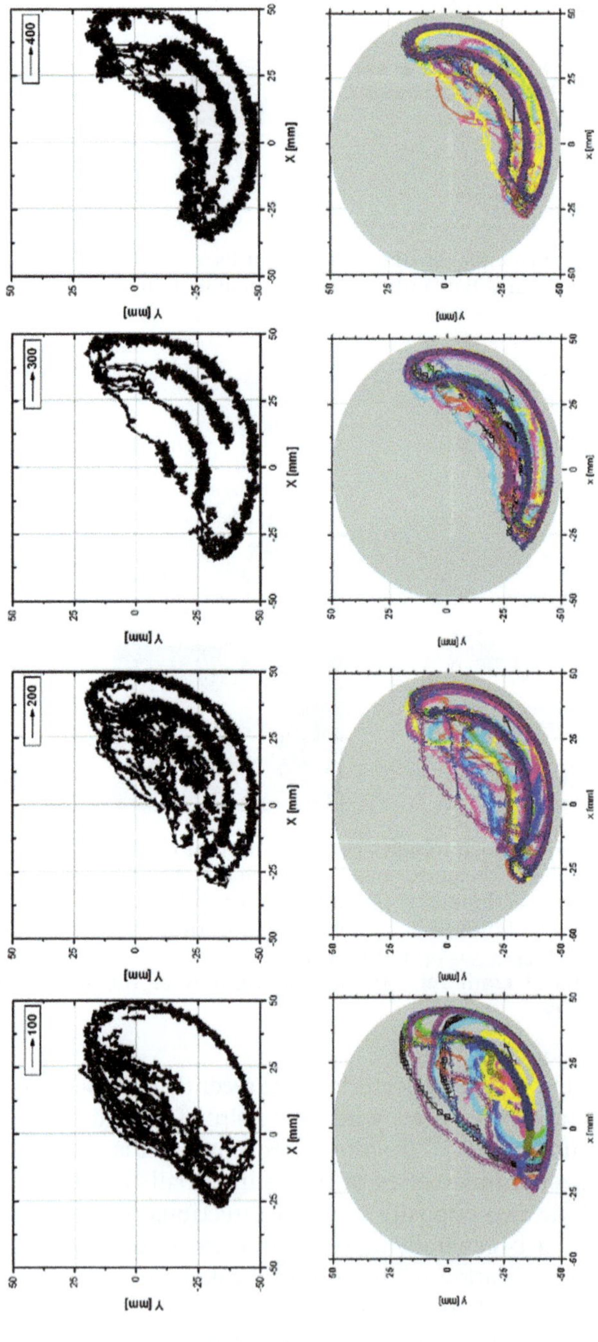

Figure 10.4 Grinding body trajectories in a planetary ball mill monitored by video analysis (top) and simulated using DEM (bottom; in each case 10 mm steel balls and without mill feed).
By courtesy of Sandra Breitung-Faes (Technical University Braunschweig).

to the movement of the planetary disc solely) depends on various variables, *e.g.* operating frequency, the perimeter ratio of the two disks, distance between the two central axis, and the gear ratio, which depends on the ratio of the rotation frequencies rpm of the two disks.[54,55] The main wear mechanisms in a PBM are friction and impact, resulting from the centrifugal trajectory and from the collision with the milling beakers wall or between different milling balls, respectively.[54]

10.3.1.2 Mixer and Vibration Ball Mills

Mixer ball mills (MBMs) and vibration ball mills (VBMs) are operated in completely different modes compared to PBMs. The initial force is generated by oscillation at variable frequencies (ν_{osc}). The difference between MBM and VBM is the degree of freedom of the oscillation and the plane wherein the oscillations takes place. The milling beaker in a MBM is mounted on a swing arm and oscillates in a horizontal plane on a circular arc (Figure 10.5). The energy density is mainly influenced by the radius of the swing arm, r_{arm}. Although the movement follows the circular arc, the tangential displacement (Δl_{tan}) is small compared to the horizontal displacement (Δl_{hor}) of the milling beaker on its trajectory. As a result, the milling beaker, including its grinding bodies, follows a straight line in the first approximation. Taking the laboratory MBMs MM301 and MM400 (Retsch GmbH) as an example, the r_{arm} and amplitude (distance A and A′ in Figure 10.5) are 138 and 20 mm, respectively. Based on those data, Δl_{hor}, Δl_{tan}, and 2θ are calculated to be 10 mm, 0.363 mm, and 89.8 rad ($\triangleq$ 0.5°), respectively. The milling balls are accelerated in a similar direction as the containment and due to their inertia they keep moving, although the oscillation reaches its reversal point (A and A′ in Figure 10.5) resulting in collision between the grinding media and the top or bottom ends of the milling containment. Ultimately, both wear mechanisms, friction and impact, contribute to the energy entry in the milling beaker and transfer to the mill feed.

In contrast to MBMs the oscillation in VBMs can take place at different planes in a Cartesian coordinate system (Figure 10.2 and Figure 10.3A).[56]

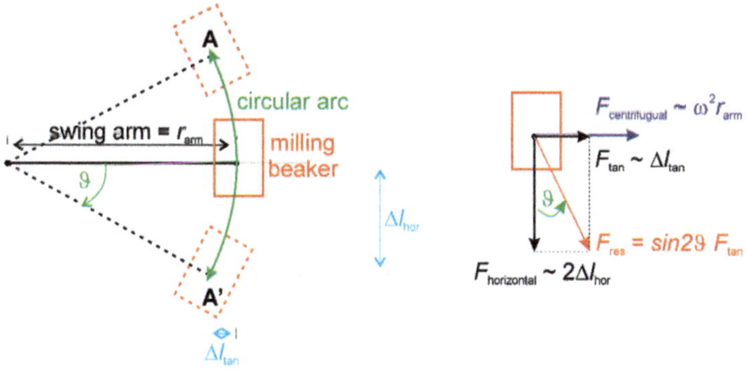

Figure 10.5 Schematic illustration of the milling beaker trajectory for a mixer ball mill (top view; left) and the force triangle for the moving parts of the milling beaker (right).

The trajectories are similar to those observed in case of a MBM, although the degree of freedom regarding the oscillation is increased by at least a factor of two (two-dimensional oscillation). Displacement in the *x*-, *y*-, and *z*-plane would increase the degree of freedom threefold compared to an oscillation in the horizontal plane. Owing to different directional vectors including their tangential displacement vectors the trajectory of the milling containment including the grinding bodies is complicated. The wear mechanisms are similar to MBM and PBM.

10.3.1.3 *Effect of the Ball Mill type on the Synthesis*

The literature reports many examples of chemical reactions performed in one of these ball mills.[10,48,57–59] However, studies comparing the effectiveness of the ball mill types in a similar reaction are seldom. Up to now, such comparative studies have been disclosed for the $Zn/ZnCl_2$-mediated Pinacol coupling of aromatic ketones and aldehydes,[60] Suzuki–Miyaura cross-coupling with $KF–Al_2O_3$,[29] Knoevenagel condensation of aromatic aldehydes,[61] the oxidative homo-coupling of anilines,[62] and the CuI-catalyzed Glaser reaction of phenylacetylene.[63] Most of the publications compare accomplishment of the respective reactions in a MBM and a PBM.[29,61–63] Experimental results for a Suzuki–Miyaura cross-coupling reaction (Scheme 10.2) indicate that there is no difference between the reactions carried out with respect to yield and selectivity if reaction time and scale are on a comparable level. The only difference is the changed operating frequency (Figure 10.6).[29] Whereas the PBM is often operated at its maximal rpm of 800 min^{-1} (equal to $\nu_{rot} = 13.3$ Hz), reaction at similar frequency in a MBM resulted in significantly lower yields.[29] Owing to the different torques of the machines the peripheral velocities (ν_p) and *g*-forces are not comparable. Thus, the energy density at the same frequency is different, resulting in different stress energies transferred to the mill feed (Eq. 10.1):[64,65]

$$E_{stress,feed} = d_{MB}^3 \nu_p^2 \rho_{MB} K^{-1} \text{ with } K = f(E_{Y,i}) \qquad (10.1)$$

where

$E_{stress,feed}$ (J) is the stress energy transferred to the feed material,
d_{MB} (m): diameter of the milling balls,
ν_p (m s^{-1}): peripheral velocity,
ρ_{MB} (kg m^{-3}): density of the milling balls,
$E_{Y,i}$ (Pa): Young's modulus.

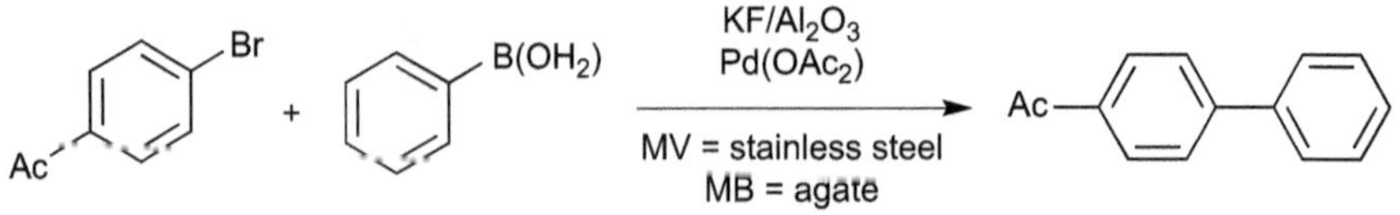

Scheme 10.2 Suzuki–Miyaura cross coupling used as model reaction for the investigation of parameter effects.

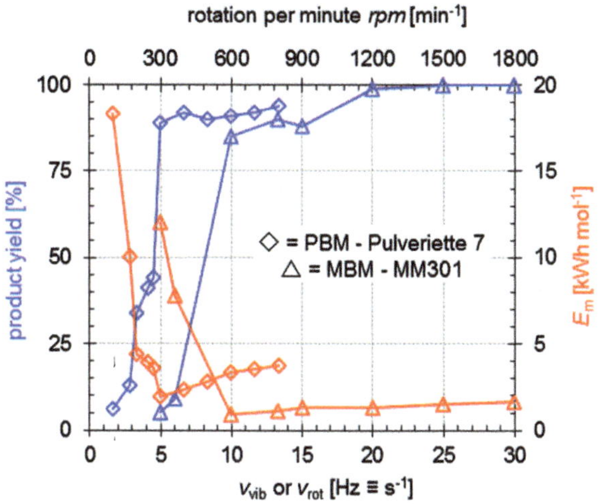

Figure 10.6 Correlation between operating frequency of two different types of ball mills and yield of product as well as energy intensity E_m (Eq. 10.16, see Section 10.5).[29]
Reproduced with permission of the Royal Society of Chemistry.

The higher centrifugal force in the case of a PBM generates *g*-forces that multiply the force generated by the propulsion drive system (*cf.* Figure 10.5). This effect results in the higher operating frequencies of a MBM in comparison to a PBM under the prerequisite that the yield of a chemical reaction is the same.

Comparison of a MBM, a VBM, and a *mortar grinder* was undertaken by Wang and Liu using the pinacol coupling of aromatic aldehydes and ketones as their model reaction.[60] The results indicate that the selectivity of the reaction is lower for accomplishment under the conditions of high-energy ball milling (MBM, VBM) compared to the experiments in the mortar grinder. Obviously, the increased energy density in the ball mills initiate decomposition of the vicinal diols to secondary alcohols. Such "association–dissociation equilibrium"[60] has been verified for the grinding of C_{60} with mortar and pestle yielding the dumbbell-shaped dimer C_{120}.[66]

10.3.2 Grinding Material

The choice of material for milling beakers and grinding media (milling balls) is important for the performance of organic reactions. Material properties like density (ρ_{MB}), Young's modulus ($E_{Y,MB}$), hardness, or abrasion behaviour significantly influence the outcome of a chemical reaction performed in a ball mill as well for particle refinement (*cf.* Eq. 10.1).[7,51,67] Based on Eq. (10.2) the kinetic energy E_{kin} of a system is determined by the moving mass m and the relative velocity v:[55]

$$E_{kin} = 0.5mv^2 \tag{10.2}$$

For rotation (i = rot) or oscillation (i = osc) processes Eq. (10.2) can be transformed into Eq. (10.3) with substitution of m and v by the moment of inertia I and the angular velocity ω or frequency ν ($\omega = 2\pi\nu$), respectively:[10]

$$E_{\mathrm{kin},i} \propto I\omega_i^2 \propto I\nu_i^2 \tag{10.3}$$

where I (kg m^2) is the moment of inertia, ω (s^{-1}) is the angular velocity and ν (s^{-1}) is the rotation or oscillation frequency.

On substitution of ν by the peripheral velocity (v_p) of a punctual mass that rotates at the distance r from the rotation axis and the general definition for the moment of inertia of a complex body, Eq. (10.3) can be transformed into Eq. (10.4), demonstrating the importance of the grinding body density for the kinetic energy of the system:[54,68]

$$E_{\mathrm{kin,rot}} = 6mv_\mathrm{p}^2 \tag{10.4}$$

where v_p (m s^{-1}) is the peripheral velocity and m (kg) is the mass of the rotating body.

The kinetic energy that originates from the torque of the propulsion unit of the ball mill is transformed through friction and impact processes into either chemical or thermal energy. If the mechanical action of the milling balls directly induce a chemical reaction, a *mechanochemical reaction* takes place.[17] On the other hand, frictional heat induces an increase of the local or bulk temperature of the mill feed and can speed up chemical reactions by providing the required activation energy. Although the kinetic energy provided by the propulsion system cannot be used completely as (thermal) activation energy due to dissipative losses, a correlation between the mass of the grinding bodies ($m_\mathrm{MB} \sim \rho_\mathrm{MB}$) and the chemical yield has been described for some organic reactions in ball mills.[10,62,69–71] In several cases a significant increase of yield was observed when the material is changed, *e.g.* from light-weight natural mineral agate ($\rho = 2.7$ g cm^{-3}) to the heavier ZrO_2 ($\rho = 5.7$ g cm^{-3}).[62,70] However, there are also examples from the field of organic synthesis in which no influence was observed, which seems to indicate a product decomposition due to high energy.[60,66,72,73]

Table 10.1 provides an overview of the materials for milling balls and milling vessels applicable for mechanochemical syntheses. Beside material density, the *chemical resistance* is also important for organic syntheses. From the available materials, tungsten carbide and stabilized zirconia are the materials of choice. Although they are very expensive, they proved to be chemically inert towards most reagents and in case of zirconia the abrasive resistance is also high. Other materials available for grinding bodies or milling containment have a high porosity (agate or steel) and, therefore, memory effects might occur if the tools are not properly cleaned after the reaction.

During ball milling, *abrasion* of particles from the surface of the milling tools may occur and can interact with the mill feed (Table 10.1). For metal-catalyzed reactions this might be a problem due to contamination with other catalytically active elements. However, it is also possible to take advantage of

Table 10.1 Overview of materials applicable for mechanochemical syntheses in ball mills.

Material	Vickers hardness	Density (ρ) (g cm^{-3})	Energy entry	Possible impurities due to wear
Stainless steel	550	7.8	High	Fe, Cr
Hardened steel	750	7.9	High	Fe, Cr, C
Tungsten carbide	1200	14.8	Very high	WC, Co
Agate	1000	2.7	Very low	SiO_2
Corundum	1750	3.9	Low	Al_2O_3[c]
Zirconia[a]	1200	5.9	Moderate	ZrO_2, MgO/Y_2O_3[b]
PTFE	Elastic	2.1	Very low[c]	F, C

[a]Magnesia-stabilized zirconia = MSZ. Yttrium-stabilized zirconia = YSZ.
[b]The deterioration of milling balls made from those materials is generally low compared to the other materials listed.
[c]In the case of polymer-coated steel balls the energy entry is high.

the abrasive behaviour and generate catalytic species *in situ*. Thus, substitution of Cu (co)-catalyst(s) in several reactions by milling vessel and milling balls made from Cu has been reported.[69]

Deterioration of the milling tools is generally a problem since the milling balls continuously lose weight,[69] which leads to a decrease of the moving masses with consequences for the kinetic energy. Since the abrasive potential not only depends on the material properties of the milling balls or containment surface but also on the hardness and particle size distribution of the mill feed,[62,74] it is important to monitor the weight loss. If the deterioration exceeds a specific value or if control experiments reveal that parameters like yield or selectivity tend to change significantly, the grinding media have to be replaced.

10.3.3 Number of Milling Balls

As indicated with Eq. (10.2) the mass of the grinding bodies influences E_{kin} of the moving bodies in a ball mill and thus the energy that can be transformed into chemical or thermal energy. Changing the mass of the grinding bodies by use of different materials is one possibility to influence the energy density. On the other hand, the number of the milling balls n_{MB} and their size d_{MB} are important parameters (*cf.* Eq. 10.1).[65,75,76]

Generally, one has to differentiate between mechanochemical syntheses in ball mills carried out in *single-ball mode* ($n_{MB} = 1$) or *multi-ball mode* ($n_{MB} \geq 2$). The possibility to run a ball mill with less than two milling balls depends on the type of mill used for the experiments. Whereas it is possible to operate a MBM or a VBM using only one milling ball, similar experiments would fail in planetary ball mills, since the wear stress would increase significantly. According to Figure 10.4 multiple milling balls are necessary to give the required trajectories inside the vessel, which are responsible for the energy transfer and the wear effects. With respect to organic syntheses in PBMs this parameter has been evaluated for various reaction examples such as Suzuki cross-coupling (Scheme 10.2),[72] oxidative homo-coupling,[62,63] as well as the oxidation of γ-terpinene to *p*-cymene[77] and of β-pinene to nopinone (Scheme 10.3A–B).[73] The correlations between product yield and n_{MB} for reactions carried out in a PBM with $d_{MB} = 15$ mm are shown in summarized in Figure 10.7.

The results indicate a strong linear relationship between those two variables, which is reasonable since Eq. (10.1) proves the direct proportionality of E_{kin} and the mass of the moving grinding bodies. While the number of grinding bodies in the milling vial increases the transferred energy is higher

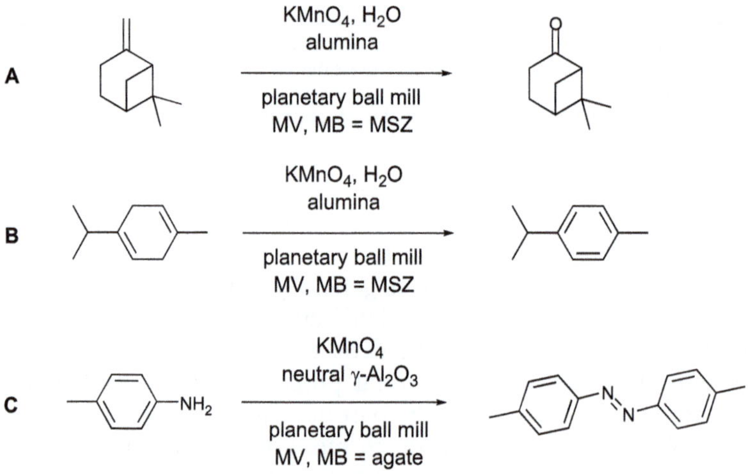

Scheme 10.3 Oxidative model reactions used for the investigation of parameter effects.

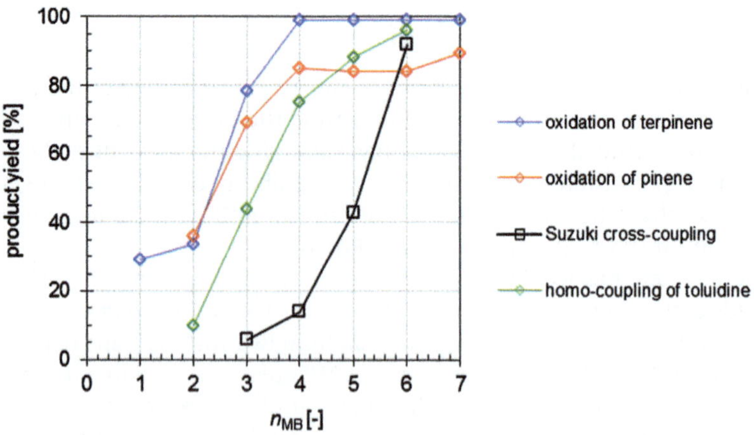

Figure 10.7 Correlation of product yield with the number of milling balls n_{MB} for several model reactions in a PBM. (For references to the individual data see the text.)

and, therefore, more energy is provided, increasing the bulk temperature and along with that the rate constant of the reaction. The fact that the relation between n_{MB} and yield is not similar for all reactions is due to the different requirements regarding activation energy. Accordingly, a discrete amount of energy has to be provided before conversion starts. For reactions like the cross- [72] and homo-coupling procedure[62] higher activation energies are needed compared to the two oxidation procedures.

10.3.4 Size of Milling Balls

Another important feature that is related to n_{MB} is the diameter of the grinding bodies d_{MB}. Generally, the milling ball diameter depends on the volume of the milling vessel and milling balls for materials listed in Table 10.1 are available with sizes from 0.25 to 30 mm. The stress energy transferred to the milling feed ($E_{stress.feed}$) showed a cubic proportionality to d_{MB} for stirred media mills operated under wet conditions, as indicated in Eq. (10.1).[65,75,76] However, it is questionable if this relation is transferrable to the conditions of organic synthesis in ball mills. The most significant differences are the mill type and the type of processing: wet milling for particle refinement and dry milling for organic syntheses. Thus, a more general consideration has to be made regarding the influence of d_{MB} on E_{kin} or $E_{stress.feed}$. In general the stress energy transferred to the milling feed ($E_{stress.feed}$) is proportional to the impact energy of the milling balls (E_{impact}) and the collision frequency ($\nu_{collision}$) (Eq. 10.5):[55]

$$E_{stress,feed} \propto E_{impact} \nu_{collision} \tag{10.5}$$

where $E_{stress,feed}$ (J) is the stress energy transferred to the feed material, E_{impact} (J) is the impact energy of colliding grinding bodies and $\nu_{collision}$ (Hz) is the collision frequency.

Regarding E_{impact} the direct proportionality to E_{kin} and therefore to the mass of the moving bodies in the milling beaker is obvious (Eq. 10.4). This part of $E_{stress.feed}$ is mainly determined by the density and the number of the milling balls. The collision frequency is expressed as the number of collisions in a given time range (Eq. 10.6). Decreasing n_{MB}, the ball-to-powder ratio decreases, alongside with a decreased collision probability and thus decreasing yield (Figure 10.7):[54]

$$\nu_{collision} \propto \frac{\text{number of collisions}}{\Delta t} \tag{10.6}$$

In general, the stress energy (E) is considered as an energy distribution summing up all stress energies, SE, from all single stress events (Eq. 10.7):

$$E = \sum_{i=1}^{N} SE_i = \overline{SE} \cdot SF \cdot t_{reaction} \tag{10.7}$$

where E (J) is the stress energy, SE (J) is the stress energy for a single event and SF (Hz) is the stress frequency $\equiv \nu_{collision}$.

The number of stress events, SN, itself is the product of $t_{reaction}$ and the stress frequency, SF.[55] SF and SN correlate with n_{MB}, which is inversely proportional to d_{MB}, assuming that the overall mass of the milling balls, $m_{MB,total}$, is constant. Both parameters, d_{MB} and n_{MB}, influence the outcome of the individual comminution process or chemical reaction in ball mills. The probability for a collision is much higher for smaller grinding bodies and increases proportionally with n_{MB}. Therefore, it is convenient to use smaller grinding bodies for fine milling.[19,20,31,51]

When performing organic syntheses in ball mills one has to differentiate the nature of the substrates. With respect to the milling of solely organic substrates the SN seemed to be the dominant effect. Hence, smaller milling balls are favourable over larger ones, which has been demonstrated recently for the Knoevenagel condensation of vanillin with barbituric acid (Scheme 10.4) and different reaction scales (Figure 10.8).[78,79]

Analysis of the data from Figure 10.8 for the experiments at low scale (PBM P7) allows for some additional insights. First of all the question might arise as why at the large scale experiments (PBM P6) no milling balls with $d_{MB} < 10$ mm were applied. Even during experiments at the smaller scale (PBM P7) it was observed that working with $d_{MB} = 2$ mm results in an enclosure of the milling balls in the substrate–product mixture and no further movement was possible (Figure 10.9). Especially at high rpm such observations were made. On the other hand, the application of 20 mm balls in the low-scale PBM (PBM P7; like in PBM P6) is critical, since in that case this PBM in combination with 45 mL beakers had to be operated with $n_{MB} < 3$. This would result in severe abrasion (high impact energies) and should be avoided when using PBMs, in contrast to MBMs. The latter may be operated with one or two milling balls only.

If inorganic (for instance solid oxidants and catalysts) are milled together with organic substances larger milling balls are necessary since higher individual SEs are required to break the inorganic particles. However, this effect might be compensated by the effect of increased SN (Table 10.2), unless the overall mass is constant.[10,63,77] Since the surface of the milling balls ($A_{s,MB}$) (Eq. 10.8) increases with decreasing volume of the single bodies (Σm_{MB} = constant; Table 10.2), it can be concluded that the surface of the grinding bodies has no significant influence on the chemical reaction:

$$A_{s,MB} = \pi n_{MB} d_{MB}^2 \quad (10.8)$$

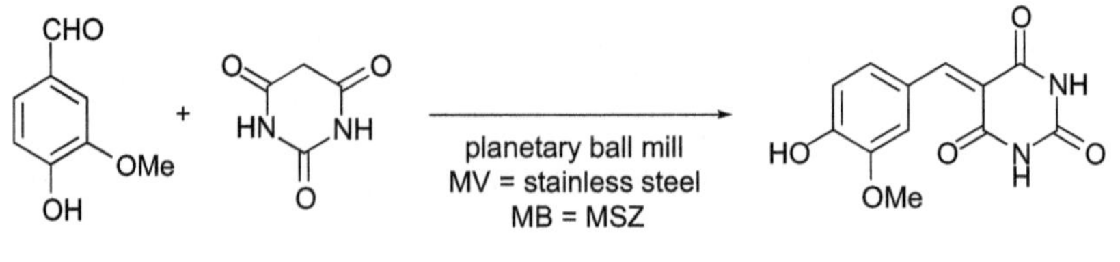

Scheme 10.4 Knoevenagel condensation used as model reaction for the investigation of parameter effects.

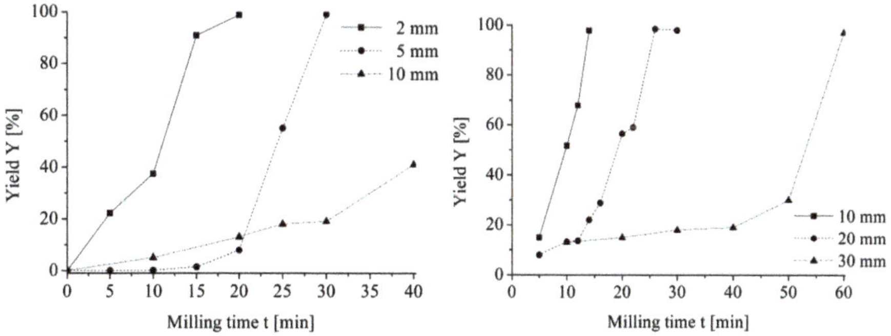

Figure 10.8 Influence of milling time, $t_{reaction}$, and milling ball diameter, d_{MB}, on the yield of the Knoevenagel condensation of vanillin and barbituric acid (equimolar ratio). Left: PBM Fritsch P7 premium line ($V_{MV} = 0.045$ L, $n = 20$ mmol, rpm $= 800$ min^{-1}). Right: PBM Fritsch P6 ($V_{MV} = 0.25$ L, $n = 100$ mmol, rpm $= 650$ min^{-1}). For further experimental conditions see Table 1 in the respective reference.[79]
Reproduced with permission of the Royal Society of Chemistry.

Figure 10.9 Hindered milling ball movement by enclosure of the milling balls in the substrate–product mixture (Scheme 10.4). Reaction conditions: PBM Fritsch P7 premium line ($V_{MV} = 0.045$ L, $d_{MB} = 2$ mm, $n = 20$ mmol, rpm $= 1000$ min^{-1}).[79]
Reproduced with permission of the Royal Society of Chemistry.

Table 10.2 Effect of changing d_{MB} and n_{MB} for the dehydrogenation of γ-terpinene to *p*-cymene (Scheme 10.3B).[77,a]

d_{MB} (mm)	n_{MB} (–)	Σm_{MB} (g)	$A_{s,MB}$ (cm^2)	Yield (%)[b]	Yield (%)[c]
5	120	45.8	94.2	99	72
10	15	45.4	47.1	99	79
15	5	46.3	35.3	99	75

[a]Conditions: 2 mmol γ-terpinene, 6 mmol $KMnO_4$, 0.1 g H_2O, 3.8 g milling auxiliary; milling beakers (MSZ, $V_{MV} = 45$ mL) and milling balls made from MSZ; PBM Fritsch P7 classic line, $t = 5$ min, rpm $= 800$ min^{-1}.
[b]Milling auxiliary = acidic α-Al_2O_3.
[c]Milling auxiliary = quartz sand.

It is more convenient to view the activation of this particular reaction as a thermal process in which no *mechanochemical reaction* took place as defined by IUPAC.[17] Moreover, the frictional forces generate heat, providing the form of energy needed for activation of the process. However, if d_{MB} is reduced while maintaining n_{MB} ($\Sigma m_{MB} \neq$ constant), the impact energy (E_{impact}) is lowered and so is the yield. This effect has been demonstrated for $Pd(OAc)_2$-catalyzed Suzuki–Miyaura reaction using KF/Al_2O_3 as base and grinding auxiliary: decreasing d_{MB} from 15 to 10 mm with $n_{MB} = 6$ reduces the yield from 92% to 6%.[72]

10.3.5 Filling Degree

The filling degree of a milling vessel (Φ) is an important measure since it influences the milling balls' trajectories in a ball mill (Figures 10.3, 10.4 and 10.9). If the milling vessel is completely filled with milling balls and substrates no movement will be possible in general. As a consequence no energy can be transferred to the system. Generally the total filling degree of the vessel (Φ_{total}) is calculated following Eq. (10.9) by adding up the volumetric filling degrees for the milling balls ($\Phi_{MB,material}$) and the grinding stock ($\Phi_{GS,packing}$). The values for the milling balls can be easily calculated considering basic parameters as outlined in Eq. (10.10). In addition to $\Phi_{MB,material}$ the milling balls' filling degree considering the milling balls' packing volume ($\Phi_{MB,packing}$) might be of interest, especially for scale-up purposes. Following Eq. (10.11) the filling degree for the grinding stock (substrates in the case of mechanochemical syntheses) can be approximated using basic material parameters as well:

$$\Phi_{total} = \Phi_{MB,material} + \Phi_{Gs,packing} \tag{10.9}$$

$$\begin{aligned}\Phi_{MB,material} &= \frac{V_{MB,material}}{V_{MV}} = \frac{\pi d_{MB}^3 \cdot n_{MB}}{6V_{MV}} \\ &\cong \Phi_{MB,packing}(1 - \varepsilon)\end{aligned} \tag{10.10}$$

$$\Phi_{GS,packing} \cong \frac{\sum_i^n \frac{m_{GS,i}}{\rho_{GS,i}}}{V_{MV}} \tag{10.11}$$

where

Φ_{total} (–) is the total filling degree,
$\Phi_{MB,material}$ (–): volumetric filling degree of the milling balls,
$\Phi_{GS;packing}$ (–): volumetric filling degree of the grinding stock,
$V_{MB,material}$ (cm^3): overall milling ball volume,
V_{MV} (cm^3): total milling vessel volume,
$\Phi_{MB,packing}$ (–): milling balls' filling degree regarding packing volume,
ε (–): porosity of the milling balls' packing,
$m_{GS,i}$ (g): mass of individual substrate,
$\rho_{GS,i}$ ($g\ cm^{-3}$): density of individual substrate,

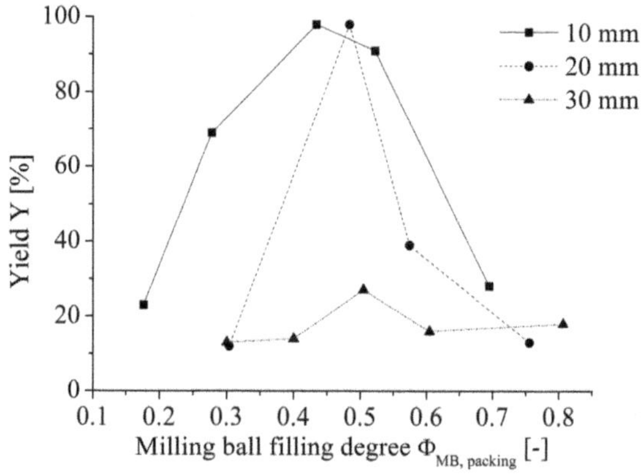

Figure 10.10 Influence of milling ball filling degree, $\Phi_{MB,packing}$, and milling ball diameter, d_{MB}, on the yield of the Knoevenagel condensation of vanillin and barbituric acid (equimolar ratio; Scheme 10.4). PBM Fritsch P6 ($V_{MV} = 0.25$ L, $n = 100$ mmol, rpm $= 650$ min^{-1}, $t_{reaction} = 12$ min). For further experimental conditions see Table 1 in the respective reference 79.
Reproduced with permission of the Royal Society of Chemistry.

Figure 10.10 visualizes the influence of $\Phi_{MB,packing}$ on the yield of the model reaction for different d_{MB}. Experiments have been conducted in the large PBM in 250 mL milling beakers at a constant reaction time of 12 min and rpm $= 650$ min^{-1}, which are the optimal conditions already outlined for $\Phi_{MB,packing} = 0.28$–0.30 in Figure 10.10. The results strikingly indicate the importance of this parameter, since independent from d_{MB} the curves pass through a maximum, which is in the range of $0.43 \leq \Phi_{MB,packing} \leq 0.50$.[79] According to Eqs (10.1) and (10.7) the overall stress energy depends on *SE* and *SF*. If $\Phi_{MB,packing}$ is < 0.4 the number of stress events, *SN*, and therefore also *SF* and the medium *SE* are reduced, which leads to the observed effect regarding the yield. Although the movement of the milling balls inside the beakers is not distorted, for an efficient energy transfer the grinding bodies need space for acceleration and movement through the milling beaker. In case of $\Phi_{MB,packing} > 0.5$ the free space for the trajectories is restricted or completely blocked and the milling balls cannot be accelerated, make their way through the vessel, and cause impacts (Figure 10.4).[9,20,31,53] Manufacturers of PBMs generally advise their customers to leave at least 33% of V_{VM} for the trajectories of the milling balls and the geometrical volume, $V_{MB,material}$, of the milling balls should not exceed 33%.[31] Based on parameters like d_{MB}, n_{MB} and ρ_{MB}, the value of $\Phi_{MB,material}$ can be estimated (Eq. 10.10). Recalculation of the optimal values for $\Phi_{MB,packing}$ from Figure 10.10 results in $\Phi_{MB,material}$ of 0.26, 0.27, and 0.28 for $d_{MB} = 10$, 20, and 30 mm, respectively. This is roughly in accordance with the manufacturers' advice.

Beside the volume of milling balls the volume of reagents also influences Φ_{total}. Changing this value would consequently result in different conditions

Table 10.3 Experimental results[a] for variation of the batch size, $n_{initial,i}$, and the effect on the filling degrees, Φ_i (−),[b] and the time to reach quantitative yield, $t_{97\%}$, for the reaction of vanillin and barbituric acid in a PBM (Scheme 10.4).[79]

$n_{initial,i}$ (mmol)	$\Phi_{MB,material}$	$\Phi_{MB,packing}$	$\Phi_{GS,packing}$	Φ_{total}	$t_{97\%}$ (min)
100	0.25	0.45	0.30	0.55	30
125	0.25	0.45	0.38	0.63	28
150	0.25	0.45	0.45	0.70	30
175	0.25	0.45	0.53	0.78	29
200	0.25	0.45	0.60	0.85	28

[a]Reaction conditions: PBM Fritsch P6, milling vessel (V_{MV} = 0.25 L) and milling balls made from MSZ (d_{MB} = 20 mm), equimolar ratio of substrates, rpm = 550 min^{-1}.
[b]Calculation according to the equations given in parentheses: $\Phi_{MB,material}$ (Eq. 10.10), $\Phi_{MB,packing}$ (Eq. 10.10), $\Phi_{GS,packing}$ (Eq. 10.11), and Φ_{total} (Eq. 10.9).

regarding the ratio of substrate amount and milling balls. One possibility to measure this ratio is the so-called "ball-to-powder-ratio", which is often applied to ascribe experimental conditions processing inorganics in ball mills.[9,30] This value is often expressed as a mass ratio, which makes this parameter inapplicable to describing the general relationship due to large differences in material or bulk densities. Thus, a volume-based measure like $\Phi_{GS,packing}$ (Eq. 10.11) should be preferred. Table 10.3 summarizes the influence of variations in $n_{initial,i}$. The total filling degree, Φ_{total}, ranges from 0.55 to 0.85. Despite the high values for Φ_{total} the times taken to reach quantitative yield were 30 min, with negligible variations. It seemed that the filling degree with respect to the substrates has only a minor influence, unless the optimal value for the milling balls has been chosen. Even a residual free volume for milling ball movement of 15% ($1 - \Phi_{total} = 1 - 0.85$) of V_{MV} is apparently enough to achieve high conversion.

10.4 Process Parameters

Variables described within this section are parameters used for controlling the energy entry during the process. The values whose influence on chemical reaction will be discussed here are (i) the operating frequency (ν, in some cases also f; in the case of PBM also rpm), (ii) the reaction time (t) and (iii) the temperature (T). Whereas the operating frequency is directly correlated to the kinetic energy of the grinding media and, thus, to the energy that can be dissipated as heat to the mill charge (originated from frictional forces), the parameters t and T assign the progression of a chemical reactions. Consequently, the chemical kinetics of a desired reaction also contribute to the performance of a solid-state reaction carried out in a ball mill.

10.4.1 Operating Frequency

The energy entry in ball mills is facilitated by the acceleration, trajectories, and impact of milling balls, resulting in wear phenomena that are able to induce chemical transformation, particle refinement, or heat dissipation to the mill feed. The impact intensity of the milling balls correlates strictly with the impulse and the kinetic energy. As previous considerations regarding the kinetic energy of a rotating or oscillating system (E_{kin}) (Eq. 10.3) as well as the stress energy transferred to the feed material ($E_{stress,feed}$) (Eq. 10.1) indicate, the correlation with the rotation, ν_{rot} ((Hz) or rpm (min^{-1})), or oscillation frequency, ν_{osc} ($Hz \equiv s^{-1} \equiv 1/60\ min^{-1}$), is of a quadratic nature.[10,29,75] In combination with the moving mass the operating frequency determines the energy amount that can be transferred into heat or chemical energy. Therefore, a strong correlation is reasonable for the influence of ν on resulting variables describing the characteristics of a product or the performance of a chemical reaction. Beside reaction time (t), it has been shown that the influence of ν on a chemical reaction in ball mills is the most important.[72] Several examples described in the literature indicate this mathematical correlation between, for example, yield or conversion and the frequency a ball mill is operated with. Generally, higher frequencies result in an increase of those variables. Examples regarding organic synthesis are very widespread, resembling many fields of synthesis chemistry: cross-coupling,[29,72] redox chemistry,[73,77,80] synthesis of heterocycles,[81] organocatalytic enantioselective aldol-type reactions,[82] condensation reactions,[61,78,79,83,84] polymer grafting,[85] reactions of carbohydrates,[86] and further examples are known dealing with these correlations.[87] The correlations are illustrated for the Knoevenagel condensation outlined in Scheme 10.4 at different reaction scales (Figure 10.11).[79] It should be mentioned that similar operating frequencies for different ball mills should not allow us to conclude that the stress energies are similar (see also Section 10.3.1).

Many authors report that the target variable (yield, conversion or selectivity) reaches a static relation or they observe a decrease after passing

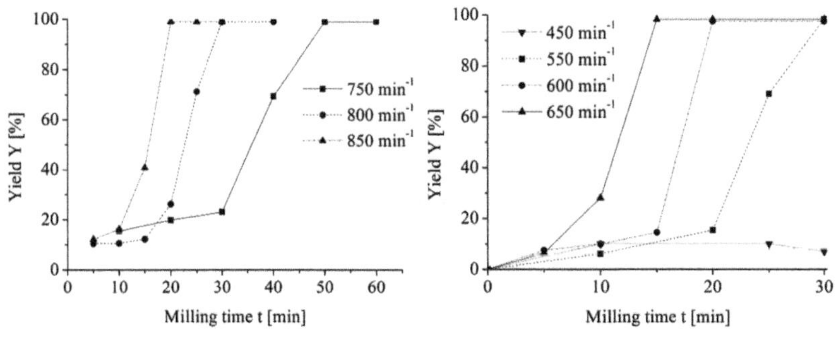

Figure 10.11 Influence of milling time, $t_{reaction}$, and rotation frequency, rpm, on the yield of the Knoevenagel condensation of vanillin and barbituric acid (equimolar ratio; Scheme 10.4). Left: PBM Fritsch P7 premium line (V_{MV} = 0.045 L, n = 20 mmol, d_{MB} = 10 mm). Right: PBM Fritsch P6 (V_{MV} = 0.25 L, n = 100 mmol, d_{MB} = 20 mm). For further experimental conditions see Table 1 in the respective reference 79.
Reproduced with permission of the Royal Society of Chemistry.

through a maximum.[29,88] As this is typical for a quadratic function it is reasonable from mathematical point of view, but it is also explainable from chemical viewpoint. High energy impacts may distort the chemical equilibrium, favouring unwanted side reactions, resulting in a decrease of the mentioned variables. On the other hand, this high energy impact may also open reaction routes to side or consecutive reaction products, lowering the performance of a specific process.[89,90] This effect is also possible for a change of the ball mill type to apparatus that realize higher energy densities.[66]

10.4.2 Milling Time

In addition to operating frequency, statistical analysis of solid-state reactions in ball mills identifies the reaction time (t) as the main variable determining the performance of a chemical reaction.[10,72,73] This is readily understandable on looking at the physics on the one hand and at chemical kinetics on the other hand. Direct proportionality between energy (E) and time (t) is a physical fact (Eq. 10.12). Hence, prolonged ball milling goes along with an increase of collisions between the moving parts of the mill and of the grinding beads with the mill chamber or feed material. More energy is distributed to the mill charge, although the energy density is kept constant:

$$E = Pt \tag{10.12}$$

Chemical kinetics, however, also prove the importance of the time for reactions carried out in ball mills. The reaction rate (r) of a specific chemical reaction indicates that more moles can be converted at increased reaction time, t (Eq. 10.13):

$$r_i = \frac{1}{\nu_i}\frac{\mathrm{d}n_i}{\mathrm{d}t} = k(T)\prod_j n_{\mathrm{educt},j}^{\nu_j} \tag{10.13}$$

where

r_i is the reaction rate for product i,
n_i (mol): molar amount,
t (s): reaction time,
$k(T)$: temperature-dependent rate constant;,
ν: stoichiometric coefficient,
index i = product; index j = educt(s).

The exponential correlation between r, the rate constant (k), and temperature (T) expressed by the Arrhenius (Eq. 10.14) or Eyring–Polanyi equation (Eq. 10.15) limits the infinite increase of conversion or yield:

$$k = A\mathrm{e}^{-\frac{E_\mathrm{a}+0.1\Delta(Vp)}{RT}} \tag{10.14}$$

$$k = \frac{k_\mathrm{B}T}{h}\mathrm{e}^{-\frac{\Delta G^\ddagger}{RT}} = \frac{k_\mathrm{B}T}{h}\mathrm{e}^{-\frac{\Delta H^\ddagger}{RT}+\frac{\Delta S^\ddagger}{R}} \tag{10.15}$$

However, since the micro- and macro-kinetics as well as the reaction order of organic syntheses carried out in ball mills differ significantly, no general correlation between performance-based variables and t can be drawn. In general, increased processing time goes along with an increase of parameters like conversion or yield, whereas the influence on selectivity is difficult to assess. Consecutive, side, or parallel reactions can be promoted or suppressed, depending on the reaction type. The promoting effect of t on the product yield has been described for the Suzuki–Miyaura reaction,[72] oxidation reactions,[73,77,87] synthesis of heterocycles,[81] enantioselective aldol-type reactions,[82] condensation reactions (see for instance Figure 10.11),[61,79] polymer grafting,[80] functional group manipulation,[86] solvent-free peptide synthesis,[88] and fullerene oxygenation.[85]

Recently, Tumanov *et al.* described the sequential product formation followed by PXRD (powder X-ray diffraction) for the synthesis of a coordination compound using an apparatus (similar to a Bridgman anvil) that allows the distribution of pulsed portions of wear energy (pressure) to the feed.[91] The authors were able to follow a mechanochemical reaction step-by-step, by changing the pressure the reactions mixture is exposed to. Since the energy is distributed solely as mechanical energy, considerations regarding the wear energy can be drawn allowing us to draw conclusions on the activation energy of such processes. The methodology has been adopted recently using *in situ* PXRD and Raman to follow chemical reactions in ball mills.[92–94]

10.4.3 Combined Assessment of Frequency and Time

The two previous sections point out the importance of operating frequency (ν) as a regulative for the energy entry and of reaction time (t) determining the reaction rate. Both variables contribute to the progress of chemical reactions carried out in ball mills. Thus, assessment of the interrelationship between those two process parameters might be of interest. Schneider *et al.* compared the influence of technological as well as process variables for their effect on the yield of the cross-coupling product of the Suzuki–Miyaura reaction of phenylboronic acid and *p*-bromoacetophenone (Scheme 10.2).[72] The use of methods for design of experiments and statistical analysis identified ν and t as the main influencing variables followed in decreasing order by the milling material characterized by the material density (ρ) and the number of milling balls (n_{MB}). Application of an artificial neuronal network for data assessment from ZnO synthesis in a PBM proves the influence of ν and identifies the ball-to-powder ratio as a further parameter.[95] Based on those results investigations on the assessment of both variables using designed factorial plans and statistical analysis were extended to two further reactions: oxidative cleavage of β-pinene affording nopinone (Scheme 10.3A)[10,73] and the dry iodination of aniline yielding several iodinated anilines (Scheme 10.5).

Figure 10.12 shows the correlation between ν_{rot} ($\equiv$ rpm), t, and percentage-yield of nopinone.[10] Results were based on a 3^2-factorial plan while the variations in ν_{rot} and t were as follows: 400, 600, and 800 min^{-1} and 10, 20,

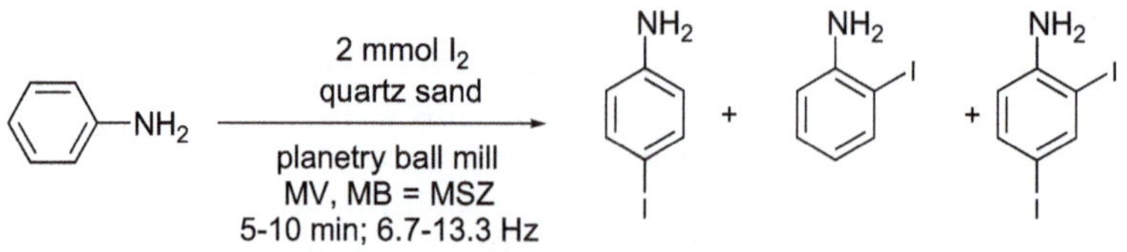

Scheme 10.5 Dry iodination of aniline in a planetary ball mill.

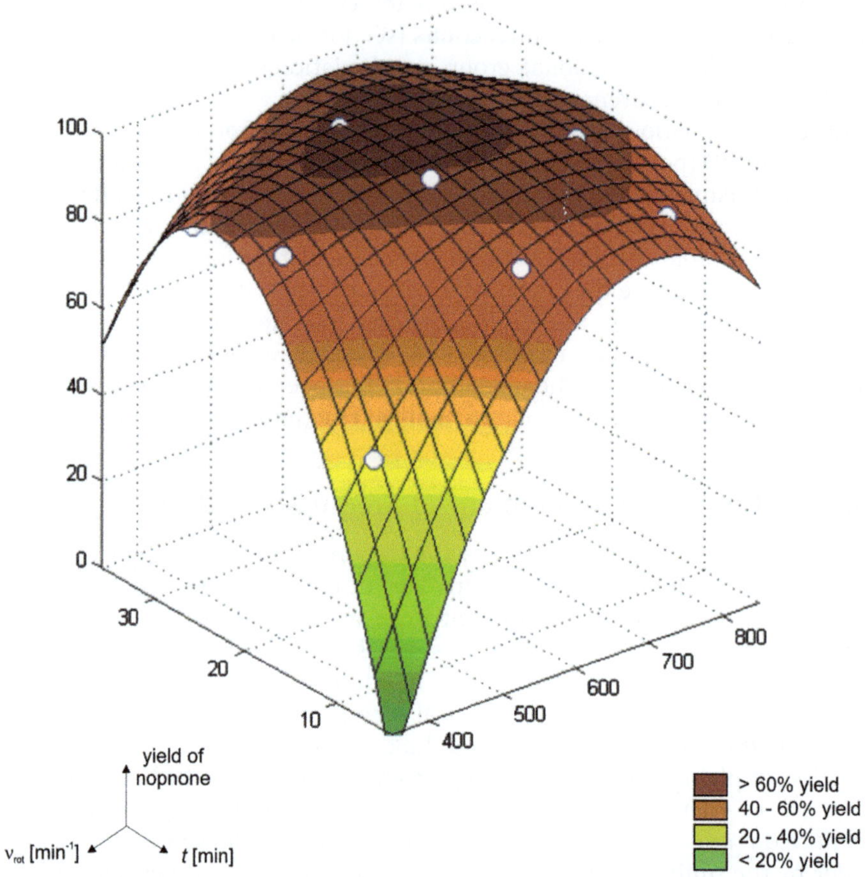

Figure 10.12 Correlation between rotation frequency ν_{rot}, milling time t, and yield of nopinone from oxidation of β-pinene (Scheme 10.3A) in a planetary ball mill.[10]
Reproduced with permission of the Royal Chemical Society.

and 30 min, respectively. Reactions were carried out using milling tools made from MSZ and applying quartz sand as the milling auxiliary.[73] Correlation in Figure 10.12 indicates a global maximum for the yield at medium ν_{rot} and $t = 20$ min. Obviously, at lower operating frequencies the energy entry is too low to induce high reaction rates. Selectivity for the target product nopinone remained at a high level for $\nu_{rot} \geq 600$ min^{-1} and $t \leq 20$ min. At higher frequencies and longer reaction time consecutive reactions become predominant, lowering the yield. The square relationship between ν_{rot} and the yield proves the relation between the operating frequency and the kinetic energy of the system (Eq. 10.1).

The correlations shown in Figure 10.12 for the synthesis of nopinone seemed very simple compared to the complex relationship of the variables in case of aniline iodination (Scheme 10.5). The results from a similar factorial plan varying ν_{rot} and t are summarized in Figure 10.13. Reactions were carried out in a PBM using a molar ratio of aniline to I_2 of 0.5 and quartz sand as the milling auxiliary. Chemically, the correlation is more complex, since the reaction did not initially afford a single product but, instead, a mixture of isomers from iodination at the *p*- and *o*-positions. According to classical rules of electrophilic aromatic substitution the +M-effect of the

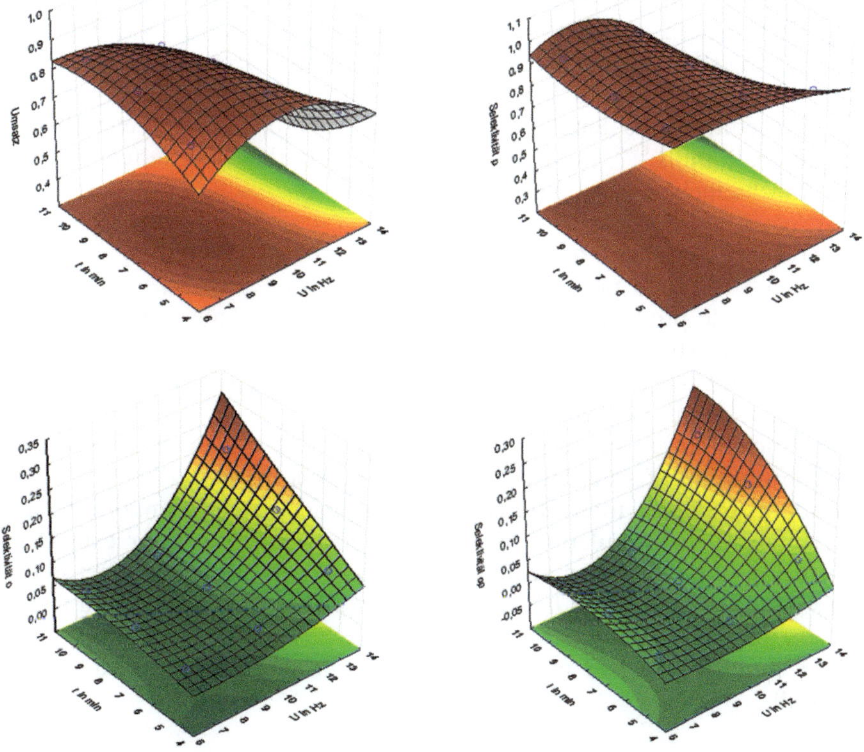

Figure 10.13 Correlation between rotation frequency ν_{rot}, milling time t, and the following variables (clockwise from top left): conversion of aniline, selectivity for *para*, 2,4-, and *ortho* product (Scheme 10.5). Conditions: 1 mmol aniline, 2 mmol I_2, 4 g quartz sand; milling beaker (MSZ, 45 mL), milling balls (MSZ, 6×15 mm).

NH_2-group favours those positions and for structural reasons *p*-substitution is strongly favoured, which is also proven by the selectivity data for *p*-iodoaniline (Figure 10.13, top right) in comparison with *o*-iodoaniline (bottom left). Consecutive reaction of the mono-iodinated products afforded 2,4-diodoaniline as further product. The progress in selectivity for 2,4-di-iodoaniline (bottom right) is very similar to that of the *o*-substituted product, although the values differ by approximately 5%.

The correlation between aniline conversion (Figure 10.13, top left) and ν_{rot} again proves the square relation of those variables. However, compared to data shown in Figure 10.12 the conversion strongly decreases at high ν_{rot} and t for the iodination of aniline in a PBM. Two reasons account for this observation:

i. The sublimation temperature of iodine is low and increased bulk temperature due to prolonged milling at high frequencies results in sublimation of I_2. Thus, no reagent is available for the reaction itself, leading to lower conversion of aniline.
ii. Decomposition of reaction products by dehalogenation to the aromatic core structure can occur under high-energy conditions. Owing to the bond length of the C–I bond the dissociation energy is rather low, forcing this reverse reaction. Similar behaviour accounts for the lower reactivity of aryl iodides compared to bromides in cross-coupling procedures like Suzuki–Miyaura and Mizoroki–Heck accomplished in ball mills.[89,90] Komatsu *et al.* reported the same effect for the mechanochemical dimerization of C_{60}.[66]

Increasing conversion of aniline goes along with an increase in selectivity for the *p*-substituted product, whereas the opposite trend favours the formation of side products. Therefore, by operating the ball mill at medium frequency and short reaction times, high yields of *p*-iodoaniline can be observed. High-energy ball milling conditions (high ν_{rot}) favour side and/or consecutive reactions (Figures 10.12 and 10.13).[10,73]

Statistical analysis of the results for both model reactions revealed that the contribution of single variables to the effects discussed before is predominant. Complex interactions of the variables are less important and contribute to the simulation of the effect matrices to a lower extent. In conclusion, both variables are independent of each other. This fact is obvious, since the operating frequency influences the wear energy and therefore the temperature evolution in the milling beaker, whereas t affects the energy input of the ball mill, wherein P is the gross power input of the ball mill.

10.4.4 Reaction Temperature

In addition to reaction time and technological parameter as well as operating frequency, the reaction temperature (T) is a further important variable, showing a direct effect on the rate constant (k) (Eqs 10.14 and 10.15).

Although this parameter significantly influences chemical kinetics and therefore the reaction rate, commercial ball mills for laboratory-scale applications (PBM, MBM) do not allow temperature regulation. Individual solutions are available; either the instalment of a heating mantel for a milling beaker of a MBM[47,71] or construction of a PBM allowing temperature regulation of the complete milling chamber.[51] Switching from laboratory to pilot or industrial scale the type of ball mill often changes to stirred media mills. These types of ball mills have a static milling chamber, thus easily providing the opportunity for heating or cooling the mill feed.[8,47]

Owing to these technological hurdles, most organic reactions carried out in ball mills are operated at room temperature, while during processing the temperature of the milling beaker, and as a result that of the mill feed, increases. The final temperature strongly depends on ρ_{MB}. Typical equilibrium temperatures after prolonged milling for $\geq$1 h are 40–80 °C.[29,54] Utilization of high-density materials like steel or WC can lead to a T of 120 °C, measured at the surface of the milling beaker and the mill charge. Calculation of the reaction conditions in the impact zone results in maximal temperatures of up to 3000 K.[51,96,97] Efforts to determine the actual temperature of steel milling balls by opening the milling beakers and transferring the balls into a calorimeter result in values of 200 °C.[98] However, these experiments were conducted by operating the ball mill at a low Φ, which influences the final temperature. Owing to continuous release of frictional heat, the bulk temperature increases with increasing t, ν, ρ_{MB}, and Φ^{-1}. For several applications, especially for temperature-sensitive materials, the employment of a milling auxiliary may be desirable for homogeneous heat distribution and reduction of local hot spots in the impact zone of the milling beakers.[63,99] Alternatively, a cycle-mode is often used: ball milling for a discrete time interval alternates with pauses, allowing the milling beaker and feed material to cool down.[81,82]

Since liquefaction of the mill charge during ball milling may disturb the efficiency of energy transfer and the chemical reaction itself,[100] conditions have to be set-up keeping all reagents in the solid or solidified state. This can either be realized through the application of high-melting inorganic milling auxiliaries or by (continuous) cooling of the milling beakers. One possibility is cryogenic milling, whereby the feed material is frozen with liquid nitrogen and afterwards comminuted for a short time to avoid unfreezing. However, the energy distributed by frictional and impact forces is so high that the milling time is reduced to several minutes only before room temperature is reached. Special equipment for MBMs allows continuous cooling with liquid nitrogen. Performing reactions under such conditions reduces molecular motion almost completely and thus most chemical reactions afforded significantly lower yields.[61,63,101] Unfreezing of the finely ground reaction mixtures can lead to post reactions, which may be very fast and result in high conversions,[63] since the surface energy of the particles is very high.

10.5 Energy Efficiency and Comparison to Other Methods

Assessment of the energy efficiency of a dedicated synthesis technique as well as its comparison to other standard methods applied in organic synthesis helps to prove the capability of the new method. Results from ball milling experiments will be discussed in the light of other methods of energy entry. Based on such experimental data energy efficiencies can be calculated, allowing evaluation of the investigated methods.

Although organic synthesis carried out in ball mills is a relatively new field compared to other advanced reaction technologies,[102] reviews have been published that compare this method with synthesis technologies like microwave-assisted heating or sonochemistry.[26,28,48,103–107] In most cases authors evaluate a specific synthesis or synthetic procedure disclosed by different authors. The individual reaction protocols might be different, complicating the assessment of the heterogeneous data sets. The most obvious difficulty is the fact that reactions in ball mills are generally carried out under solvent-free reaction conditions, in most cases as solid-state reactions.[26,38,57] Comparison of chemical differences between reactions in ball mills and solution-based protocols have been undertaken for the oxidation of *p*-toluidine (Scheme 10.3C) with $KMnO_4$ as oxidant.[62] Figure 10.14 shows the conversion and the selectivity for azo-compound resulting from different experimental procedures. Apart from accomplishments in ball mills, reactions in microwaves as well as those using classical heating or sonochemistry have been carried out in acetonitrile as solvent to facilitate proper mixing, energy entry, and energy distribution. From a selectivity point of view all methods are comparable, whereas for solid-sate reactions mainly 1,2-diphenyldiazene oxide (the azoxy compound) and *p*-nitrotoluene are formed as side products. Oxidation in solution facilitated moreover oxidation of the methyl group to *p*-nitro- or *p*-aminobenzoic acid as parallel reaction channels.[108] Conversion for ultrasound processing, classical heating, and performance in a PBM are on the same high level, although the reaction rate is highest for the last example (short t and high batch size; for details see literature).[62] The similarity of the results from mechano- and sonochemical processing displays the correlation between these two methods of energy entry, which has also been demonstrated by the studies of Friščić on the LAG-synthesis of co-crystals.[43] The homogenizing effect of the acoustic field as well as the action of the milling balls are advantageous for the formation of highly dispersed systems with well distributed particles.[43,109] The lower energy density for the MBM compared to PBM at similar operating frequencies[29] is responsible for the lower conversion. For the microwave-assisted examples the low solubility of the oxidant results in a heterogeneous reaction mixture hampering the mass transport and, thus, the reaction rate.

Comparison of solvent-free reaction procedures explored with different methods of energy entry might also be difficult, since these protocols can either be liquid phase or solid state reactions, but investigation might be helpful for exploration of chemical differences. With respect to catalytic reactions in ball mills, comparison of TON as well as TOF revealed strong similarities with solvent-free accomplishment in microwave apparatus. However, it has to be mentioned that similar performance in a microwave field or under mechanochemical conditions is only possible if proper mixing is applied. Otherwise the product yield is lower, as demonstrated in Figure 10.14 for aniline oxidation[62] and also for other types of reactions like Knoevenagel condensation[61] or Suzuki–Miyaura cross-coupling.[29]

Beside comparison with other technologies of energy entry, ball milling procedures are often compared to simple *stirring in a flask* [82,110–112] and with the comminution of reagents using mortar and pestle.[26,28,29,58,59] The last method in particular is outmoded due to low robustness.[29] Although many reactions might be realized with this technique, the application of mortar and pestle for advanced organic synthesis procedure is questionable (Section 10.1.2). However, grindstone chemistry undoubtedly has one big advantage: its energy efficiency, since the energy provided to the reaction mixture is *truly renewable*. For all other methods the activation energy is generated from energy transformation starting, in most cases, with electrical energy. The energy needed for the production of a defined quantity of product (n_{product}) is defined as molar energy (E_{m}, Eq. 10.16):[10,29]

$$E_m = \frac{E_{\mathrm{grid}}}{\sum_i n_{\mathrm{product},i}} \tag{10.16}$$

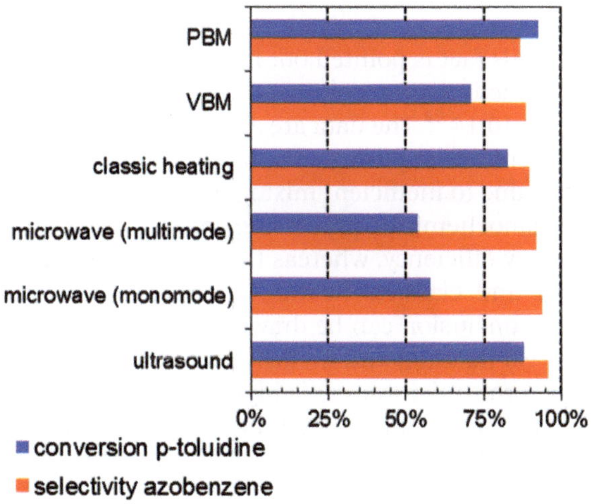

Figure 10.14 Comparison of conversion and selectivity for the $KMnO_4$-mediated oxidation of *p*-toluidine (Scheme 10.3C) carried out with different methods of energy entry.[62]

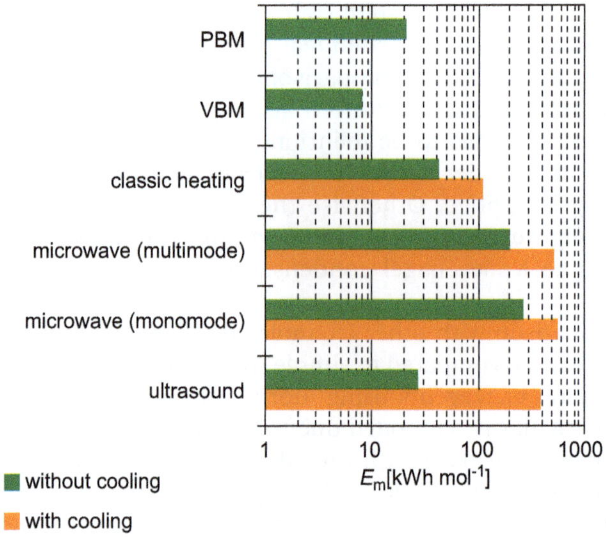

Figure 10.15 Comparison of the molar energy E_m (Eq. 10.16) for the $KMnO_4$-mediated oxidation of *p*-toluidine (Scheme 10.3C) carried out with different methods of energy entry.[62]

The index i in Eq. (10.16) represents the number of milling beakers the ball mill can be operated with, *i.e.* for MBMs $i=2$ and for PBMs i varies between 1 and 4 (Figure 10.2). The value E_m is also defined as the energy intensity of a desired reaction or process.[11,113,114] The energy efficiency of ball milling procedures expressed by E_m has been proven for several examples. Especially in comparison to microwave-assisted synthesis, the application of ball mills is more sustainable, if similar reaction systems are compared.[29,61,62] This fact is pointed out in Figure 10.15, which summarizes E_m for different reaction protocols regarding the examples discusses in context with Figure 10.14.[62] The data are also interesting in that the product yield for the different accomplishments differs by not as much as reported for other reactions due to inefficient mixing in the heterogeneous phase.[29,61] Sono- and mechanochemical processing result in comparable values regarding the energy efficiency, whereas the energy demand for microwave-assisted synthesis and classical heating is approximately one magnitude higher.[29,61,62] The conclusion can be drawn that the energy loss due to heat dissipation (classical heating) and inefficient energy transfer from electrical to desired energy form (microwaves) is higher for these methods. Including the energy demand for cooling of the liquid-phase reactions decreases the energy efficiency, which is manifested by an increase of E_m in Figure 10.15.[62]

10.6 Conclusion and Outlook

The concepts of sustainable chemistry and process intensification are the main driving forces leading to application of ball mills as reactors in organic synthesis. Although theoretical models predict extreme reaction conditions in the impact zones of milling balls, several publications indicate that this type of synthesis is possible and often results in spectacular and unintended results regarding chemical reactivity or selectivity in comparison to classical solution-based synthesis. The omission of solvents in the reaction step is certainly one of the many advantages of ball milling, making this technology a tool with fast growing fields of application, beyond solely organic synthesis and particle refinement.

Beside chemical aspects of reactions in ball mills, technological parameters influence the reactions considerably. Whereas the type of ball mill – planetary ball mill, mixer ball mill, vibration ball mill, stirred media mill – seems to be relevant to the outcome of the reaction, variables like the milling material, number and size of milling balls, as well as the filling degree of the milling vessels are important. The choice of milling material is not only important for the energy entry, since the material density determines the transferred wear energy, but also chemical aspects should be considered, *i.e.* the chemical resistance and the mechanical stability of the material. The filling degree of the milling beakers is important with respect to the number of collisions and impacts during ball milling.

Although technological parameters are important for several reasons, experiments in ball mills with respect to organic synthesis are significantly influenced by process parameters like operating frequency, reaction time, and reaction temperature. Correlation of these variables to the stress energy (frequency), kinetics (temperature), or reaction rate (time) displays their importance in the successful accomplishment of reactions in a ball mill.

In summary, ball mills haven proven to be powerful tools for the performance of organic reactions in the solid state, allowing us to conduct various types of organic transformations. Compared to chemistry in solution, the parameter set for ball milling is enlarged, but offers various possibilities to manipulate the outcome of a chemical reaction (Scheme 10.1). Assessment of the different influencing variables contributes to our understanding of reactions carried out under these prerequisites.

Abbreviations and Symbols

Abbreviations

DEM	descrete element method
GS [index]	grinding stock, mill feed
LAG	liquid-assisted grinding
MB [index]	milling ball(s)
MBM	mixer ball mill
MSZ	magnesia-stabilized zirconia
MV [index]	milling vessel(s)
osz [index]	oscillation
PBM	planetary ball mill

PXRD	powder X-ray diffraction
rot [index]	rotation
VBM	vibration ball mill
YSZ	yttrium-stabilized zirconia

Symbols

ε (–)	porosity of the milling balls' packing
ν (s^{-1})	rotation or oscillation frequency
$\nu_{collision}$ (Hz)	collision frequency
ρ (kg m^{-3})	density
Φ (–)	filling degree
d_{MB} (mm)	diameter of milling balls
E (J)	stress energy
E_{grid} (J)	electrical energy consumption
E_{impact} (J)	impact energy of colliding grinding bodies
E_{kin} (J)	kinetic energy
E_m (J mol^{-1})	molar energy
$E_{stress,feed}$ (J)	stress energy transferred to the feed material
$E_{Y,i}$ (Pa)	Young's modulus
f (Hz)	frequency
m (kg)	mass
n_i (mol)	molar quantity of substance i
n_{MB} (–)	number of milling balls
rpm (min^{-1})	rotation per minute, rotation frequency
SE (J)	stress energy for a single event
SF (Hz)	stress frequency $\equiv \nu_{collision}$
SN (–)	number of stress events
T (°C)	temperature
$t_{reaction}$ (min)	reaction time $\equiv$ milling time
v_p (m s^{-1})	peripheral velocity
V (m^3)	volume

References

1. G. Centi and R. A. v. Santen, *Catalysis for Renewables*, Wiley-VCH Verlag GmbH, Weinheim, 2007.
2. F. Cavani, G. Centi, S. Perathoner and F. Trifiró, *Sustainable Industrial Chemistry*, Wiley-VCH Verlag GmbH, Weinheim, 2009.
3. G. Rothenberg, *Catalysis – Concepts and Green Applications*, Wiley-VCH Verlag GmbH, Weinheim, 2008.
4. C. Wiles and P. Watts, *Micro-Reaction Technology in Organic Synthesis*, CRC Press, Boca Raton, 2011.
5. R. S. Oakes, A. A. Clifford and C. M. Rayner, *J. Chem. Soc., Perkin Trans. I*, 2001, 917–941.
6. C. O. Kappe and D. Dallinger, *Mol. Divers.*, 2009, **13**, 71–193.

7. G. Heinicke, *Tribochemistry*, Akademie-Verlag, Berlin, 1984.
8. P. Baláž, *Mechanochemistry in Nanoscience and Minerals Engineering*, Springer-Verlag, Berlin, 2008.
9. C. Suryanarayana, *Progr. Mat. Sci.*, 2001, **46**, 1–184.
10. A. Stolle, T. Szuppa, S. E. S. Leonhardt and B. Ondruschka, *Chem. Soc. Rev.*, 2011, **40**, 2317–2329.
11. A. D. Curzons, D. J. C. Constable, D. N. Mortimer and V. L. Cunningham, *Green Chem.*, 2001, **3**, 1–6.
12. A. Lapkin and D. Constable, *Green Chemistry Metrics – Measuring and Monitoring Sustainable Processes*, John Wiley & Sons, Ltd, Chichester, 2008.
13. T. E. Fischer, *Annu. Rev. Mater. Sci.*, 1988, **18**, 303–323.
14. V. V. Boldyrev, *Solid State Ion.*, 1993, **63–65**, 537–543.
15. M. K. Beyer and H. Claus-Schaumann, *Chem. Rev.*, 2005, **105**, 2921–2947.
16. J. F. Fernández-Bertrán, *Pure Appl. Chem.*, 1999, **71**, 581–586.
17. A. D. McNaught and A. Wilkinson (ed.), *IUPAC Compendium of Chemical Technology*, 2nd edn (the "Gold Book"), Blackwell Scientific Publications, Oxford, 1997.
18. G. Kaupp, *CrystEngComm*, 2009, **11**, 388–403.
19. C. C. Koch, *Nanostruct. Mater.*, 1993, **2**, 109–129.
20. P. Baláž and E. Dutkova, *Miner. Eng.*, 2009, **22**, 681–694.
21. M. Casas-Cabanas, G. Binotto, D. Larcher, A. Lecup, V. Giordani and J.-M. Tarascon, *Chem. Mater.*, 2009, **21**, 1939–1947.
22. D. Braga and F. Grepioni, *Angew. Chem., Int. Ed.*, 2004, **43**, 4002–4011.
23. A. V. Trask and W. Jones, *Top. Curr. Chem.*, 2005, **254**, 41–70.
24. Z. V. Todres, *Organic Mechanochemistry and its Practical Applications*, Taylor & Francis, Boca Raton, FL, 2006.
25. V. V. Boldyrev, *Russ. Chem. Rev.*, 2006, **75**, 177–189.
26. K. Tanaka, *Solvent-free Organic Synthesis*, 2nd edn, Wiley-VCH Verlag GmbH, Weinheim, 2008.
27. S. L. James, C. J. Adams, C. Bolm, D. Braga, P. Collier, T. Friščić, F. Grepioni, K. D. M. Harris, G. Hyett, W. Jones, A. Krebs, J. Mack, L. Maini, A. G. Orpen, I. P. Parkin, W. C. Shearouse, J. W. Steed and D. C. Waddell, *Chem. Soc. Rev.*, 2012, **41**, 413–447.
28. K. Tanaka and F. Toda, *Chem. Rev.*, 2000, **100**, 1025–1074.
29. F. Schneider, T. Szuppa, A. Stolle, B. Ondruschka and H. Hopf, *Green Chem.*, 2009, **11**, 1894–1899.
30. L. Takacs, *Progr. Mater. Sci.*, 2001, **47**, 355–414.
31. P. Baláz, *Mechanochemistry in Nanoscience and Minerals Engineering, Springer-Verlag, Berlin*, 2008, pp. 103–132.
32. P. A. May and J. S. Moore, *Chem. Soc. Rev.*, 2013, **42**, 7497–7506.
33. V. Sepelak, A. Duvel, M. Wilkening, K.-D. Becker and P. Heitjans, *Chem. Soc. Rev.*, 2013, **42**, 7507–7520.
34. S.-E. Zhu, F. Li and G.-W. Wang, *Chem. Soc. Rev.*, 2013, **42**, 7535–7570.

35. D. Braga, L. Maini and F. Grepioni, *Chem. Soc. Rev.*, 2013, **42**, 7638–7648.
36. G.-W. Wang, *Chem. Soc. Rev.*, 2013, **42**, 7668–7700.
37. E. Boldyreva, *Chem. Soc. Rev.*, 2013, **42**, 7719–7738.
38. G. Rothenberg, A. P. Downie, C. L. Raston and J. L. Scott, *J. Am. Chem. Soc.*, 2001, **123**, 8701–8708.
39. T. Welton, *Green Chem.*, 2006, **8**, 13.
40. T. Friščić, A. V. Trask, W. Jones and W. D. S. Motherwell, *Angew. Chem., Int. Ed.*, 2006, **45**, 7456–7550.
41. C. Maheshwari, A. Jayasankar, N. A. Khan, G. E. Amidon and N. Rodríguez-Hornedo, *CrystEngComm*, 2009, **11**, 493–500.
42. T. Friščić and L. Fábián, *CrystEngComm*, 2009, **11**, 743–745.
43. T. Friščić, S. L. Childs, S. A. A. Rizvi and W. Jones, *CrystEngComm*, 2009, **11**, 418–426.
44. T. Friščić, D. G. Reid, I. Halasz, R. S. Stein, R. E. Dinnebier and M. J. Duer, *Angew. Chem., Int. Ed.*, 2010, **49**, 712–715.
45. F. C. Strobridge, N. Judäs and T. Friščić, *CrystEngComm*, 2010, **12**, 2409–2418.
46. F. Bernhardt, R. Trotzki, T. Szuppa, A. Stolle and B. Ondruschka, *Beilstein J. Org. Chem.*, 2010, **6**, No. 7.
47. G. Kaupp, *CrystEngComm*, 2006, **8**, 794–804.
48. A. Bruckmann, A. Krebs and C. Bolm, *Green Chem.*, 2008, **10**, 1131–1141.
49. C. F. Burmeister and A. Kwade, *Chem. Soc. Rev.*, 2013, **42**, 7660–7667.
50. R. G. Blair, K. Chagoya, S. Biltek, S. Jackson, A. Sinclair, A. Taraboletti and D. T. Restrepo, *Faraday Discuss.*, 2014, DOI: 10.1039/C4FD00007B.
51. F. K. Urakaev, in *High-energy Ball Milling – Mechanochemical Processing of Nanopowders*, ed. M. Sopicka-Lizer, Woodhead Publishing Limited, Cambridge, 2010, pp. 9–44.
52. G. Kakuk, A. Csanády, I. Sajo, K. Papp, P. Németh, H. Hargitai and A. Sztaniszláv, *Mater. Sci. Forum*, 2008, **589**, 397–402.
53. S. Rosenkranz, S. Breitung-Faes and A. Kwade, *Powder Technol.*, 2011, **212**, 224–230.
54. X. Jiang, M. A. Trunov, M. Schoenitz, R. N. Dave and E. L. Dreizin, *J. Alloys Compd.*, 2009, **478**, 246–251.
55. G. Kakuk, I. Zsoldos, Á. Csanády and I. Oldal, *Rev. Adv. Mater. Sci.*, 2009, **22**, 21–38.
56. D. V. N. Prasad and J. Theuerkauf, *Chem. Eng. Technol.*, 2009, **32**, 1102–1106.
57. B. Rodríguez, A. Bruckmann, T. Rantanen and C. Bolm, *Adv. Synth. Catal.*, 2007, **349**, 2213–2233.
58. G. Kaupp, *Top. Curr. Chem.*, 2005, **254**, 95–183.
59. G. Kaupp, *J. Phys. Org. Chem.*, 2008, **21**, 630–643.
60. G.-W. Wang and L. Liu, *Chin. Chem. Lett.*, 2004, **15**, 587–590.
61. R. Trotzki, M. M. Hoffmann and B. Ondruschka, *Green Chem.*, 2008, **10**, 767–772.

62. R. Thorwirth, F. Bernhardt, A. Stolle, B. Ondruschka and J. Asghari, *Chem.-Eur. J*, 2010, **16**, 13236–13242.
63. R. Schmidt, R. Thorwirth, T. Szuppa, A. Stolle, B. Ondruschka and H. Hopf, *Chem.-Eur. J*, 2011, **17**, 8129–8138.
64. A. Kwade, *Powder Technol.*, 1999, **105**, 14–20.
65. S. Breitung-Faes and A. Kwade, *Powder Technol.*, 2011, **212**, 383–389.
66. K. Komatsu, G.-W. Wang, Y. Murata, T. Tanaka and K. Fujiwara, *J. Org. Chem.*, 1998, **63**, 9358–9366.
67. R. Paudert, H. Harenz, R. Pöthig, G. Heinicke, L. Dünkel and H. Schumann, *Chem. Techn.*, 1978, **30**, 470.
68. G. Heinicke, *Tribochemistry*, Akademie-Verlag, Berlin, 1984, p. 111.
69. D. A. Fulmer, W. C. Shearhouse, S. T. Mendonza and J. Mack, *Green Chem.*, 2009, **11**, 1821–1825.
70. R. Thorwirth, A. Stolle and B. Ondruschka, *Green Chem.*, 2010, **12**, 985–991.
71. M. R. Naimi-Jamal, J. Mokhtari, M. G. Dekamin and G. Kaupp, *Eur. J. Org. Chem.*, 2009, 3567–3572.
72. F. Schneider, A. Stolle, B. Ondruschka and H. Hopf, *Org. Process Res. Dev.*, 2009, **13**, 44–48.
73. T. Szuppa, A. Stolle, B. Ondruschka and W. Hopfe, *ChemSusChem*, 2010, **3**, 1181–1191.
74. A. Kwade and J. Schwedes, *Chem. Ing. Tech.*, 2004, **68**, 809–812.
75. M. Becker, A. Kwade and J. Schwedes, *Int. J. Miner. Process.*, 2001, **61**, 189–208.
76. A. Kwade, *Int. J. Miner. Process.*, 2004, **74-S1**, S93–S101.
77. T. Szuppa, A. Stolle, B. Ondruschka and W. Hopfe, *Green Chem.*, 2010, **12**, 1288–1294.
78. C. F. Burmeister, A. Stolle, R. Schmidt, K. Jacob, S. Breitung-Faes and A. Kwade, *Chem. Eng. Technol.*, 2014, **37**, 857–864.
79. A. Stolle, R. Schmidt and K. Jacob, *Faraday Discuss.*, 2014, DOI: 10.1039/C3FD00144J.
80. W. Qiu and T. Hirotsu, *Macromol. Chem. Phys.*, 2005, **206**, 2470–2482.
81. E. M. C. Gérad, H. Sahin, A. Encinas and S. Bräse, *Synlett*, 2008, 7202–7204.
82. B. Rodríguez, A. Bruckmann and C. Bolm, *Chem.-Eur. J.*, 2007, **13**, 4710–4722.
83. Z. Zhang, Y.-W. Dong, G.-W. Wang and K. Komatsu, *Synlett*, 2004, 61–64.
84. S. Wada and H. Suzuki, *Tetrahedron Lett.*, 2003, **44**, 399–401.
85. H. Watanabe, E. Matsui, Y. Ishiyama and M. Senna, *Tetrahedron Lett.*, 2007, **48**, 8132–8137.
86. P. R. Patil and K. P. R. Kartha, *J. Carbohydr. Chem.*, 2008, **27**, 279–293.
87. R. Schmidt, A. Stolle and B. Ondruschka, *Green Chem.*, 2012, **14**, 1673–1679.
88. V. Declerck, P. Nun, J. Martinez and F. Lamaty, *Angew. Chem., Int. Ed.*, 2009, **48**, 9318–9321.
89. F. Schneider and B. Ondruschka, *ChemSusChem*, 2008, **1**, 622–625.

90. L. M. Klingensmith and N. E. Leadbeater, *Tetrahedron Lett.*, 2003, **44**, 765–768.
91. I. A. Tumanov, A. F. Achkasov, E. V. Boldyreva and V. V. Boldyrev, *CrystEngComm*, 2011, **13**, 2213–2216.
92. I. Halasz, A. Puškarić, S. A. J. Kimber, P. J. Beldon, A. M. Belenguer, F. Adams, V. Honkimäki, R. E. Dinnebier, B. Patel, W. Jones, V. Štrukil and T. Friščić, *Angew. Chem., Int. Ed.*, 2013, **52**, 11538–11541.
93. T. Friščić, I. Halasz, P. J. Beldon, A. M. Belenguer, F. Adams, S. A. J. Kimber, V. Honkimäki and R. E. Dinnebier, *Nat. Chem.*, 2013, **5**, 66–73.
94. D. Gracin, V. Štrukil, T. Friščić, I. Halasz and K. Užarević, *Angew. Chem., Int. Ed.*, 2014, **53**, 6193–6197.
95. O. M. Lemine, M. A. Louly and A. M. Al-Ahmari, *Int. J. Phys. Sci.*, 2010, **5**, 2721–2729.
96. F. K. Urakaev, *Powder Technol.*, 2000, **107**, 93–107.
97. F. Dachille and R. Roy, *Nature*, 1960, **186**, 34.
98. L. Takacs and J. S. McHenry, *J. Mater. Sci.*, 2006, **41**, 5246–5349.
99. R. Thorwirth, A. Stolle, B. Ondruschka, A. Wild and U. S. Schubert, *Chem. Commun.*, 2011, 4370–4372.
100. G. Kaupp, *CrystEngComm*, 2003, **5**, 117–133.
101. D. C. Waddell and J. Mack, *Green Chem.*, 2009, **11**, 79–82.
102. M. O'Brien, R. Denton and S. V. Ley, *Synthesis*, 2011, 1157–1192.
103. F. Alonso, I. P. Beletskaya and M. Yus, *Tetrahedron*, 2004, **61**, 11771–11835.
104. F. Alonso, I. P. Beletskaya and M. Yus, *Tetrahedron*, 2008, **64**, 3047–3101.
105. M. A. P. Martins, C. P. Frizzo, D. N. Moreira, L. Buriol and P. Machado, *Chem. Rev.*, 2009, **109**, 4140–4182.
106. A. Stolle and B. Ondruschka, *Pure Appl. Chem.*, 2011, **83**, 1343–1349.
107. W. C. Shearouse, D. C. Waddell and J. Mack, *Curr. Opin. Drug Discovery Dev.*, 2009, **12**, 772–783.
108. S. Paul, P. Nanda and R. Gupta, *Synlett*, 2004, 531–533.
109. V. V. Boldyrev, *Ultrason. Sonochem.*, 1995, **2**, S143–S145.
110. E. Tullberg, D. Peters and T. Frejd, *J. Organomet. Chem.*, 2004, **689**, 3778–3781.
111. C.-C. Hsu, N.-C. Chen, C.-C. Lai, Y.-H. Liu, S.-M. Peng and S.-H. Chiu, *Angew. Chem., Int. Ed.*, 2008, **47**, 7475–7478.
112. A. M. Flock, C. M. M. Reucher and C. Bolm, *Chem.–Eur. J*, 2010, **16**, 3918–3921.
113. C. J. Cleveland and C. G. Morris, *Dictionary of Energy*, Elsevier, Amsterdam, 2006.
114. G. Centi and S. Perathoner, in *Sustainable Industrial Processes*, ed. F. Cavani, G. Centi, S. Perathoner and F. Trifiró, Wiley-VCH Verlag GmbH, Weinheim, 2009, pp. 279–318.

Subject Index